The Evolution of Parental Care

MONOGRAPHS IN
BEHAVIOR AND ECOLOGY

Edited by John R. Krebs and
Tim Clutton-Brock

The Evolution of Parental Care

T. H. CLUTTON-BROCK

With original drawings by Dafila Scott

Princeton University Press
Princeton, New Jersey

Published by Princeton University Press,
41 William Street,
Princeton, New Jersey

In the United Kingdom:
Princeton University Press, Oxford

Library of Congress Cataloging-in-Publication Data

Clutton-Brock, T. H.
The evolution of parental care / T.H. Clutton-Brock.
p.   cm. — (Monographs in behavior and ecology)
Includes bibliographical references and index.
ISBN 0-691-08730-X (alk. paper)
    0-691-02516-9 (pbk.)
1. Parental behavior in animals.   2. Behavioral evolution.
I. Title.   II. Series.
QL762.C48   1991
591.56—dc20   90-8832

This book has been composed in Lasercomp Times
Roman and Univers 689

Princeton University Press books are printed on acid-free
paper, and meet the guidelines for permanence and
durability of the Committee on Production Guidelines for
Book Longevity of the Council on Library Resources

Printed in the United States of America by
Princeton University Press, Princeton, New Jersey

10 9 8 7 6 5 4 3 2 1
10 9 8 7 6 5 4 3 2 1
(Pbk.)

This book is dedicated to Robert Hinde

Contents

List of Drawings at Chapter Openings

(all drawings by Dafila Scott)

Acknowledgments

I am deeply grateful to many colleagues for generous advice, comments or criticism, or for access to unpublished literature. They include Steve Albon, Pat Bateson, Brian Charlesworth, Andrew Cockburn, Bernard Crespi, Nick Davies, Hugh Drummond, Marco Festa Bianchet, Charles Godfray, Alan Grafen, Paul Harvey, Robert Hinde, Mariko Hiraiwa-Hasegawa, John Krebs, Hans Kruuk, John Lazarus, Nigel Leader Williams, Kate Lessells, Mark McGinley, Rob Magrath, Geoff Parker, Linda Partridge, Marion Petrie, Michael Reiss, Mark Ridley, Rick Shine, Judy Stamps, Steve Stearns, Kevin Teather, Fritz Trillmich, Amanda Vincent, and Ken Yasukawa.

Hannah Clarke, Camilla Lawson, Owen Price, and Alison Rosser provided invaluable help with the figures, while Pat Reay turned innumerable drafts of dubious handwriting into carefully organized text with unfailing good humor. Sigal Earn generously helped with the indexing.

Lastly, my greatest debt is to my wife, Dafila Scott, who not only drew the plates at the beginning of each chapter, but looked after our children while I wrote about parental care.

For permission to reproduce material from published papers, I am indebted to the authors (whose names appear next to the relevant figures) and to: Academic Press Ltd. (Figs. 4.6, 8.3, 8.6); *American Zoologist* (Figs. 7.2, 7.3, 8.1, 9.11, 10.2, 11.1); Baillière Tindall (Figs. 2.9, 2.10, 8.4, 8.5, 9.6, 9.8, 10.5, 10.7, 11.2); Blackwell Scientific Publications (Figs. 2.6, 3.5); E. J. Brill, Publishers (Figs. 7.4, 12.10); Elsevier Trends Journals (Figs. 6.6, 7.1); Harvard University Press (Figs. 2.4, 2.5, 2.7); Longman Group, U.K. (Fig.5.2); Macmillan Magazines Ltd. (Figs. 3.7, 6.2, 9.2, 9.7, 12.1, 12.8, 13.1, 13.2, 13.6, 13.9, 13.12); Museum of Zoology, University of Michigan (Fig. 5.3); The National Research Council of Canada (Fig. 13.5); *Oecologia* (Figs. 4.4, 4.5); *Oikos* (Figs. 4.2, 8.2); Oxford University Press (Fig. 6.3); Paul Parey, Berlin (Fig. 3.8); The Royal Society, London (Fig. 13.2); The Society for the Study of Evolution (Figs. 2.2, 3.2, 3.6); Springer-Verlag, London (Figs. 2.3, 10.8, 12.7); University of Chicago Press (Figs. 3.3, 4.3, 4.10, 4.11, 6.1, 9.5, 10.1, 10.4, 12.2, 12.5, 12.9, 13.8); Wiley & Sons (Figs. 4.9, 7.6); The Zoological Society of London (Figs. 4.7, 6.4, 6.5, 6.7, 6.8).

The Evolution of Parental Care

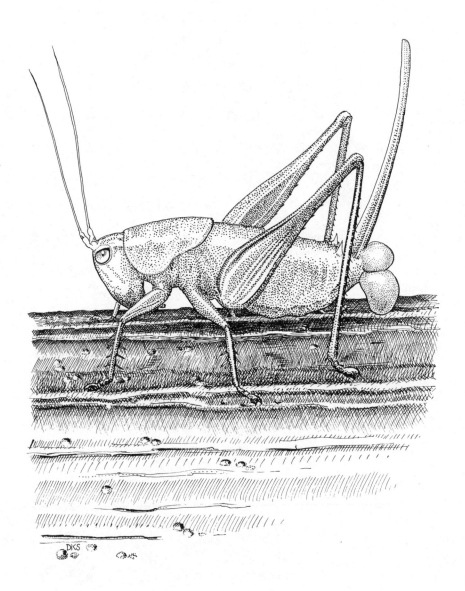

1

Parental Care and Competition for Mates

1.1   Parental Care and Sexual Selection

Few areas of evolutionary biology have progressed as rapidly over the past two decades as our understanding of animal breeding systems. In this area, an understanding of the evolution of parental care is of central importance since many of the most striking differences in the reproductive behavior of males and females are associated with variation in their involvement in parental care in its broadest sense, including any form of parental expenditure on producing large eggs or bearing young (Bateman 1948; Williams 1966; Trivers 1972). In particular, parental care plays an important part in determining the intensity of competition for mates within both sexes (Trivers 1972), which, in turn, affects selection pressures operating on many different aspects of behavior, physiology, and anatomy (Clutton-Brock, in press).

Surprisingly, the importance of parental care in determining the pattern of mating competition was not explicitly recognized by Darwin. As Bateman noted in 1948, Darwin "took it as a matter of general observation that males were more eager to pair with any female, whereas the female, though passive, exerted choice. He was at a loss, however, to explain this sex difference, though it is obviously of great importance for an understanding of intersexual selection."

Bateman showed that variance in reproductive success in *Drosophila* was greater among males than females because male success increased proportionately with number of matings while female success did not. He then went on to discuss the reasons underlying this difference, linking them directly to the evolution of anisogamy:

> In most animals the fertility of the female is limited by egg production which causes a severe strain on their nutrition. . . . In the male, however, fertility is seldom likely to be limited by sperm production but rather by the number of inseminations or the number of females available to him. . . .
>
> The primary cause of intra-masculine selection would thus seem to be that females produce much fewer gametes than males. Consequently, there is competition between *male* gametes for *female* gametes.

Bateman's argument attributed the greater intensity of reproductive competition among males to the smaller size of male gametes and (by implication) to the faster rate of gamete production by males. Where neither sex cares for eggs or young, as in *Drosophila*, differences in gamete size may be responsible for sex differences in reproductive rate and in competition for the opposite sex (see Baylis 1981), but involvement of either or both sexes in parental care can easily reverse these differences (see Section 7.2).

The relevance of parental care to mating competition and sexual selection was first appreciated by Williams (1966b):

> It is a common observation that males show a greater readiness for reproduction than females. This is understandable as a consequence of the greater physiological sacrifice made by females for the production of each surviving offspring. A male mammal's essential role may end with copulation, which involves a negligible expenditure of energy and materials on his part, and only a momentary lapse of attention from matters of direct concern to his safety and well-being. The situation is markedly different for the female, for which copulation may mean a commitment to a prolonged burden, in both the mechanical and physiological sense, and its many attendant stresses and dangers. Consequently the male, having little to lose in his primary reproductive role, shows an aggressive and immediate willingness to mate with as many females as may be available. (p. 163).

Six years later, in an influential paper that laid the foundations of current thinking about the evolutionary consequences of parental care, Trivers (1972) advanced the argument in three main ways. Following Darwin and Fisher, he emphasized the need to consider all ways in which parents contributed to the fitness of their offspring, combining these in a new term, "parental investment," defined as "any investment by the parent in an individual offspring that increases the offspring's chance of surviving (and hence reproductive success) at the cost of the parent's ability to invest in other offspring."

Second, he stressed that the currency of all forms of parental investment is its cost to the parent's ability to invest in other offspring:

> I measure the size of a parental investment by reference to its negative effect on the parent's ability to invest in other offspring: a large parental investment is one that strongly decreases the parent's ability to produce other offspring. There is no necessary correlation between the size of parental investment in an offspring and its benefit for the young. . . . Decrease in reproductive success resulting from the negative effect of parental investment on *nonparental* forms of reproductive effort (such as sexual competition for mates) is excluded from the measurement of parental investment.

And third, he argued that relative parental investment by males and females controls the degree of intrasexual competition and the intensity of sexual selection:

> *What governs the operation of sexual selection is the relative parental investment of the sexes in their offspring.* Competition for mates usually characterizes males because males usually invest almost nothing in their offspring. . . . Where male parental investment strongly exceeds that of the female (regardless of which sex invests more in the sex cells) one would expect females to compete among themselves for males and for males to be selective about whom they accept as a mate.

1.2   Parental Care and Competition for Mates

Where females care for the young and males play little or no part in parental investment, as in most mammals, Trivers's argument provides a satisfactory explanation of competition between males for females. However, it is less satisfactory where males are heavily involved in parental care for, here, it is often impossible to determine which sex invests most in its progeny. Take, for example, the three-spined stickleback *Gasterosteus aculeatus*, where females lay in nests built by males who subsequently guard and fan the eggs (Wootton 1984). Do females invest more heavily than males because the energetic costs of egg production are high, or do males invest more than females as a consequence of their expenditure on guarding and fanning and their reduced food intake or increased liability to predators? In practice, questions of this kind are usually impossible to answer, and it is seldom possible to use estimates of parental investment by males and females to predict the pattern of competition for mates. As a result, there is a danger that we lapse into the circularity of assuming that females invest more heavily than males because males compete more intensely for mates.

Is it possible to reconstitute Trivers's proposals to generate more testable predictions concerning the direction of mating competition? In fish, Baylis (1981) attributes both the increased intensity of reproductive competition between mates and the evolution of paternal care to the faster rate of gamete production by males, which allows them to breed more rapidly than females. However, though sex differences in gamete size and breeding rate may have influenced the original evolution of parental care, once parental care has evolved, sex differences in care can exert an important influence on sexual selection and on the pattern of competition for mates. For example, in some species where males are heavily involved in parental care, females compete for males more intensely than vice versa, yet males still have a higher rate of gamete production (see Chapters 7 and 8).

Once parental care has evolved, it is not the rate of gamete production by the two sexes that will determine the pattern of mating competition but their potential reproductive rates, measured in terms of the number of independent offspring that parents can produce per unit time, calculated over periods when both sexes are reproductively active. Where members of one sex can breed more quickly, some individuals will be disproportionately succeessful, with the result that others will experience an Operational Sex Ratio (Emlen and Oring 1977) biased toward the faster sex and will be forced to compete intensely for mates. In many ectotherms where males are responsible for parental care they can still achieve faster reproductive rates than females. In three-spined sticklebacks, for example, females can lay a clutch of eggs at least once a week while males guard clutches for around 15–20 days before hatching. However, males can guard up to six clutches at a time, giving them a maximum rate of one clutch every 2.5 days—a rate twice as high as females (Wootton 1984). Where males are able to "process" clutches of eggs more rapidly than females can produce them, males are likely to be the primary competitors despite their involvement in parental care (see Chapter 7).

Where males are responsible for parental care, females are most likely to be the main competitors where the rate at which males can process clutches or eggs is relatively low. In fish, this is most likely to be the case where males brood eggs internally. For example, in some pipefish (Syngnathidae) females produce eggs more rapidly than males can brood them, and females are larger and more colorful than males and more active during courtship (Berglund, Rosenqvist, and Svensson 1986, 1989). Similarly, in some cardinal fish where males brood the eggs in their mouths and females spawn several times in each season, females are primarily responsible for defense of spawning sites and may guard brooding males (Garnaud 1950, 1962; Usaki 1977). In birds where males are the primary care-givers, female competition and full role reversal are confined to the smaller shorebirds (Jenni 1974; Erckmann 1983). Here, too, the rate at which males can process eggs is low because of the allometry of egg size (see Chapter 4).

The relative rate at which males and females can produce independent offspring is not the only factor that will affect the Operational Sex Ratio and the pattern of competition for mates. Especially where males and females have similar rates of reproduction, variation in the costs of competition to the two sexes or in the net benefits of retaining potential mating partners may generate exceptions to the general rule. So, too, may fluctuations in the adult sex ratio caused either by sex differences in lifespan or by local concentrations of one sex. For example, in some fish where successful males hold territories used simultaneously by several females, the latter may compete for the male's attention during the breeding peak (Fricke 1980; Thresher 1984). Finally, it is important to emphasize that even where one sex is the predominant competitor, direct competition for access to mates may be well developed in the other sex, too.

1.3 The Evolution of Parental Care

Despite the importance of parental care in determining patterns of reproductive competition, the evolution of parental care has not attracted the same degree of attention as other aspects of animal breeding systems. In particular, little attempt has been made to link work on different animal groups into a unified theoretical framework or to integrate the results of detailed studies of variation in parental care within species to our knowledge of interspecific trends.

This book explores five principal questions about the evolution of parental care in its broadest sense:

What are the principal benefits and costs of parental care?

All functional explanations of the extent and distribution of parental care assume that the benefits and costs of parental care are substantial, but attempts to measure them have produced varied results. Chapter 2 describes different forms of parental care and briefly reviews evidence that they increase the fitness of offspring, while Chapter 3 examines attempts to measure the costs of reproduction.

Why does the extent of parental care vary so widely between species?

Chapter 4 examines the adaptive significance of variation in the size of eggs and neonates; Chapter 5 reviews the distribution of viviparity and other forms of "bearing"; and Chapter 6 examines the reasons for variation in the duration of incubation, gestation, and lactation periods.

Why do only females care for eggs and young in some animals, only males in others, and both parents in a few?

The extent to which males and females care for their young varies widely within as well as among most major animal groups. Since the relative costs of parental care to males and females commonly differ between ectotherms and endotherms, it is convenient to treat these two groups separately. Chapter 7 examines the distribution of parental care and the extent to which males and females are involved among invertebrates, fish, amphibia, and reptiles, while Chapter 8 reviews the distribution of care in birds and mammals.

To what extent is parental care adjusted to variation in its benefits to offspring and its costs to parents?

Parents might be expected to adjust their expenditure in relation to its costs to their own fitness and its benefits to that of their offspring. Chapters 9 and 10 review the empirical evidence that they do so. In sexually reproducing animals, the evolutionary interests of parents and offspring are likely to differ, and offspring may attempt to extract a higher level of expenditure from their parents than the latter are selected to give. Chapter 11 reviews predictions about the extent and distribution of parent-offspring conflict and examines the evidence that offspring are able to influence the extent of parental expenditure.

How do parents divide their expenditure between their sons and daughters?

In many sexually dimorphic animals, young males and females require different amounts of resources. Chapter 12 reviews the empirical evidence that parents invest to different extents in sons and daughters. Another way in which parents might manipulate their expenditure on offspring is by varying the sex ratio in relation to the availability of resources. Chapter 13 briefly reviews evidence of sex ratio variation in invertebrates, birds, and mammals.

1.4 Definitions of Parental Care and Parental Investment

The current terminology used in studies of the evolution of parental care is often diffuse and misleading. Since it is important to distinguish clearly between the various currencies in which the costs and benefits of parental care are measured, I use three different terms.

Parental care. Any form of parental behavior that appears likely to increase the fitness of a parent's offspring. This is a descriptive term and carries no implications about costs in terms of energy or fitness. In its broadest sense, parental care includes the preparation of nests and burrows, the production of large, heavily yolked eggs, the care of eggs or young inside or outside the parent's body, the provisioning of young before and after birth, and the care of offspring after nutritional independence (see Chapter 2). In its narrowest sense, it refers only to the care of eggs or young when they are detached from the parent's body.

It is useful to recognize two contrasting categories of care, though many forms may lie between these extremes: *depreciable care*, such as the provision of food, where the benefits of parental expenditure decline as brood size increases; and *nondepreciable* care, such as parental vigilance, where individual benefits do not decline with increasing brood size (see Altmann, Wagner, and Lennington 1977; Lazarus and Inglis 1978, 1986: Wittenberger 1979a). (I have adopted Altmann's terms since an unfortunate terminology has developed. While Wittenberger [1979a] refers to care of the first kind as "shareable" and to care of the second kind as "nonshareable," Lazarus and Inglis [1978, 1986] refer to the first as "unshared" and the second as "shared"!)

Parental expenditure and relative parental expenditure. The expenditure of parental resources (including time and energy) on parental care of one or more offspring. "Parental effort" is sometimes used in this context, but the disadvantage here is that other authors use it to refer to the fitness costs of care (see below).

In some cases, parental expenditure is calculated as a proportion of the mother's resources expended on one or more of its offspring (e.g., Hirschfield and Tinkle 1975). I refer to measures of this kind as relative parental expenditure. It is often assumed that relative parental expenditure reflects the fitness costs of care, and measures of this kind are sometimes treated as if they were measures of parental investment (see below). However, this need not be the case for the proportion of resources that a parent can afford to expend, for a given fitness cost is likely to vary with its age and size as well as with many other biological and environmental factors (see Chapters 9 and 10).

Parental investment. The extent to which parental care of individual offspring reduces the parent's residual reproductive value (RRV). The definition of parental investment is the source of considerable discussion. Today, it is generally used in a less restrictive way than Trivers's original definition (see section 1.1) to refer to any characteristics or actions of parents that increase the fitness of their offspring at a cost to any component of the parent's fitness (see Alexander and Borgia 1979; Gwynne 1984a; Knapton 1984; Thornhill and Gwynne 1986), including any costs of parental care to the parent's subsequent mating success as well as to its survival or fecundity. Costs to the parent's growth also need to be considered, for these may affect the parent's subsequent mating success or fecundity, as well as its future survival (Gross and Sargent 1985). In addition, it is logical to include costs to the fitness of other offspring or any reduction in the parent's ability to increase the fitness of other relatives by behaving nepotistically. Parental investment is usually (though not always) used to refer to the fitness costs of parental care of *individual* offspring, while the total costs of caring for *all* progeny are designated parental effort (PE), which, with mating effort (ME), is a part of the organism's reproductive effort (RE) (Low 1978; Alexander and Borgia 1979).

Not surprisingly, often there are problems in deciding what should and what should not be included in parental investment. In particular, male expenditure on territorial defense or nuptial gifts can be hard to classify since they may both increase mating success (and thus be regarded as a form of mating effort, ME) and improve the survival or quality of offspring (Zeh and Smith 1985). Though the distinction may be difficult to draw in practice, the most logical course may be to include as parental investment any costs beyond those necessary to ensure mating success. However, as Gwynne (1984a) points out, where females acquire a substantial amount of energy or other resources from the nuptial gifts of males, access to males may limit their reproductive output, and females may consequently compete for mating access while males become the choosy sex (see Gwynne 1983, 1984a,b). Thus the consequences of heavy expenditure on nuptial gifts may be similar whether or not we regard it as a form of parental investment.

Finally, it is important to remember that variation in parental investment need not be closely related to variation in the amount of resources received by offspring (Trivers 1972; Evans 1990). For example, reduction in food availability may be associated both with increased costs of rearing young to the parent and with reductions in parental expenditure. Depending on their scarcity, environmental changes may either increase or reduce the effects of parental expenditure on offspring fitness. Consequently, though it is sometimes convenient to assume that variation in the cost of care to the parents reflects differences in the benefits of care to their offspring, this may not always be the case.

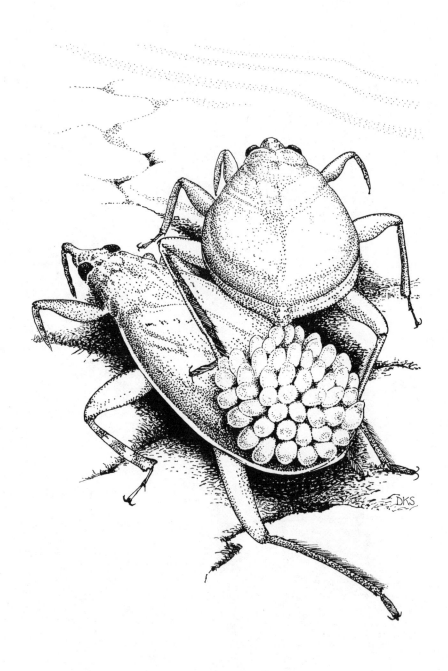

2

Forms of
Parental Care

2.1   Introduction

This chapter describes different forms of parental care and briefly examines the evidence that they contribute to the fitness of progeny. In its broadest sense, parental care includes the preparation of nests and burrows (Section 2.2), the production of large, heavily yolked eggs (Section 2.3), the care of eggs or young outside or inside the parent's body (Sections 2.4, 2.5), the provisioning of young before and after birth (Sections 2.6, 2.7), and the care of offspring after nutritional independence (Section 2.8).

As yet, much of the evidence of the benefits of parental care relies on correlations between parental expenditure and offspring fitness. It is important to appreciate that the causal relationships underlying these associations are often uncertain. In at least some cases, correlations between parental quality and parental expenditure may generate correlations between the latter and offspring fitness where causal relationships either do not exist or are weak. For example, correlations between egg size, growth, and offspring fitness (Section 2.3) could either be caused by direct effects of egg or juvenile size on offspring fitness, or by phenotypic or genetic correlations between some independent variable, such as the mother's body size, and different components of reproductive performance. In the last case, egg size need not have any causal effect on juvenile growth or viability. Similarly, it is usually assumed that correlations between the social rank of female primates and their reproductive success show that high rank increases the fitness of her offspring (see Sections 2.8 and 2.9). However, an alternative interpretation is again that some other variable, such as a female's body size, affects both her rank and her breeding success and that the effects of rank on breeding success are either minimal or nonexistent. Objections of this kind apply to much of the evidence for the benefits of parental care and emphasize the need for experiments.

2.2   Preparation of Nests, Burrows, or Territories

The preparation of holes, burrows, or nests by either or both parents is common in invertebrates and vertebrates. For example, in bush crickets (*Anurogryllus muticus*), males call from burrows they have constructed (West and

Alexander 1963). When a receptive female arrives, the male mates with her then vacates the burrow, which she uses to lay eggs in. Similarly, in many invertebrates as well as in vertebrates, males defend patches of resources where females feed before mating (see Thornhill and Alcock 1983; Rubenstein 1986). In both cases, it is often unclear whether the costs of this behavior should be regarded as a form of parental effort (PE) or mating effort (ME), for females typically refuse to mate with males that do not possess burrows or territories (see Section 1.4).

2.3   Production of Gametes

In many animals, gamete production by females represents the principal form of parental expenditure. Gamete size is commonly related to the survival, growth, and eventual breeding success of offspring, though this is not always the case (Karlsson and Wiklund 1984). For example, in *Daphnia magna*, females reared at high food availability produce larger eggs with more fat reserves, and young from these eggs are more resistant to starvation than young hatched from the smaller eggs of females reared on low levels of food availability (Tessier, Henry, et al. 1983; Goulden and Henry 1984). In aphids, large-born young are more likely to survive periods of food stress or encounters with predators than small-born young (Dixon 1985). Similarly in birds, chick survival commonly increases with egg size (Figure 2.1; Parsons 1970; Schifferli

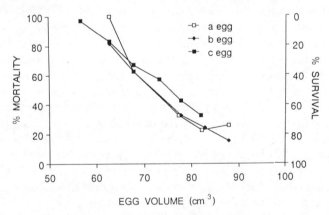

Figure 2.1   The relationship between egg volume and post-hatching mortality for each egg of the laying sequence in herring gulls (from Parsons 1970).

1973; Lundberg and Vaisanen 1979; Ankney 1980; Boersma, Wheelwright, et. al 1982; Galbraith 1988). In several fish, egg size is also related to early growth (Hempel 1965; Bagenal 1969): in Coho salmon (*Oncorhynchus kisutch*), for example, there is a close correlation between the size of eggs, early growth, and adult size (Figure 2.2a,b). Adult size is, in turn, associated with increased fecundity (Figure 2.2c), larger egg size (Figure 2.2d), higher nest survival (Figure 2.2e), and increased reproductive output.

Similar correlations between growth at different stages, juvenile viability, and adult breeding success have been found in a wide variety of plants and animals. Minor differences in egg size may have a disproportionate effect on fitness, especially where juveniles compete directly for resources (e.g., Harvey and Corbet 1985; Prout and McChesney 1985; Sinervo 1990). However, large egg size can have disadvantages, including longer incubation times and higher instantaneous rates of mortality before hatching (Balon 1984; Sargent, Taylor, and Gross 1987; see also Karlsson and Wiklund 1985).

Males may contribute to the production of gametes by females in at least four different ways. First, they may defend resources used by females before or during egg production (Lack 1966; Thornhill and Alcock 1983). Second, they may present females with "gifts" of food, feces, or glandular products that are used in the formation of eggs (Nisbet 1973; Thornhill 1981, 1986; Austad and Thornhill 1986; Steele 1986). Third, males may pass nutrients directly into the female's genital tract in spermatophores or other ejaculatory products (stomatopods: Deecaraman and Subramoniam 1983; Lepidoptera: Rutowski, Newton, and Schaefer 1983). And fourth, males may permit themselves to be eaten by the female (Buskirk, Frohlich, and Ross 1984), though it is not yet clear whether this represents an adaptive strategy on the part of males or one of the costs of mating that they try to avoid (Gould 1984; Liske and Davis 1987).

A substantial amount of nutrients may be passed by males to females in these ways (Boggs and Gilbert 1979; Rutowski 1982). In some Lepidoptera, single spermatophores represent 10% of the male's body weight (Rutowski 1982), while in one katydid, *Requena verticalis*, males produce spermatophores, consisting of an ampulla containing sperm and a nutritional, sperm-free mass called the spermatophylax, which represent up to 10% of their weight (Gwynne 1983, 1988; Gwynne, Bowen, and Codd 1984). After mating, the female eats the spermatophylax while insemination occurs, subsequently transferring nutrients for use in egg production. Consumption of the spermatophylax increases the size of the eggs that the female subsequently lays, and offspring born from large eggs have a better chance of surviving the winter and mature earlier than those born from small eggs (Gwynne 1988). In contrast, courtship feeding in red-backed shrikes (*Lanius colluro*) increases clutch size but not egg size (Carlson 1989).

2.4   Care of Eggs

The care of fertilized eggs by one or both parents is found in many different groups of animals. Care of eggs is necessary in virtually all endotherms, while in ectotherms it may also be necessary to control environmental factors includ-

Figure 2.2 Correlations between egg size and subsequent growth and breeding success in Coho salmon (from Holtby and Healey 1986, van den Berghe 1984, and van den Berghe and Gross 1989):
(a) between egg size and fry size;
(b) between freshwater and ocean growth;
(c) between female size and fecundity.

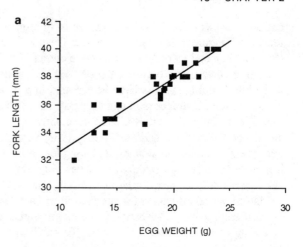

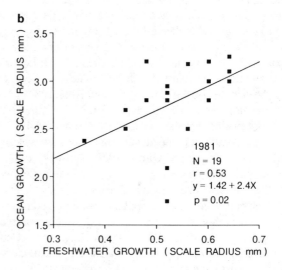

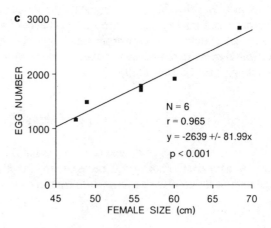

Figure 2.2
(continued).
(d) between
female size and
egg size;
(e) between
female size and
nest survival;
and
(f) between
female size and
the total number
of offspring
surviving to
maturity.

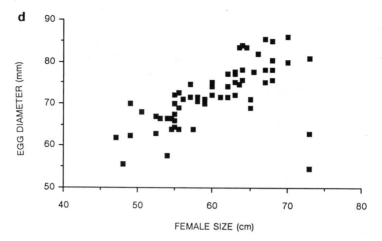

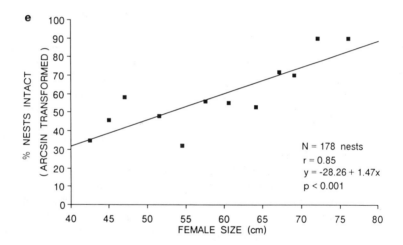

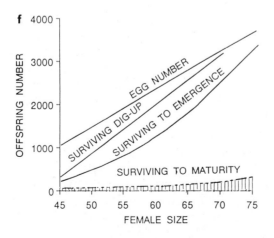

ing humidity, oxygen levels, and salinity or to reduce the risks of predation, parasitism or egg dumping by members of the same or other species (see Wilson 1971; Salthe and Mecham 1974; Eberhard 1975; Forester 1979; Perrone and Zaret 1979; Smith 1980; Hinton 1981; Simon 1983). For example, if guarding fathers are removed immediately after laying in the Puerto Rican frog, *Eleutherodactylus coqui*, the frequency of predation increases and few or no eggs survive to hatching (see Figure 2.3). In the dusky salamander, *Des-*

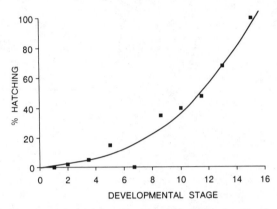

Figure 2.3   Percentage of eggs of *Eleutherodactylus coqui* that survived to hatch following removal of the male at different stages of egg development, corresponding to 1–2 days each (from Townsend 1986).

mognathus ochrophaeus, the removal of parents increases the susceptibility of the eggs to fungal infection as well as to predators (Forester 1979). Similar effects have been demonstrated in other amphibia (e.g., Tilley 1972; Simon 1983) as well as in insects (Eberhard 1975; Nafus and Schreiner 1988; Scott 1990), decapods (Diesel 1989) and fish (Dominey 1981).

Four types of egg care are common: (1) Care of eggs laid directly on the substrate or attached to it or in nests or burrows, as in many invertebrates (Hinton 1981), fish (Perrone and Zaret 1979), amphibia (Salthe and Mecham 1974) and birds (Lack 1968). (2) Care of eggs attached externally to or carried by either or both parents, as in sea spiders, some Hemiptera, some fish and some amphibia (Hinton 1981; Hoar, 1969; R. L. Smith 1976; McDiarmid 1978; Wells 1981). (3) The retention of fertilized eggs within the female's reproductive tract, often associated with ovoviviparity (where young develop and hatch internally but are not provisioned from sources other than the egg) or viviparity (where young develop internally and are provisioned by the parent; see Chapter 5). In many species, eggs are retained within the mother's body cavity for part or all of their development, usually in the ovary or the oviduct or in specialized brood pouches (Hinton 1981). For example, in the cockroach *Nauphoeta*, eggs (ootheca) are extruded, rotated through 90 degrees, then retracted into a specialized brood sac, where development proceeds until the larvae are ready to hatch, when they are extruded (Hinton 1981). In ovoviviparous animals, young are retained within the mother's body cavity after hatching, but are not nourished from sources other than yolk (for insects, see Hinton 1981; fish: Hoar 1969; reptiles: Yaron 1985; amphibia: Salthe and Mecham 1974). However, all degrees of parental nourishment exist, and the distinction between ovoviviparity and viviparity is often difficult to draw. (4) Care of eggs in the parent's body cavity but not in the reproductive tract. One category includes care of eggs in the mouth (as in some cichlid fish and in some amphibia) or in specialized brood pouches or chambers (as in some sea horses and pipefish of the family Syngnathidae) or in the stomach (as in females of the Australian myobatrachid frog, *Rheobatrachus silus* (Corben Ingram, and Tyler 1974).

2.5 Care of Young without Provisioning from Sources outside the Egg

Care of young without provisioning occurs in a variety of invertebrates, fish, reptiles, and amphibia as well as in some birds. In some invertebrates, either or both parents care for aggregations of young, while in others, larvae shelter beneath one parent (see Figure 2.4) or cling to some part of the parent's body (Hinton 1981).

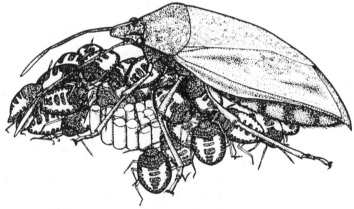

Figure 2.4 A female of the cacao stinkbug (*Antiteuchus limbativentris*) of Trinidad is shown guarding her first instar nymphs 4 days after they hatched (from Wilson 1971; see Eberhard 1975). In this species, females guard eggs and nymphs. When guarding mothers were removed, no eggs survived to produce nymphs. (Drawing by Sarah Landry.)

Care of young (either externally or internally) commonly plays an important role in protecting juveniles against predators or parasites (see Shine 1985). For example, in some Hemiptera, parents (usually females) drive off intruders and can play an active role in shepherding their broods together during local migrations between feeding sites (Wilson 1971; see Figure 2.4). The importance of parental care in guarding eggs or juveniles against predators has been demonstrated experimentally in a number of species, including bluegill sunfish (Dominey 1981). In other cases, the principal function of parental care can be to regulate the environment of the nest or burrows. For example, in the staphylinid beetle *Bledius*, which lives in intertidal mud, the mother keeps the burrow ventilated by renewed burrowing activity, and if she is removed, the brood perishes for lack of oxygen (Bro Larsen 1952). Similarly, in mole crickets that nest on moist vegetable matter, the mother's presence can be necessary to prevent the invasion of the nest by fungi (West and Alexander 1963).

2.6   Provisioning Young before Hatching or Birth

In some insects, either or both parents may provision brood chambers where eggs are laid so that food is available for the emerging young. Well-documented examples include studies of parasitoid wasps and scarab beetles (Wilson 1971). In several scarabs, both sexes cooperate in provisioning: for example, in the genus *Lethrus*, pairs form in spring and both partners cooperate in excavating the nest (see Figure 2.5). The male, who has two long tusks, clears a "courtyard" around the entrance, which he subsequently defends. Males and females play different roles in provisioning the burrow, the male gathering fresh leaves and buds and taking them to the female, who presses them into elliptical balls fitted to the interior of brood cells where the eggs are laid (Schreiner 1906; Wilson 1971). In other insects, parents prepare specialized food sources for the larvae before egg laying, including adult feces and partially digested food (Hinton 1981).

In viviparous animals, young are nourished from parental sources before birth. This may involve a wide variety of mechanisms, but five main types of maternal provisioning exist:

Where young are provisioned from the mother's own soma, as in viviparous mites where young feed on their mother, finally bursting out and dispersing (W. D. Hamilton 1967).

Where young are provisioned by ingesting eggs (oophagy) or other embryos (adelphagy). For example, in the porbeagle shark, *Lamna cornubica*, the young absorb their egg yolks early in development and subsequently depend on swallowing immature eggs and degenerating ovarian tissues that pass down the oviduct packaged in a secretion of the shell gland (Shann 1923). In the viviparous salamander, *Salamandra atra*, only 1–2 eggs per oviduct are fertilized normally and the remaining 20–30 degenerate into a sort of yolk that is eaten by the developing larvae after their own yolk is used up (Salthe and Mecham 1974).

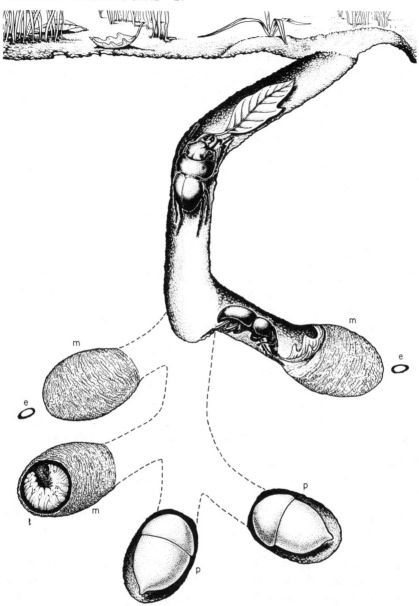

Figure 2.5   A pair of *Lethrus apterus* are shown provisioning their nest. The male is pulling a freshly cut leaf backward through the entrance gallery, while the female packs another leaf into the brood cell currently being provisioned: (m) leaf masses; (e) eggs; (l) larva; and (p) pupae, which are enclosed in acorn-shaped cocoons. After a cell has been provisioned, the pair pack soil into the main tunnel to that level, as indicated by the limits of the dashed lines (drawing by Sarah Landry, based on von Lengerken 1939, from Wilson 1971).

Where the young are provisioned from specialized secretions produced by the ovary, the oviduct or related sources, as in a wide variety of viviparous animals including insects, fish, reptiles, and amphibia (Hoar 1969; Imms 1970; Salthe and Mecham 1974; Wourms, 1981; Yaron 1985). For example, in the tsetse fly, *Glossina palpalis*, a single larva is maintained in the mother's uterus until the end of its second instar and fed with a milky secretion that is absorbed through the mouth (Buxton 1955). In aphids, the larger embryos developing in the parent's body cavity already have embryos developing within them so that three generations can develop simultaneously (Dixon 1985). The origin and nature of nutritional secretions vary widely, as does the method of transfer to the offspring.

Where the young feed from secretions produced within specialized brood chamber(s) where the growing young are maintained. For example, in some pipefish and sea horses as well as in some frogs, young are nourished from blood vessels surrounding specialized brood chambers (Salthe and Mecham 1974; Kronester-Frei 1975; Haresign and Shumway 1981).

Where young are nourished from the mother's blood supply via placentas or pseudoplacentas, as in some fish, amphibia, and in all therian mammals (Eisenberg 1981; Wourms 1981; Yaron 1985).

Provisioning before birth commonly has a substantial influence on neonatal survival as well as on reproductive performance in adulthood. For example, red-deer calves born below average birth weight are more likely to die in the first winter (Guinness, Clutton-Brock, and Albon 1978) and, if they do survive, bear light young throughout their reproductive lives and lose a high proportion of their progeny before maturity (see Figure 2.6). A variety of other studies have produced evidence of similar effects (Allden 1970; Cundiff 1972; Legates 1972; Hartsock and Graves 1976; Thorne, Dean, and Hepworth 1976).

2.7 Provisioning Young after Hatching or Birth

In many animals, young are provisioned after hatching or birth. Four main types of provisioning occur: *With the adult diet.* In "ambrosia" beetles of the family Scolytidae adults cultivate fungi on palisades lining the gallery walls of their burrows (Batra 1963). In some species, larvae are reared in short chambers ("cradles") leading off the main galleries, and the mouth of each cradle is kept closed by a plug of fungus that is eaten by the young and renewed by the mother until the young are mature (Wilson 1971). In the Jamaican bromeliad crab *Metopaulias depressus*, females feed their growing larvae with small prey items, and experimental removal of parents causes an increase in larval mortality and a reduction in growth rates (Diesel 1989). Similarly, in many insectivorous or carnivorous birds, young are fed on the adult diet or something approximating closely to it (Newton 1979).

Figure 2.6 Correlations between birth weight, offspring birth weight, and offspring survival in red-deer calves (from Albon, Clutton-Brock and Guinness 1987). The upper graph shows the mean birth weight of offspring born to different cohorts of mothers plotted against the mean birth weight of the mothers' cohorts. The lower plot shows the mean survival of calves born to members of different cohorts plotted against the mean birth weight of the cohort. Numbers beside each point show the year in which the cohort was born.

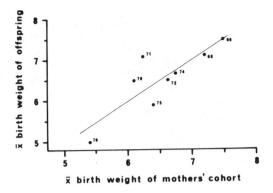

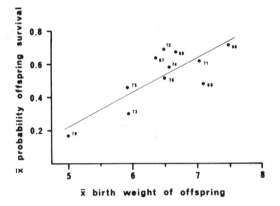

$\bar{x}$ birth weight of offspring

With a specialized diet suited to the requirements of the young. For example, in some passerine birds where adults feed on grains, pulli are initially fed predominantly with insects (Lack 1966).

With preprepared or partly digested food. In a number of animals, including crickets (Wilson 1971), spiders (Kaston 1965), birds (Lack 1966), and carnivorous mammals (Ewer 1973), young are fed with partially digested food regurgitated by the parent. In some wood-boring beetles belonging to the family Passalidae, young are fed with wood that has been chewed and either partly digested or converted into frass (Tallamy and Wood 1986). Similarly, in naked mole rats (*Heterocephalus glaber*), young are fed with the feces of adult colony members (Jarvis 1981).

With specialized products or secretions. Young are fed with specialized secretions or other products in a substantial number of invertebrates, including members of the Coleoptera and Orthoptera. For example, in burying beetles of the genus *Nicrophorus*, pairs of beetles construct a roughly spherical ball from the corpse of a small vertebrate that they install in a burrow, sealing them-

Figure 2.7   A female of the burying beetle (*Nicrophorus vespillo*) with her larvae. The larvae, which in this case are in the third instar, rest in a depression dug by the female on top of the carrion ball. Both parents feed them by regurgitation. (Original drawing by Sarah Landry; based on photographs by Erna Pukowski; from Wilson 1971.)

selves in with the nest (Pukowski 1933; Scott and Traniello 1990). The female then eats out a crater-shaped depression in the ball in which she lays her eggs (see Figure 2.7). When the young hatch, the female (and, in some species, the male too) feed the young with a dark liquid from their mouths. When the larvae are only five or six hours old they begin to feed on the ball itself but continue to be fed by their parents at irregular intervals and are totally dependent on them for a short time after each moult.

IN SOME fish, young are fed cutaneously from the mother's blood supply (Wootton, 1979) while in others, like the cichlid, *Symphylodon discus*, the young feed on an ectodermal mucus produced by both parents (Ward and Barlow 1967; Nicoll and Bern 1971; Noakes and Barlow 1973; Robertson 1973). In a few birds, including penguins, pigeons and doves, and flamingoes, young are fed with secretions from the crop or esophagus (Lang 1963; Prevost and Vilter 1963; Vandeputte-Poma 1980): in pigeons and doves the walls of the crop thicken during incubation and produce a cream-colored secretion ("crop milk") made up of cells from the outer layers of mucosal epithelium which is regurgitated into the throats of the nestlings (Beams and Meyer 1931; Chadwick 1977). Immediately after hatching, the parents do not feed and the squabs are fed exclusively with this secretion while, later, the adults begin to feed again and the young receive a mixture of grain and "crop milk." Finally, juveniles are fed on true milk in all mammals, though the quality of the milk varies widely (see Chapter 6).

In virtually all species where young are fed by their parents, they do not survive if parents are removed, though where both parents are involved the removal of one is not necessarily fatal (see Chapter 8). Both across and within species, there is usually a close relationship between feeding rate and the growth rate and survival of young (Figure 2.8; see also Lack 1966; Geist 1971;

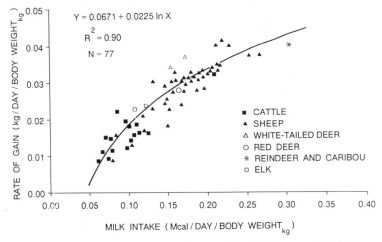

Figure 2.8 Milk intake compared to growth rates during the first month of life in several ungulates (from Robbins 1983).

Drent and Daan 1980; Clutton-Brock, Albon, and Guinness 1982; Prentice and Whitehead 1987). Early growth may also affect reproductive success in adulthood. In mammals, for example, adult size is commonly related to breeding success and is usually well correlated with early growth, which is affected by birth weight and the mother's milk yield (Gunn 1964a,b; Fraser and Morley Jones 1975; Schinkel and Short 1961; Russell 1976; Sadleir 1969; Clutton-Brock, Albon, and Guinness 1982; Gosling, Baker, and Wright 1984).

2.8   Care of Offspring after Nutritional Independence

Especially in the longer-lived vertebrates, parental care commonly extends beyond the age at which young are capable of feeding themselves. For example, in Kloss's gibbon (*Hylobates klossii*), both parents may help to defend a territory for their offspring until the latter acquire a mate and can defend it themselves (Tilson 1981), while in three-toed sloths (*Bradypus infuscatus*), mothers assist their offspring in acquiring a territory by vacating part of their own former range (Montgomery and Sunquist 1978).

In many social vertebrates, parents help to protect their young from competition with older conspecifics (Harcourt and Stewart 1987). For example, in most geese and swans, juveniles do not reach adult weight in their first year of life, and offspring usually associate with their parents for at least six months after fledging (Kear 1970). During this time, parents help to protect them from competition with conspecifics, allowing them to gain access to resources from which they would otherwise be excluded by unrelated adults (see Figure 2.9).

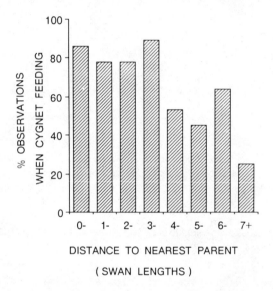

Figure 2.9   Time spent feeding by Bewick's swan cygnets at different distances to the nearest parent (from Scott 1980).

In social primates, parental protection of this kind may even include the suppression of likely competitors: for example, in several macaque species where females remain in their natal troop while males disperse after reaching adolescence, mature females attack or threaten the daughters of unrelated females more frequently than their sons (Dittus 1979; Silk 1983; Simpson and Simpson 1985). Complex social strategies are involved. When subordinate members of a matriline are attacked or threatened by unrelated females in rhesus macaques, their older relatives may retaliate by attacking a younger member of the same matriline as the original aggressor (Harcourt 1989a,b). The benefits of extended care can be large: in many social primates the reproductive performance of females is affected by their social rank, which is commonly established by parental support in social interactions early in life (Berman 1980; Horrocks and Hunte 1983a; Simpson and Simpson 1985; Harcourt 1989a,b).

2.9   Social Assistance to Mature Offspring

In some social mammals, mature daughters adopt home ranges overlapping their mother's range or remain in her social group throughout the rest of her life. This is often associated with situations where the social rank of females increases with their age, as in red deer (Thouless and Guinness 1986), or where subgroups of related females compete for resources within a larger troop, as in macaques and baboons (see Harcourt 1987). In both circumstances, daughters are likely to benefit from the presence of their mother and other relatives because this allows them to gain access to resources from which they would otherwise be excluded. For example, in vervet monkeys (*Cercopithecus aethiops*), females with mothers still present in the group have higher reproductive success than those without (see Figure 2.10).

2.10   Summary

Forms of parental care in its broad sense include the preparation of nests and territories; the production of large, heavily yolked eggs; guarding, brooding or bearing eggs and young; provisioning offspring before and after birth; and supporting them after nutritional independence. Patterns of parental care vary widely, often between closely related species.

Empirical studies of the effects of parental care show that these are often substantial. Much of the evidence of the effects of parental care is correlational rather than experimental and is open to the objection that some confounding factor independently affects the level of expenditure and the fitness of offspring. However, both experimental and correlational evidence indicate that the benefits of parental care are often large and that the level of expenditure commonly affects the reproductive performance of offspring throughout their lives as well as their survival to weaning age. As a result, estimates of the benefits of parental care based on single components of offspring fitness will tend to underestimate its real effects.

Although parental care evidently affects offspring fitness, in no single case do we yet know the form of the relationship between parental investment and the fitness of offspring or how this varies with environmental conditions. There is thus a large gap between empirical studies of parental care and models of its evolution, which mostly rely on assumptions concerning the form of relationships between parental investment and offspring fitness. As a result,

Figure 2.10   Effects of maternal presence in the same troop in vervet monkeys (from Fairbanks and McGuire 1986). Histograms compare (a) the frequency of aggression received; (b) pregnancy rate; (c) production of surviving infants by young adult females whose mothers were present or not present in the same troop.

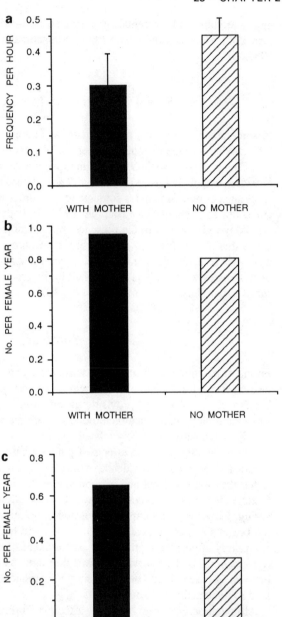

few theoretical predictions concerning the extent or distribution of parental care have a firm empirical basis. In this they differ from predictions concerning other areas of behavior, such as foraging behavior, where the consequences of variation in food choice or intake rate can be measured more reliably (see Stephens and Krebs 1987).

3

The Costs of Breeding

3.1 Introduction

Birth is painful, old age is painful, sickness is painful, death is painful. —From *The Teachings of the Compassionate Buddha*

Virtually all functional arguments concerning the evolution of parental care rely on the assumption that caring for offspring reduces the survival or reproductive success of parents. Surprisingly, some attempts to measure the costs of breeding have been unable to detect any change in the survival or reproductive success of parents (see below), and both the extent of reproductive costs and the reliability of attempts to measure them are a controversial subject (see Bell 1980; Reznick 1985; Partridge and Harvey 1988; Partridge 1989a,b).

This chapter reviews the results of attempts to measure the fitness costs of reproduction. Unfortunately, we know relatively little about the relative costs of different components of reproduction, though it is commonly assumed that the costs of breeding stem from the most energetically costly activities (in most cases, feeding young and lactation). This is a dangerous assumption, as recent work on *Drosophila* shows. Here, mating alone reduced the lifespan of females whose rate of egg production was held constant (Fowler and Partridge 1989). In addition, comparatively little is known of the relative costs of mate competition, mating, and parental care to males.

Section 3.2 briefly reviews evidence of the energetic costs of breeding, while Sections 3.3–3.7 review three different types of evidence concerning the costs of reproduction to the survival or future reproductive success of parents: (1) correlations between the reproductive output of individuals and their subsequent survival or breeding success; (2) experiments that manipulate reproductive output, either in the laboratory or in the field, and investigate effects on parental survival and subsequent breeding success; (3) negative genetic correlations between reproductive output at different stages of the life history or between reproductive output and subsequent survival. The final section synthesizes evidence of the costs of breeding derived from measures of different kinds.

3.2   Parental Expenditure

Most aspects of parental care involve substantial expenditure of energy or other resources by the parents, though the energetic costs of some aspects of parental care have received more attention than others.

EGG PRODUCTION

The energetic costs of egg production are usually substantial (Hinton 1981; Wootton 1979). For example, in reptiles, between 5% and 20% of the annual energy budget is usually spent on egg laying (Congdon, Dunham, and Tinkle 1982). In birds, the daily energetic costs of egg laying are around 29–35% BMR while daily protein requirements increase by between 86% and 230% (Robbins 1983). In some species, females increase the proportion of insects and other protein-rich foods in their diets during egg laying and may also mobilize a proportion of body protein (Robbins 1983).

EGG CARE

The costs of egg care probably vary widely. In ectotherms that do not incubate their eggs, the main costs are likely to be caused by reduction in feeding opportunities or, in males, by reduced mating opportunities. In a number of species where males guard eggs, their condition declines during the period of guarding, though they may compensate by eating a proportion of the eggs (De Martini 1987). For example, in the Puerto Rican frog, *Eleutherodactylus coqui*, males guard eggs throughout their development, their food intake declines, and their body weight falls by around 20% (Townsend 1986; see also Forester 1979; Simon 1983), while in the gastric brooding frog *Rheobatrachus* there is total inhibition of feeding and of the secretion of gastric juices during incubation (Corben, Ingram, and Tyler 1974). In some species, this period of starvation may reduce the parent's reproductive performance the following season (Hairston 1983), and, where parental care affects the parent's growth rate and subsequent size, this can have important consequences for fecundity or mating success throughout the rest of its lifespan (Gross and Sargent 1985; Berglund and Rosenqvist 1986).

In almost all endotherms, the costs of egg care include the costs of maintaining egg temperature. In birds, estimates of the amount of heat transferred to the eggs ranges from 10% to 30% BMR in passerines (King 1973), while in petrels and penguins energy expenditure on incubation represents around 30–40% BMR (Croxall 1982). The amount of heat that must be transferred to the eggs is relatively high in smaller species compared to large ones, since smaller eggs lose heat more rapidly on account of their high surface-to-volume ratio (Ricklefs 1974) and increases with clutch size (Coleman and Whittall 1988). Other factors affecting the energetic costs of incubation include the ambient temperature and the structure or location of the nest (Robbins 1983). Effects of incubation on the parent's food intake should be added to the costs of incubation itself, for the feeding time of incubating females is severely reduced in many species (Robbins 1983).

PROVISIONING YOUNG BEFORE BIRTH IN VIVIPAROUS SPECIES

In most viviparous species, parental expenditure on provisioning young before birth rises disproportionately in the later stages of gestation. Among different species of mammals, the total costs of maintaining the fetus(es) and of production increase almost isometrically with body size, though, relative to the female's size, daily expenditure declines since the relatively smaller neonates of larger species are produced over longer periods of time (K. L. Blaxter 1964; Payne and Wheeler 1968; Gordon 1989). For pregnant rodents, the increase in mean daily calorie intake over nonreproductive rates ranges from 18% to 25% (Gittleman and Thompson, in press).

PROVISIONING YOUNG AFTER HATCHING OR BIRTH

In birds and mammals, feeding young has heavy energetic costs that typically exceed those of egg production or gestation by a substantial margin (Drent and Daan 1980; Oftedal 1985). In five bird species where energy expenditure on feeding nestlings has been measured, this rises at peak demand to around 4.0 BMR (see Figure 3.1). This is similar to the cost of heavy labor in humans. In mammals, energy expenditure at peak lactation (including the use of stored energy) commonly lies between 2.5 and 5 times that for nonreproductive females, and mean calorific intake rises by up to 200%, depending partly on litter size (Randolph, Randolph, et al. 1977; Millar 1978; Gittleman and Thompson, in press). Lactating females commonly lose weight, in some cases very rapidly (see Costa, Le Boeuf, et al. 1986) and may show marked physiological changes, including hypertrophy of the liver, kidneys and digestive organs (Widdowson 1976a,b; Hanwell and Peaker 1977). The proportion of daytime spent feeding may increase by as much as 30% with an associated reduction in other activities (e.g., Altmann 1983; Clutton-Brock, Albon, and Guinness 1982). However, a variety of mechanisms exist that reduce the costs of feeding young in mammalian mothers. These include the timing of breeding seasons (Wasser and Barash 1982; Blurton–Jones, 1986), the storage of fat reserves prior to conception (Frisch 1984), reduction in resting metabolic rate (Prentice and Whitehead 1987), and a variety of changes in behavior (Brodie 1989).

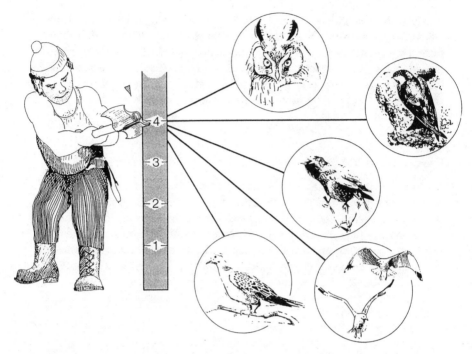

Figure 3.1   Maximum sustained working level of parents tending nestlings in five birds (long-eared owl, house martin, starling, glaucous winged gull, and turtle dove), expressed as metabolizable energy per day (DME) in multiples of BMR (basal metabolic rate) (from Drent and Daan 1980).

CARE OF OFFSPRING AFTER NUTRITIONAL INDEPENDENCE

Little attempt has been made to measure the energetic costs of caring for juveniles and subadults, but several studies indicate that they may be appreciable. In Bewick's swans, pairs with six-month-old cygnets spend less time feeding and gain weight more slowly during the winter than pairs without cygnets (Scott 1980). Similar costs are likely to be associated with care of older offspring, for the proximity of mature offspring may reduce the mother's feeding efficiency or food availability. For example, in red deer, the frequency of competition for feeding sites increases with the size of the mother's matrilineal group (Thouless 1987).

3.3   Correlations between Components of Fitness

A substantial number of studies have shown that producing gametes and rearing offspring is associated with reductions in the parent's subsequent survival or breeding success. In many animal groups, interspecific differences in fecundity are negatively related to longevity across species (Figures 3.2 and 3.3),

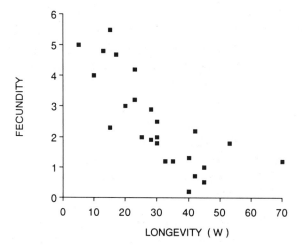

Figure 3.2   The relationship between longevity (in days) and the rate of reproduction in different species of Cladocera. Plotted points represent, as nearly as possible, midpoints of the ranges of values given by an author or a number of authors (from Bell 1984b). The measure of fecundity is the rate of egg production per day of adult life per unit adult volume.

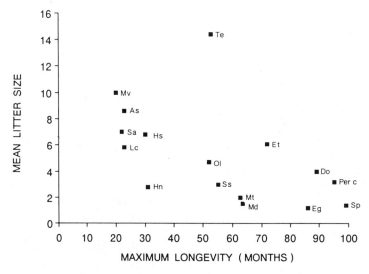

Figure 3.3   Mean litter size plotted against maximum longevity for a sample of insectivores, rodents, and marsupials. Mv = *Marmosa robinsoni*; As = *Antechinus stuarti*; Sa = *Sorex araneus*; Hs = *Hemicentetes semispinosus*; Lc = *Lagurus ordii*; Hn = *Hemicentetes nigriceps*; Ol = *Onychomys leucogaster*; Ss = *Setifer setosus*; Et = *Echinops telfairi*; Mt = *Microgale talazaci*; Md = *Microgale dobsoni*; Do = *Dipodomys ordii*; Perc = *Peromyscus californicus*; Eg = *Echinosorex gymnura*; Sp = *Solenodon paradoxus*; Te = *Tenrec ecaudatus* (from Eisenberg 1981).

and clutch size is negatively correlated to egg weight (crustacea: Allan 1984; insects: Hinton, 1981 fish: Wootton, 1979; reptiles: Congdon, Dunham, and Tinkle 1982). However, across stable populations there is an inevitable inverse correlation between fecundity and survival so that these trends constitute only weak evidence for evolutionary trade-offs (Sutherland, Grafen, and Harvey 1986; Gustaffson and Sutherland 1988).

There is also extensive evidence that parental care is associated with a reduction in subsequent survival or breeding success. In a variety of reptiles, carrying eggs affects the female's mobility (Shine 1980, 1988a; Brodie 1989). For example, in some viviparous skinks, gravid females lose about 25% of their nongravid running speed and are consequently more vulnerable to predators (Shine 1980). In many birds, mortality of females during the breeding season is higher than that of males as a result of predation during incubation (Lack 1966, 1968), while in some species of fish where males care for the eggs, they suffer higher predation rates than females (see Figure 3.4). Differences in

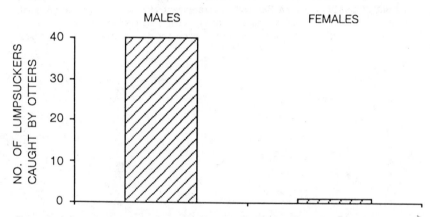

Figure 3.4 Number of male and female lumpsuckers (*Cyclopterus lumpus*) caught by otters (H. Kruuk, unpubl. data). In this species, males guard adhesive, demersal eggs in shallow water.

clutch size are sometimes related to variation in parental survival or subsequent breeding success. In house martins (*Delichon urbica*), a lower proportion of females that rear two broods return the following year compared to those that rear a single brood, though their mates are unaffected (Bryant 1979; see Figure 3.5). In willow tits (*Parus montanus*), parental survival declines with increasing fledgling production in the previous season (Figure 3.6: Ekman and Askenmo 1986). In this case, both sexes were affected and the survival of breeding males was lower than that of nonbreeders (Ekman and Askenmo 1986).

In mammals, lactation can affect subsequent fecundity by several mechanisms. In addition to the direct effects of energy expenditure on subsequent

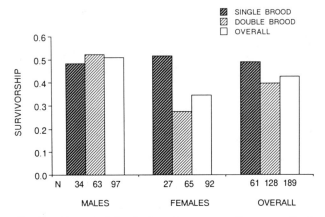

Figure 3.5   Survivorship to the following breeding season of house martins in relation to the number of broods produced (from Bryant 1978).

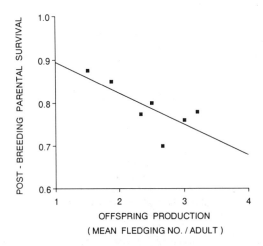

Figure 3.6   Relation between the postbreeding survival rate of willow tit parents (end of breeding to middle of September) and the average fledgling production per adult. Survival rates are based on Jolly-Seber estimates, and offspring production values are means for each year (from Ekman and Askenmo 1986).

condition and fecundity (e.g., Albon, Mitchell, and Staines 1983), lactation can increase the mother's parasite load (Festa-Bianchet 1989). Moreover, the stimulation of the mother's nipples alone generates hormonal changes that prolong the period of postlactational anestrus (Short 1976a, 1983).

   Few studies have been able to separate the costs of incubation or gestation from those of feeding the young after birth or hatching, though it is often assumed that the costs of lactation exceed those of gestation. However, in red deer, it is possible to compare the survival and fecundity of females that failed to produce calves (true yelds) with those that produced calves but lost them

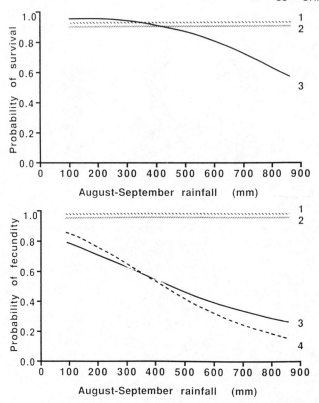

Figure 3.7   Variation in the costs of breeding in red deer (from Clutton-Brock, Albon, and Guinness 1989). (a) Differences in survival among true yeld (1), summer yeld (2), and milk hinds (3) on the Isle of Rhum; (b) differences in fecundity during the subsequent breeding season among true yeld (1), summer yeld (2), and milk hinds. In this analysis, milk hinds are subdivided into those that lost their calf during its first winter (3) and those that reared it to at least one year of age (4). In both analyses, values are plotted against summer rainfall, since this affects adult survival and fecundity: wet summers are associated with low-standing crop of grass at the onset of winter (Albon and Clutton-Brock 1988). The figure shows that costs of lactation are only apparent after wet summers, when food availability in autumn and winter is low.

shortly after birth (summer yelds) and those that reared them successfully (milk hinds). As expected, differences in survival and fecundity between milk hinds and summer yelds were substantially larger than those between summer yelds and true yelds (see Figure 3.7), indicating that the fitness costs of lactation are substantially greater than those of gestation.

Parental care after weaning or fledging can also be costly. In red-necked wallabies (*Macropus rufogriseus*), mothers that associate closely with their offspring are more likely to lose their next infant (C. N. Johnson 1985, 1987; see Figure 3.8), and the breeding success of females declines with increasing

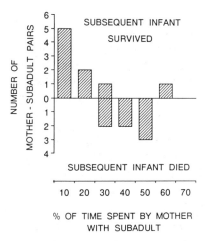

Figure 3.8 Frequency of association between female red-necked wallabies and their subadult offspring during the six months following their weaning in relation to the survival of their subsequent infants (from Johnson 1986).

matriline size in many social species (van Schaik 1983; Clutton-Brock, Albon, and Guinness 1982). Parental care may also reduce the fitness of the female's *previous* offspring, as parental expenditure on the next offspring frequently terminates investment in the previous one (e.g., Clutton-Brock, Guinness, and Albon 1983).

In males, the extension of paternal care beyond fertilization usually constrains fecundity or mating rate. For example, in the frog *Eleutherodactylus*, guarding males do not call and can rarely attract additional mates so that they lose further mating opportunities during the three-week period of development (Townsend 1986, 1989). The extent of these costs depends on the breeding system and will usually be greater in endotherms than ectotherms (see Chapter 7). In many species it probably represents the largest cost of parental care to males (Gross and Sargent 1985).

Several studies show that the costs of parental care vary with environmental conditions. For example, in red deer, differences in subsequent fecundity between milk and yeld hinds are found only when population density is high and vary with the mother's dominance rank (Clutton-Brock, Guinness, and Albon 1983, unpublished).

3.4 The Limitations of Phenotypic Correlations

Not all studies have found negative correlations between reproductive success and subsequent survival or fecundity (Bell 1980; Linden and Pape Møller 1989). Under favorable conditions, reproductive success is often either unrelated or *positively* related to subsequent survival and fecundity (Bell 1984b; Bell and Koufopanou 1986). Several field studies have even found positive associations between the reproductive success of parents and their subsequent survival or breeding success. In song sparrows (*Melospiza melodia*), J. M. Smith (1981) found that, compared to those that subsequently died, surviving females had previously shown *higher* reproductive success, and that young which recruited to the breeding population tended to come from clutches of *above* average size (see also Hogstedt 1981; Harvey, Stenning, and Campbell 1985). Of nine studies of the relationship between reproductive rate and subsequent longevity reviewed by Bell and Koufopanou, three found a significant negative association, two found no relationship, and four found a significant positive association. Of eleven studies of the relationship between egg production early in life and fecundity later in life, only one found a significant negative association, four found little or no evidence of a correlation and six found a positive association (Bell and Koufopanou 1986).

It is sometimes suggested that positive correlations between reproductive success and subsequent survival or fecundity indicate that reproduction has little or no cost (J. M. Smith 1981; Bell 1984a,b). This may occasionally be the case, for animals may minimize the costs of breeding by limiting reproduction to periods when its energetic costs can be met by increased food intake or from accumulated food reserves (Tuomi, Haka, and Haukioja 1983). However, a more likely explanation is that populations contain "good" and "bad" quality individuals and that these differences between individuals affect both their fecundity and longevity, obscuring the costs of reproduction (Reznick 1985; Reznick, Perry, and Travis 1986; van Noordwijk and de Jong 1986). For example, the successful song sparrows in Smith's sample might have had superior territories and consequently shown both high fecundity and high longevity. Alternatively, they might have been larger or healthier birds and shown superior survival and reproductive performance for this reason. It is significant that positive correlations between reproductive success and subsequent survival or fecundity often disappear in harsh environments. For example, in *Artemia*, Browne (1982) showed that a positive association between egg production and survival disappeared when food was short, while Calow (1977) found that mated female water boatmen, *Corixa*, had the same longevity as virgins when maintained under favorable conditions but had shorter lifespans when starved.

The list of confounding variables that may obscure the costs of reproduction is a long one, and their effects are not confined to field studies. Despite the most stringent attempts to maintain cultures under identical environmental conditions, environmental differences usually exist that can affect all components of reproduction. In *Drosophila*, differences of this kind can account for as much variation in viability as controlled treatments of temperature or food type (Dobzhansky and Levine 1955). Effects of parental phenotype or early environment on offspring phenotype are another common source of positive associations beween fitness components. For example, in Scottish red deer, calves born in warm springs have higher birth weights and grow faster, showing higher fecundity, superior reproductive performance, and greater longevity as adults (Albon and Clutton-Brock 1988).

Environmental factors can also induce *negative* correlations between components of reproductive success (see Bell 1984a,b; Bell and Koufopanou 1986). In cladocerans, variation in temperature creates positive correlations between age at first laying and lifespan, but negative correlations between both variables and the duration of the adult instar. In *Daphnia*, variation in fecundity falls with rising population density, at first slowly and then rapidly, while longevity initially increases and then falls, creating negative correlations between longevity and fecundity in the lower ranges of population density and positive ones at high density.

The dangers of interpreting positive correlations between fitness components as evidence that the costs of breeding are low is well illustrated by a simple economic parallel (van Noordwijk, pers. comm.). People might be expected to spend their money either on their cars or on their houses. If so, negative correlations might be expected between the value of people's houses and that of their cars. Evidence of positive correlations between the value of houses and cars does not mean that trade-offs do not occur—only that people's access to resources varies widely.

So far, few attempts to measure reproductive costs by phenotypic comparisons have attempted to control for the most obvious confounding variables (including individual differences in age, environmental quality, and body size), so it is unsurprising that many have produced equivocal and misleading results. Where attempts have been made to control for confounding factors, negative associations between reproduction and survival are usually found in natural populations (see Altmann, Altmann, and Hausfater 1978; Clutton-Brock, Guinness and Albon 1983); but even here, individual variation in phenotypic quality is likely to lead to underestimates of the real costs of breeding.

3.5   Experimental Manipulation of Parental Expenditure

The problems of phenotypic correlations underline the need for experiments that manipulate reproductive effort and observe the consequences for the parents' subsequent survival and fecundity (Partridge and Harvey 1985, 1988). One of the commonest techniques has been to manipulate reproductive effort by exposing one group of females to males and leaving another group as virgins (Bell and Koufopanou 1986). Of twenty-three studies reviewed by Bell and Koufopanou (1986), nineteen found that virgins lived longer. However, an alternative interpretation of these results is that mating itself (as against investment in gametes) is responsible for the reduction in longevity (see Dean 1981; Fowler and Partridge 1989).

Several experiments have investigated the costs of parental care as distinct from gamete production. In the tingid bug *Gargaphia*, where females normally care for eggs and nymphs until maturation, individuals that are removed from their egg masses resume egg production and produce more clutches than control females that are allowed to care for their eggs (Tallamy and Denmo 1982; Tallamy and Horton 1990). Artificial brood enlargement in natural populations of several passerine birds is associated with reductions in parental survival, subsequent fertility, juvenile growth, or juvenile survival (Askenmo 1977, 1979; Nur 1984a,b; Gustaffson and Sutherland 1988; Partridge 1989a,b). However, the reproductive variables affected and the magnitude of effects differ widely between studies. For example, brood enlargement in tree swallows (*Iridoprocne bicolor*) and house wrens (*Troglodytes aedon*) had little effect on female fitness (De Steven 1978; Finke, Milinkovich, and Thompson 1987), while in rooks (*Corvus frugilegus*), Røskaft (1985) found no effect on parental survival but a reduction in fecundity and rearing success the following year. In some cases, this is apparently because parents do not respond to enlarged brood size by increasing their work load, underlining the importance of combining manipulations of brood size with measures of changes in parental behavior.

Experimental attempts to investigate reproductive costs have their own limitations. The costs of reproduction are often delayed and so may be missed. If the environment is exceptionally favorable, reproductive costs are likely to be small or absent, while, if it is unusually harsh, they may be artificially high. In addition, they usually exclude components of reproductive cost, such as increased susceptibility to predation, which may be important. Field experiments should avoid some of these difficulties but introduce other problems. In particular, there is a danger that variables other than reproductive effort will be affected by brood manipulation. For example, artificial brood enlargement may affect the synchrony of hatching and the degree of relatedness between nestmates, and both these effects could influence the costs of parental care.

Another criticism both of phenotypic correlations and of experimental studies is that they do not necessarily reveal the pleiotropic effects of genes affecting the level of parental expenditure. Some authors have consequently argued that they are irrelevant to investigations of the evolution of life histories (Reznick 1985; Reznick, Perry, and Travis 1986). However, there are at least three reasons for adopting a more catholic position (see Winkler and Wilkinson, in press). First, current genetic variants probably represent a subset of those that have existed in the past so that constraints operating in contemporary populations need not have limited the options of past generations. Second, studies of phenotypic costs offer one way of investigating the causal sequences underlying trade-offs between life history variables. Genetic correlations rarely shed light on these and, for this reason, have not infrequently misinterpreted the significance of associations (see below). And third, interest in the evolution and adaptive significance of life histories is not limited to predicting the genetic consequences of selection on the range of genetic variation present in contemporary populations. In particular, there is considerble plasticity in most reproductive functions (Low 1978; Tuomi, Hakala, and Haukioja 1983), and questions concerning the adaptive significance of nonheritable differences constitute an important area of interest, too (see Chapters 9 and 10).

3.6 Negative Genetic Correlations between Fitness Components

Other evidence of the costs of reproduction comes from studies of heritable variation in different components of reproductive success. For example, Rose and Charlesworth (1981a) found negative genetic correlations between fecundity and subsequent survival in adult *Drosophila melanogaster*, and a number of other studies have produced similar findings.

However, *positive* genetic correlations between fecundity and longevity or between fecundity early in life and fecundity later in life are as common or commoner than negative ones (see Table 3.1). At first view, these results appear to contradict the assumption that reproductive effort depresses the parent's subsequent survival and reproductive performance, but, like phenotypic correlations, they cannot be taken at face value. In a large proportion of these studies, correlations refer to comparisons between inbred lines or between these lines and their F_1 hybrids. As a result, the deleterious effects of inbreeding may create inferior genotypes with low viability and fecundity, thus generating *positive* correlations between fitness components that would not be found in natural, outbred populations (Rose 1984a). In addition, effects of parental phenotype on offspring quality operating through gamete size or other components of investment can easily be misinterpreted as genetic correlations (see Section 2.3).

Experimental demonstrations that consistent artifical selection for one component of fitness gradually reduces average fitness at some other stage of the lifespan provide the clearest evidence of the pleiotropic effects of genes affecting reproductive success. A substantial number of selection experiments have now demonstrated negative associations between components of fitness (see Table 3.2). For example, artificial reduction in lifespan of *Drosophila melanogaster* is associated with an increase in fecundity early in life plus reduced longevity (Rose and Charlesworth 1981a,b).

Table 3.1
Genetic correlations between longevity (LY) and a component of fecundity and
between components of fecundity (from Bell and Koufopanou 1986).

Var 1	Var 2	r_G	Organism	Material	Reference
LY	FEC	>0	Goose	Small closed flock	Merritt (1962)
LY	FEC	>0	Fowl	Small closed flock	Dempster and Lowry (1952)
LY	FEC	0	Drosophila		Tantawy and Rakha (1964)
LY	FEC	>0	Mouse	Inbred lines	Roderick and Storer (1961)
LY	FEC	>0	Drosophila	Inbred and outbred	Gowen and Johnson (1946)
LY	FEC	>0	Drosophila	Inbred lines	Rose (1984a)
LY	FEC	>0	Drosophila		Temin (1966)
LY	FEC	>0	Drosophila	Outbred	Murphy, Giesel, and Manlove (1983)
LY	EFEC	0	Drosophila		Tantawy and El-Helw (1966)
LY	EFEC	>0	Drosophila	Inbreds and hybrids	Giesel (1979)
LY	EFEC	>0	Drosophila	Inbreds and hybrids	Giesel, Murphy, and Manlove (1982)
LY	EFEC	>0	Drosophila	Inbreds and hybrids	Giesel and Zettler (1980)
LY	LFEC	>0	Drosophila	Inbreds and hybrids	Giesel (1979)
LY	LFEC	0	Drosophila	Inbreds and hybrids	Giesel and Zettler (1980)
EFEC	LFEC	>0	Oncopeltus	Half-sibs from wild	Hegmann and Dingle (1982)
EFEC	LFEC	>0	Drosophila	Inbreds and hybrids	Giesel, Murphy, and Manlove (1982)
EFEC	LFEC	<0	Drosophila	Large outbred cage	Rose and Charlesworth (1981a)
EFEC	LFEC	>0	Drosophila	Inbreds and hybrids	Giesel and Zettler (1980)
EFEC	LFEC	>0	Drosophila	Inbred lines	Rose (1984a,b)
EFEC	LFEC	>0	Daphnia	Clones	Bell (1984b)
EFEC	LFEC	>0	Daphnia	Clones	Lynch (1984)

NOTE: In Tables 3.1 and 3.2 the sign given for the correlation often represents only a crude summary of a large and heterogeneous data set. "Inbreds and hybrids" means comparisons were made among inbred lines and between these lines and their F_1 hybrids.
  Abbreviations are: FEC, fecundity (as a rate, eggs per unit time); EFEC, fecundity early in life; LFEC, fecundity later in life; LY, longevity.

As measures of reproductive cost, selection experiments have some of the same limitations as phenotypic experiments. Like short-term phenotypic experiments, they will not include all the costs of reproduction, and their results are likely to vary with environmental conditions (see above). For example, in *Drosophila melanogaster*, genetic correlations change in strength and direction with changes in temperature (Giesel, Murphy, and Manlove 1982) and directional selection can also give rise to spurious positive genetic correlations (Service and Rose 1985). In addition, since directional selection can modify the pleiotropic effects of genes, the costs of heritable variation in reproductive output in contemporary populations may bear little relationship to their costs in ancestral populations.

Table 3.2

Correlated responses to selection of major components of fitness (from Bell and Koufopanou 1986).

Direct Var	Selection Response	Indirect Var	Selection Response	Gens	Organisms	Reference
EFEC	Upward	LFEC	Downward	6	Fowl	Lerner (1958)
EFEC	Upward	LFEC	Downward	5	Fowl	Erasmus (1962)
EFEC	±Upward	LY	±Upward	4	Fowl	Nordskog and Festing (1962)
EFEC	Upward	LFEC	Downward	12	Fowl	Morris (1963)
EFEC	Upward	LY	None	12	Fowl	Morris (1963)
EFEC	Upward	LY	Downward	10	Goose	Merritt (1962)
EFEC	Upward	LFEC	Downward	23	Mouse	Wallinga and Bakker (1978)
EFEC	?Upward	LY	Downward	40	*Tribolium*	Sokal (1970)
EFEC	Upward	LFEC	Downward	12	*Tribolium*	Mertz (1975)
EFEC	Upward	LY	None	12	*Tribolium*	Mertz (1975)
LFEC	Upward	EFEC	Downward		*Drosophila*	Wattiaux (1968)
EFEC	Upward	LFEC	None	3	*Drosophila*	Rose and Charlesworth (1981b)
EFEC	Upward	LY	None	3	*Drosophila*	Rose and Charlesworth (1981b)
LFEC	None	EFEC	Downward	3	*Drosophila*	Rose and Charlesworth (1981b)
LFEC	Upward	EFEC	Downward	3	*Drosophila*	Rose and Charlesworth (1981b)
EFEC	Upward	LFEC	Downward	17	*Drosophila*	Luckinbill et al. (1985)
LFEC	Upward	EFEC	Downward	17	*Drosophila*	Luckinbill et al. (1985)
LFEC	Upward	EFEC	Downward	15	*Drosophila*	Rose (1984b)
LFEC	Upward	LY	Downward	15	*Drosophila*	Rose (1984b)

NOTE: Abbreviations as in Table 3.1. "Gens" is number of generations (for domestic birds, years) for which selection was practiced.

Finally, in some species, negative correlations between genetic variation in reproductive output or survival at different stages of the life history may arise for reasons unrelated to the costs of breeding. In some invertebrates, the number of mitoses in different cell lineages is predetermined so that anatomical features commonly have a rather invariant number of cells at maturity and the number of primary germ cells may also be predetermined (Bell 1984b). For example, in the gastrotrich *Lepidodermella squammata*, females produce four or occasionally five eggs during their lives (Brunsom 1949; Sacks 1964). Where lifetime fecundity is determined in this manner, selection for early fecundity will necessarily reduce later fecundity (Bell 1984b), and negative correlations between longevity and reproductive rate are likely since the latter is usually measured as total egg production divided by lifespan less age at first breeding (Bell 1984b). Bell suggests that reproductive determinism rather than reproductive cost may account for the negative correlations between fecundity and longevity found in a number of studies of invertebrates. In future, it may be possible to distinguish between reproductive determinism and cost by examining the length of intervals between the production of successive eggs rather than the rate of egg production over the lifespan (Bell 1984b), or by manipulating resources to determine whether the effects of high initial fecundity on subsequent fecundity can be reversed later in the lifespan by supplying extra food (Bell and Koufopanou 1986). However, experiments of this type have yet to be performed.

3.7 Summary

The most obvious conclusion to be drawn from the attempts to measure the costs of reproduction is that all approaches have their limitations. Energy expended on reproduction is not necessarily related to the effects of breeding on survival or subsequent breeding success (Section 3.2). Negative relationships between phenotypic or genetic variation in reproductive output and subsequent survival or breeding success are often obscured by confounding variables associated with breeding success (Sections 3.3–3.5). And responses to experimental manipulation of phenotypic or genetic variation in reproduction may be difficult to interpret and can be strongly influenced by the environment in which they are carried out (see Section 3.6). Virtually all attempts to measure reproductive costs ignore many aspects of the parents fitness that are likely to be affected by previous breeding attempts, including the viability of subsequent offspring. And almost all studies have concentrated on finding out whether or not reproductive costs exist rather than on determining the form of relationships between reproductive output and the parents' subsequent survival or breeding success.

Studies of the evolution of parental care thus face a fundamental problem. They depend on assumptions about the magnitude and form of cost and benefit functions, yet it has so far proved impossible to measure the complete array of costs or benefits of care in any animal species or to determine the form of cost or benefit functions. Similar difficulties affect the measurement of the benefits of parental care (see Section 2.10).

These difficulties underline the need for further investigation of aspects of parental care about which little is yet known, including the extent to which reproductive costs are affected by environmental differences and the form of cost and benefit functions. In addition, the frequency with which correlations between fitness components have been misinterpreted emphasizes the need to understand their causal basis. Field experiments that investigate the phenotypic consequences of changing reproductive effort provide the most reliable evidence here, but, where experiments are not feasible, sensible phenotypic comparisons also have a role to play.

The problems of measuring the costs and benefits of parental care necessarily limit attempts to predict the form and duration of parental care (Chapters 9 and 10). This should not be allowed to discourage research, for many differences in costs and benefits are so large that predictions based on them do not require detailed measurement and some predictions are so general that they can be tested by comparisons of parental expenditure. At the moment, it is on these relatively robust predictions that empirical studies need to focus.

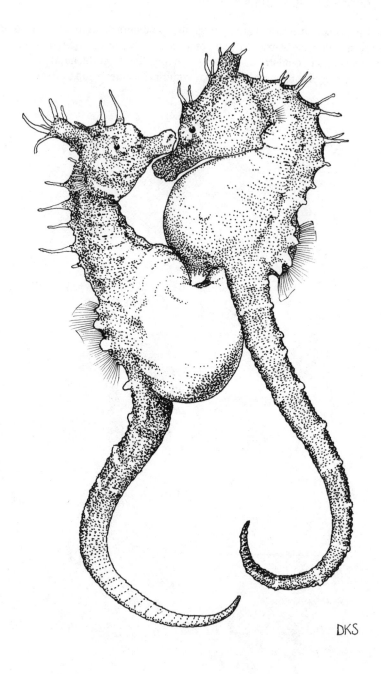

4

Propagule Size

4.1 Introduction

Instead of caring for their young after hatching, the great majority of animals provide the resources necessary for their offspring during the first hours, days, or weeks of life by producing large eggs. However, increases in the size of eggs or young reduce the number that the parent can produce, and species that have relatively large eggs or neonates typically produce small clutches or litters (insects: Hinton 1981; fish: Wootton 1979; amphibia: Salthe 1969; birds: Lack 1968).

Three main questions have been asked about egg or neonate size. First, how large should eggs be? Second, how much time should organisms spend in the egg stage and how long in subsequent stages of development? And third, should there be a single optimal egg size or an optimal range of variability? Sections 4.2 and 4.3 discuss theoretical predictions concerning the first two questions. Section 4.4 examines evidence of constraints on egg size, while Sections 4.5 and 4.6 review empirical trends in egg size. Section 4.7 examines whether or not selection is likely to favor a single optimal propagule size or an optimal range of variation.

4.2 How Large Should Propagules Be?

The basis for most subsequent attempts to investigate the optimal size of propagules was laid by Smith and Fretwell's (1974) model (see also Ware 1975; Lloyd 1987; Winkler and Wallin 1987; Clutton-Brock and Godfray, in press). Their argument assumes that parents produce a substantial number of eggs and have a fixed amount of energy that can be divided between variable numbers of offspring; that the way in which available resources are divided among offspring does not affect the fitness of the entire clutch or the parents' residual reproductive value; and that offspring fitness initially rises slowly with increasing parental expenditure, then rises rapidly before reaching a point at which increased parental expenditure has diminishing returns (see Figure 4.1a).

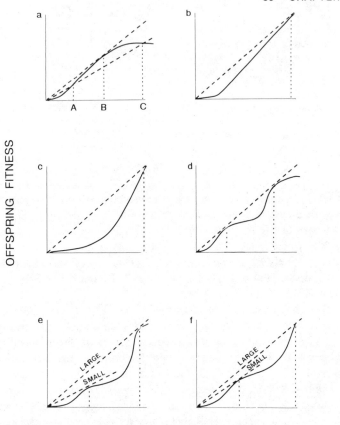

PARENTAL EXPENDITURE

Figure 4.1 Hypothetical relationships between the fitness of individual offspring and parental expenditure. The situation considered by C. C. Smith and Fretwell (1974) is shown in (a). In (b), once an initial threshold has been passed, the fitness of young increases in a linear fashion, while in (c) the increase is exponential. In (d) the fitness function is complex and two different optima are present. (e) and (f) consider optimal expenditure by large and small parents, where it is assumed that small parents can only achieve 75% of the reproductive expenditure of large parents. If offspring fitness is a complex function of parental expenditure, these differences can lead to different optima for large and small parents.

Optimal egg size under these conditions can be derived graphically (Figure 4.1a). The optimum is defined by the point at which a straight line from the origin is tangent to the offspring fitness curve. For example, in Figure 4.1a both the points at which the lower fitness function intersects the curve (A,C) will give lower values of parental fitness than the intermediate point (B).

If the fitness of offspring increases in a linear or exponential fashion with rising parental expenditure (Figure 4.1b,c), a line drawn through the origin will always meet the curve at its maximum value, and selection will favor the production of a single, large offspring. Where parents produce a single offspring at a time, trade-offs between propagule size and the parent's own survival will determine the optimal size of neonates (Stearns 1976; Winkler and Wallin 1987; Partridge and Harvey 1988), usually generating positive correlations between parental size and propagule size.

Relationships between parental expenditure and offspring fitness need not be as simple as those shown in Figure 4.1a-c, and multiple optima may occur, favoring multimodal distributions of egg or offspring size (Figure 4.1d). Such effects can exaggerate differences in offspring size among parents because some parents are unable to "reach" higher optima. For example, in Figure 4.1e and f, "small" parents can only expend three-quarters of the expenditure that "large" parents can achieve. As Chapter 13 describes, effects of this kind may generate individual differences in offspring sex ratios in some dimorphic species (see Chapter 12).

Subsequent size versus number models have extended Smith and Fretwell's predictions. Where optimal propagule size varies between environments, conditional strategies for varying propagule size in relation to parental environment might be expected (Lloyd 1984; Haig 1989). In multiparous organisms, optimal propagule size may change if offspring differ in quality (Temme 1986); if parents vary in their ability to care for their eggs (Sargent, Taylor, and Gross 1987); or if resource levels vary but propagule number is predetermined (Lloyd 1987; McGinley and Charnov 1988; Silvertown 1989). In at least some models, reduced juvenile survival can also select for increased egg size (Kolding and Fenchel 1981; Sibly and Calow 1986). For example, if egg number n declines exponentially with egg size z while juvenile survival S_j increases but reaches an asymptote at some value of egg size, reduced juvenile survival will favor an increase in egg size (see Figure 4.2).

Both the harshness of the environment and the intensity and form of competition may have important effects on propagule size. In particular, interference or contest competition for resources or mates may generate nonlinear relationships between the fitness of offspring and their priority of access to resources, raising the benefits of increased parental expenditure per offspring (see Miller 1969; Gill 1974). For example, if the resources for which juveniles are competing are defendable and normally distributed in size or quality, the relationship between offspring quality and fitness is likely to be nonlinear and increased parental expenditure may be favored (Brockelman 1975).

In some situations, egg size will not be independent of selection pressures affecting clutch size. For example, Parker and Begon (1986) show that if competition among siblings increases with brood size, depressing the average fit-

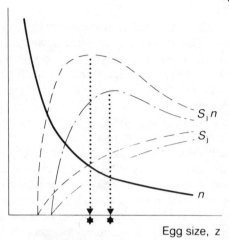

Figure 4.2 Effects of egg size on fecundity and juvenile survival (S_j) where increasing egg size reduces time spent in the juvenile stage. n = the number of surviving progeny, and S_{jn} is consequently a measure of parental fitness. If S_j is reduced, natural selection should favor bigger eggs. Asterisks show the optimal egg sizes for different values of S_j (from Kolding and Fenchel 1981).

ness of larvae in big broods, smaller clutches of larger eggs are likely to be favored (Table 4.1). This is because selection favors an optimal clutch size, and variation in the resources available for reproduction is consequently channeled to increasing egg size. In extreme cases where offspring fitness is almost completely determined by clutch size and only slightly affected by egg size, clutch size should be fixed and egg size should vary widely (Clutton-Brock and Godfray, in press). In practice, both clutch and egg size are likely to affect offspring fitness, and both will be affected by changes in total reproductive effort. As a result, clutch size and egg size are likely to be positively correlated with each other. Parker and Begon (1986) show that both sib and nonsib competition favor constancy of clutch size for all females, so that larger females, who can afford to expend more energy on reproduction, produce larger eggs. However, if competition is hierarchical, intensifying the benefits of large egg size, it will tend to favor constant egg sizes for all females and positive relationships between female size and clutch size. A subsequent model suggests that, if the resources available for reproduction are fixed at the outset and there is a significant mortality risk to the parent between clutches, larger egg and clutch sizes are likely to be favored early in the life span (Begon and Parker 1986). Trends of this kind have been documented in a number of invertebrates (e.g., Karlsson and Wiklund 1984, 1985).

In other situations, large clutch size may *enhance* the fitness of offspring— for example, because predators are satiated or the juveniles create a favorable microclimate. Relationships of this kind, too, are likely to lead to an optimal

Table 4.1
Parker and Begon's (1986) model of optimal egg and clutch size.

If larval fitness depends only on absolute egg size:
Parker and Begon's initial model considers the reproductive options of an insect where the female forages for resources which she converts into eggs, subsequently leaving her feeding habitat to find an oviposition site where she lays a single clutch of eggs before returning to the feeding site to collect resources for the next clutch. Females can alter two variables: the amount of time spent collecting resources (t) and the amount of resources spent on each egg (m). If a total amount of gametic resources M is gathered in time t, decisions as to what size of eggs to lay also determine clutch size, $M(t)/m$. d is the average time taken to find an oviposition site, oviposit and then find another feeding site and, if oviposition duration is short, corresponds to the egg-independent time cost of Parker and Courtney (1984). Maternal fitness (w) is equivalent to the gain per unit of time and is dependent on the mother's "decisions" about m and t: $w(m, t) = M(t)f(m)/m(t + d)$. To find the optimal foraging time, t_*, set $\alpha w(m,t)/\alpha t = 0$, giving

$$M'(t_*) = M(t_*)/(t_* + d); \tag{1}$$

and to find optimal egg size, m_*, set $\alpha w(m,t)/\alpha m = 0$, giving

$$f'(m_*) = f(m_*)/m_*. \tag{2}$$

The conclusions of this model are similar to those derived from C. C. Smith and Fretwell's (1974) model: egg size and foraging time will be independent of each other; larger females should lay larger clutches and have longer interclutch times.

If larval success is affected by competition among siblings:
If larval fitness is depressed by competition between siblings and the latter is positively related to clutch size, selection pressures on egg size will be affected by clutch size. If $s(n)$ is the fitness component of larvae that is depressed by clutch size n and its decline is monotonic, female fitness will be $w(m,t) = [ns(n)f(m)]/(t + d)$. Setting $\alpha w/\alpha t = 0$ gives the optimization for t_* (optimal foraging time or interclutch interval),

$$M'(t_*) = \frac{M(t_*)}{t_*+d}\left[\frac{s(n)}{s(n)+ns'(n)}\right], \tag{3}$$

in which $n = M(t_*)/m$.
Equation (3) should be compared directly with (1). Setting $\alpha w/\alpha m = 0$ gives the optimization for M_*

$$f'(m_*) = \frac{f(m_*)}{m_*}\left[\frac{s(n)+ns'(n)}{s(n)}\right], \tag{4}$$

in which $n = M(t)/m_*$.

NOTES: Equation (4) should be compared with (2). If both t and m are to be optimized, then (3) and (4) must be satisfied simultaneously with $n = M(t_*)/m_*$.
Predictions now differ from those of Smith and Fretwell's model. m_* and t_* are interdependent through their effects on clutch size. Sib competition will reduce foraging time and increase egg size. This is because it becomes optimal to spend less time gathering resources and to lay larger and fewer eggs where sib competition reduces offspring fitness. Big females would be expected to lay bigger eggs and bigger clutches.

clutch size and to positive correlations between maternal size and egg size. A model by McGinley (1989) envisages a species with a prereproductive life history in two stages: an initial stage where the probability of survival is independent of offspring size but increases with clutch size; and a subsequent stage where offspring fitness depends on egg size. Clutch size is assumed to affect offspring fitness in two ways: first, by influencing the survival of neonates immediately after hatching and; second, by affecting offspring size. Positive relationships between clutch size and offspring fitness (like the negative ones modeled by Begon and Parker) might be expected to favor constancy of clutch size so that larger females produce bigger eggs (McGinley 1989). However, where the eggs of several females are laid close together and hatch synchronously, the benefits of increased clutch size may disappear. McGinley suggests that the early life history of sea turtles may meet the assumptions of his model; if so, correlations between maternal size and egg size might be expected within solitary-nesting or asynchronous species but not in synchronous breeders showing high local densities.

Most models of optimal propagule size assume that the way in which parental resources are distributed within clutches or broods does not affect the average fitness of the entire clutch or the parent's subsequent survival or breeding success. In some situations, these assumptions, too, will be invalid. For example, if competitive interactions between nestlings attract predators but are minimized by differences in size and dominance between siblings (see Slagsvold, Sandvik, et al. 1984; Magrath 1988), selection may favor parents that produce young (or eggs) of different sizes within the same clutch (Chapter 9). Similarly, in species where offspring assist their parents' subsequent breeding attempts or compete with subsequent (or previous) sibs (see Chapter 13), the form of these interactions may affect the optimal distribution of parental expenditure among progeny.

4.3 How Much Time Should Be Spent in the Egg?

Section 4.2 considered trade-offs between egg size and clutch size. However, egg size may also be related to the amount of time spent in the next stage of development. For example, large eggs of the California newt, *Taricha torosa*, produce larvae that metamorphose at a younger age (see Figure 4.3).

Associations between egg size and the development of juveniles also occur across species. Among marine benthic invertebrates, species whose larvae feed little during their planktonic stage produce larger eggs and smaller clutches compared to species whose young depend on planktonic food (Vance 1973a,b). The large eggs of precocial birds produce downy young that can walk soon after hatching and are able to survive protracted periods without food (Lack 1968; O'Connor 1984), while mammals that produce precocial

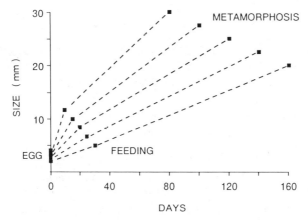

Figure 4.3   A schematic representation of the effects of egg size on size and time to first feeding and metamorphosis for embryos and larvae of the California newt, *Taricha torosa*. Each dashed line represents the developmental path of a single embryo (from Kaplan and Cooper 1984). Big eggs produce offspring that metamorphose earlier.

young show relatively long gestation periods for their body size (May and Rubenstein 1984; Harvey, Read, and Promislow 1989).

All other things being equal, natural selection might be expected to maximize the amount of time that developing organisms spend in the safest stages of development and minimize time spent in those where the instantaneous rate of mortality is highest (Williams 1966b; Shine 1978, 1989a). For example, where the instantaneous rate of mortality of eggs is low while that of juveniles is high, selection should favor prolonged development in the egg and an increase in egg size to accommodate this. Conversely, where mortality of eggs is high and mortality of juveniles is low, selection should minimize time spent in the egg and maximize the proportion of development time spent as a juvenile.

Though a number of other models have considered the trade-off between time spent in different stages of development (Vance 1973a,b; Christiansen and Fenchel 1979; Sibly and Calow 1986), the best known is a model by Shine (1978), often referred to as the "safe harbor" hypothesis. Shine assumes (1) that the overall survival of progeny from fertilization to age at reproduction depends on the instantaneous rates of egg and juvenile survival and the amount of time spent in each stage of development; (2) that larger eggs take longer to hatch but that increased time in the egg stage is subtracted from the total time necessary for development, thus reducing time spent as a juvenile; and (3) that the instantaneous rate of juvenile survival is unrelated to egg size so that offspring hatching from larger eggs show higher survival only because they spend less time in the juvenile stage. His model predicts that where the instantaneous rate of mortality of eggs is lower than that of juveniles, parents should minimize time spent in the juvenile stage by increasing egg size. Conversely, where the mortality rate of eggs is higher than that of juveniles, selection should minimize time spent in the egg stage by favoring small eggs with short incubation periods. A small shift in instantaneous egg mortality should either have no effect on optimal egg size or should shift mortality from one extreme to the other.

The assumptions of Shine's original model are simplistic and several of its predictions are not fulfilled. In particular, it is clearly unrealistic to assume that egg size does not affect juvenile survival (Chapter 2). A subsequent model by Sargent, Taylor, and Gross (1987) extends Shine's model by incorporating a *positive* relationship between egg size and the instantaneous rate of juvenile survival (see also Shine 1989a). This has the effect that the instantaneous rate of egg survival does not have to exceed the rate of juvenile survival for increases in egg size to be favored and a continuous distribution of egg size is predicted. The longer development periods associated with larger eggs impose a survival cost over the egg stage, but this can be compensated by the improved survival of larger juveniles.

A more fundamental objection to the predictions of the "safe harbor" hypothesis is that the lengths of different developmental stages appear to be regulated in different ways so that negative correlations between the length of successive stages do not always occur (Strathmann 1977; e.g., Stearns and Koella 1986). For example, the duration of some larval stages in some species is determined by fixed time limits rather than by larval size, so that there is no relationship between the size at which larvae enter a stage and its duration (e.g., Bakker 1959). In other cases, larvae hatching from larger eggs take longer to develop than smaller ones (Sargent, Taylor, and Gross 1987; Sinervo 1990), so that the duration of successive stages is positively, not negatively, correlated.

Similar differences in relationships between successive stages of development occur among species. For example, while egg size in some groups of marine invertebrates is associated with increased independence from external food sources, in others (including thoracican barnacles and pagurid hermit crabs) there is no consistent association between egg size and the time spent by larvae feeding in the plankton (Underwood 1974; Strathmann 1976). The reasons for these contrasting relationships are poorly understood (Strathmann 1976).

Among vertebrates, too, large egg or neonate size is sometimes associated with a reduction in the length of the juvenile stage while, in others, it is associated with *longer* periods of juvenile development (see Blaxter 1969; Bagenal 1971; Ware 1975; Steele 1977; Eisenberg 1981; O'Connor 1984; Nussbaum 1985; Sinervo 1990). For example, offspring of mammals that produce large neonates relative to the mother's size usually show relatively *slow* rates of development and relatively *long* periods of infant dependence compared to the offspring of species that produce relatively small neonates (Harvey and Clutton-Brock 1986; Ross 1988; Harvey, Read and Promislow 1989; Read and Harvey 1989).

These examples suggest that many different trade-offs and constraints probably affect the amount of time that organisms spend in particular stages of development (see Chapter 5). As a result, increasing egg size can probably have a variety of different effects on the duration of subsequent growth stages. However, as Shine (pers. comm.) points out, we should still expect selection to minimize the relative amount of time spent in dangerous stages of development within these constraints.

4.4   Constraints on Propagule Size

As described in Section 4.2, both positive and negative relationships between clutch size and offspring fitness may favor correlations between maternal size and egg size (Parker and Begon 1986; McGinley 1989). Positive correlations between egg size and female size are found in a variety of species of ectotherms, including fish (Figure 2.2d) and frogs (Crump and Kaplan 1979), as well as in birds and mammals (Clutton-Brock 1988), apparently confirming the predictions of these models.

Correlations between maternal body size and egg or neonate size may also be a consequence of developmental or morphological constraints (Maynard Smith, Burian, et al. 1985; McGinley 1989). For example, Congdon and Gibbons (1987) suggest that the size of the pelvic girdle limits maximum egg size in some freshwater turtles. Similar constraints on birth weight have been suggested in mammals: for example, in marmosets, Leutenegger argues that the size of the pelvic girdle constrains birth weight, requiring these species to bear twins instead of a single, larger neonate like other anthropoids (Leutenegger 1980). One difficulty faced by these arguments is that they need to explain why the size of pelvic girdles cannot get bigger if selection favors increased egg or neonate size.

Developmental or anatomical constraints are often invoked to account for the pervasive relationship between interspecific differences in egg size and body size (e.g., Leutenegger 1973; Western and Ssemakula 1982; Millar 1984). In many animal groups, egg and neonate size are allometrically related to body weight, usually increasing at around weight$^{.66-.75}$, the approximate scaling factor for metabolic rate (Table 4.2). The effect of these allometries on the energetics of parental expenditure is considerable: for example, a 10 g passerine bird typically produces an egg that is around 12% of her body weight, whereas a 100 g passerine produces an egg around 4% of her body weight. Similarly, in ungulates, a 50 kg female typically produces a neonate that weighs between 8% and 12% of her own weight (though there is wide

Table 4.2

Regression equations for interspecific relationships between the weight of single eggs or neonates on maternal weights and the proportion of variation in egg/neonate weight accounted for by maternal body weight (from Blueweiss, Fox, et al. 1978).

Taxonomic Group	Units	Equation		r^2
Crustacea	mm³/egg	$y = 0.01$	$W^{0.24}$	0.35
Fish	mm³/egg	$y = 0.59$	$W^{0.43}$	0.26
Fish and Crustacea combined	mm³/egg	$y = 0.06$	$W^{0.77}$	0.82
Reptiles	g/egg	$y = 0.41$	$W^{0.42}$	0.70
Birds	g/egg	$y = 0.26$	$W^{0.77}$	0.83
Birds	g/hatchling	$y = 0.28$	$W^{0.69}$	0.86
Mammals	g/neonate	$y = 0.097$	$W^{0.92*}$	0.94

* Lower exponents, in the range of 0.65–0.85 are usual within mammalian orders or families (Western 1979; Gittleman 1986; Kovacs and Lavigne 1986). Moreover, in mammals, combining altricial (mostly small, producing relatively small neonates) and precocial species (mostly large, producing relatively large neonates) artificially inflates the exponent, which otherwise lies around 0.8 (Martin and MacLarnon 1985).

variation in relative birth weight at this end of the scale), while species over 1,000 kg usually produce neonates that are around 4–5% of the mother's weight (Robbins and Robbins 1979). On account of the negative allometry of neonatal weight and litter weight, the proportion of a mother's daily energy budget expended on reproduction declines with increasing body size (see Chapter 6).

However, both egg weight and birth weight vary widely among species of similar body weight, indicating that, if body size does constrain the size of propagules, it does not do so very closely (Harvey and Read 1988; Harvey, Read, and Promislow 1989). Though it is sometimes suggested that other size-related variables, such as metabolic rate or fetal brain growth, may account for these discrepancies (Martin 1981; Hofman 1983), there is little evidence that this is the case (see Chapter 6).

In practice, the importance of body size constraints on egg and neonate size almost certainly varies widely. At one extreme lie cases where body size and egg size are correlated because some simple anatomical property of the mother directly constrains egg size and cannot be altered by selection pressures favoring increased egg size—as, perhaps, in the case of Congdon's turtles (see above). At the other extreme, are cases where egg size and body size covary because both represent independent adaptations to the species' environment—for example, low temperatures may independently favor large egg size, and large adult size or high adult mortality may independently favor low birth weights and small adult size (Harvey, Read, and Promislow 1989). In between lies a substantial area where both constraints and adaptive mechanisms have to be invoked to account for relationships between body size and egg size. In such cases, it is meaningless to juxtapose adaptive and nonadaptive explanations for both adaptation and constraint are necessarily involved.

One implication of the pervasive relationship between species differences in egg and body size is that the influence of environmental factors on egg size may only be apparent if the effects of body size are controlled (see Lack 1968; Harvey, Read, and Promislow 1989). Unfortunately, this is seldom simple (Harvey and Mace 1982; Clutton-Brock and Harvey 1984), for the slope and intercept of relationships between maternal weight and egg or neonate weight, as well as the proportion of variance explained by maternal weight, vary widely between groups (see Figure 4.4). Exponents of egg weight on maternal

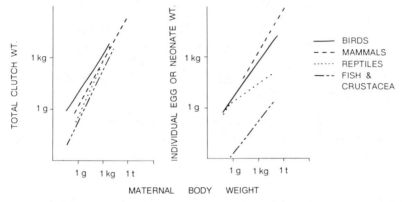

Figure 4.4 Comparison of regression lines of clutch weight on maternal weight and of egg or neonate weight on maternal body weight for birds, mammals, reptiles, and fish and Crustacea (from Blueweiss, Fox, et al. 1978).

weight are generally higher for endotherms than ectotherms and account for a larger proportion of variance in egg or neonate size (see Table 4.2). In addition, there is a widespread tendency for exponents to be higher when calculated across phylogenetically diverse groups of animals than when calculated across more closely related groups of species, because intercepts commonly vary between high-level taxonomic groups (Clutton-Brock and Harvey 1984). For example, fish are generally larger than crustaceans and typically have larger eggs for a given body weight (Figure 4.5). If the two groups are combined and a single regression coefficient is calculated, this is higher than the regression coefficient calculated for either group separately (see Table 4.2).

Parallel effects commonly occur at lower taxonomic levels. For example, if litter weight at birth is plotted on maternal weight for altricial and precocial mammals combined, a higher exponent is found than if these two groups of mammals are separated (Martin 1984). Similarly, exponents for neonatal

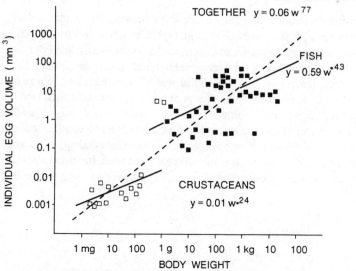

Figure 4.5 Individual egg volume plotted on maternal weight for fish, crustaceans, and for both groups combined (from Blueweiss, Fox, et al. 1978).

weight on maternal weight calculated within mammalian genera (see Table 4.3) are typically lower than exponents calculated for all mammals (Table 4.2). Although species clearly differ in the relative size of eggs and neonates, the complexities of controlling for size effects mean that all estimates of relative egg size are necessarily imprecise.

A variety of other biological factors apart from body size may constrain the size of eggs or neonates in particular cases (Maynard Smith, Burian, et al. 1985). For example, bears (Ursidae) give birth to relatively small litters of tiny, immature neonates, and litter weight is between a third and a tenth of that predicted for female mammals of comparative size (Ramsay and Dunbrack 1986). This may reflect physiological constraints on the ability of pregnant

Table 4.3
Allometric relationships between neonatal weight and adult body weight across species within genera of mammals (from Martin and MacLarnon 1985).

Species	n	Exponent	r
Domestic dog breeds	8	0.56	0.97
Felis spp	8	0.73	0.92
Panthera spp	6	0.50	0.84
Peromyscus spp	11	0.61	0.87
Cercopithecus spp	7	0.61	0.79
Macaca spp	7	0.35	0.86

females to meet the requirements of fetal metabolism while hibernating without access to food or water. Though all postnatal mammals can rely on body fat to supply energy during fasting, the mammalian fetus cannot catabolize significant quantities of free fatty acids and requires glucose as the principal oxidative substrate (Ramsay and Dunbrack 1986). A fasting female could continue to supply glucose only by using her body protein, which would eventually jeopardize her own survival. By shortening the period of gestation and giving birth to very small neonates, bears shift from transplacental to mammary nourishment of the young and can incorporate lipids into milk, thus sparing their own body protein.

4.5   Variation in Propagule Size

In many species, egg size increases and clutch size declines as environmental conditions deteriorate (Bagenal 1969; Capinera and Barbosa 1977; Fraser 1980). For example, in the terrestrial isopod, *Armadillium vulgare*, females produce few, large offspring when food availability is low, and more, smaller offspring when it is high (Brody and Lawlor 1984). Winter-breeding *Gammarus* produce fewer and larger eggs than summer-breeding ones (Kolding and Fenchel 1981). In several species of fish, egg size increases in populations where competition is intense (Blaxter 1969; Reznick and Endler 1982; Stearns 1983a) or conditions are unfavorable. For example, the Atlantic herring, *Clupea harengus*, produce few, large eggs at the beginning and end of the spawning season, when conditions are relatively unfavorable, and a larger number of smaller eggs during the peak of spawning (see Figure 4.6).

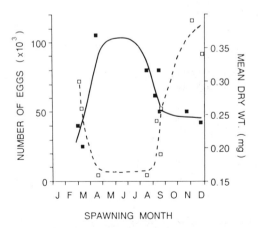

Figure 4.6   Fecundity (solid squares) and mean dry weight of ripe herring (*Clupea harengus*) eggs (open squares) in relation to spawning season (redrawn from Blaxter 1969). The vertical axis shows egg number in thousands.

Interspecific comparisons also support an association between propagule size and the severity of the environment. In several groups of invertebrates, egg size apparently increases in species living on resources that are unpredictable or difficult to exploit. Among marine benthic invertebrates, the proportion of species producing relatively large eggs and larvae that feed little on external food sources increases with latitude and depth (Thorson 1950; Vance 1973a,b). In parasitic copepods, egg size increases in species living in conditions where intraspecific competition is intense (Gotto 1962). Similarly, parasitic wasps that lay their eggs on their hosts produce fewer, relatively larger eggs than those that lay in their hosts, possibly because the larvae of exoparasites need to be larger to pierce the host's integument (Price 1973).

In fish, species breeding in fresh water, where environmental conditions are more variable and predators and parasites are more abundant, produce relatively larger eggs than those breeding in the sea (see Figure 4.7). Within sev-

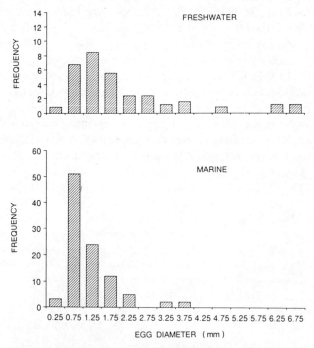

Figure 4.7 Frequency distribution of egg diameters of 101 marine fish species and 33 freshwater species from European waters (from Wootton 1979).

eral families of marine, demersal-spawning fish, the most northern species tend to have the largest eggs and the smallest clutch sizes, possibly because interspecific competition is most intense in the north (Rass 1942; Svardson

1949; Thresher 1988). Among salamanders, large egg size is associated with environments where hatchlings feed on relatively large-sized prey (Salthe 1969; Nussbaum 1985), an explanation that may also apply to some fish (Nikolsky 1963). In turtles, terrestrial species lay fewer and larger eggs than aquatic or marine species (Elgar and Heaphy, in press).

Among birds, ground-nesting species produce relatively larger eggs than species nesting on trees or cliffs (see Nice 1962; Lack 1968). For example, the brown kiwi, *Apteryx australis*, produces a 400 g egg, representing more than 14% of the female's body weight. Kiwi chicks are born fully feathered and usually receive no food from their parents. Similarly, wildfowl breeding in the high Arctic, where chicks commonly face adverse condition, produce relatively large eggs compared to those breeding in temperate areas. In several groups of mammals the relative weight of neonates is associated with the environmental conditions faced by the young after birth (May and Rubenstein 1984). For example, in phocid seals that breed on fast ice, the birth weight of pups is relatively low in species that shelter their young in caves or fissures (Kovacs and Lavigne 1986).

The predictability of the environment experienced by the young may also be important. Birds whose offspring are fed at long and irregular intervals, like many tropical seabirds, produce large, yolky eggs and juveniles that are able to survive for relatively long periods without food (Lack 1968; Boersma 1982). Environmental factors affecting the parent's access to resources, too, are sometimes associated with variation in optimal propagule size. In some groups of mammals, interspecific differences in diet quality are associated with variation in the relative size of neonates: folivorous rodents and primates tend to produce relatively smaller neonates than specialized fruit or seed eaters (Mace 1979; Harvey and Clutton-Brock 1986), and omnivorous canids produce heavier young than specialized carnivores (Gittleman 1986).

Across a variety of mammals, neonatal size relative to maternal size declines in species where life expectancy is relatively low (Harvey, Promislow, and Read 1989; Harvey, Read, and Promislow 1989). One explanation of this trend is that heavy birth weights require longer gestation periods, and these are selected against where the chance is relatively high that the mother will die before her offspring reach independence. If so, low birth weight relative to the mother's size should be expected in species where the mortality of pregnant females is unusually high.

Though relative egg size is often correlated with environmental factors, it is often difficult to identify the adaptive significance of these relationships. For example, interspecific variation in the resources available to juveniles is almost inevitably correlated with many other ecological factors—including climatic predictability, rates of predation and parasitism, and the intensity of inter- and intraspecific competition. As a result, it is usually hard to tell which is responsible for variation in egg size.

Lastly, it is important to remember that not all differences in egg size need reflect variation in parental expenditure, for egg quality varies widely. In birds, the percentage of weight made up by yolk ranges from around 15% in the gannets and boobies to over 60% in the kiwi (Carey, Rahn, and Parisi 1980; O'Connor 1984). In general, the largest yolks are found in the eggs of precocial species as well as in some seabirds where chicks are fed at long and irregular intervals (see Figure 4.8). In fish, there is an inverse relationship between

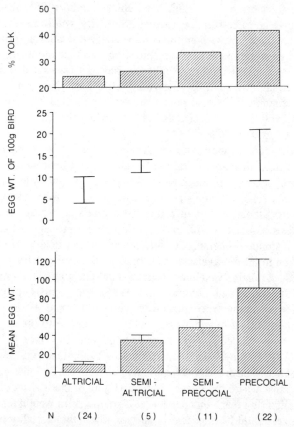

Figure 4.8   Egg weight, relative egg weight (for a 100 g bird), and percentage yolk for different bird species allocated to four categories of development (from O'Connor 1984; data from Carey, Rahn, and Parisi 1980; Rahn, Pagnelli, and Ar 1975). Altricial: passerines; semialtricial: herons, hawks, and owls; semiprecocial: gulls and terns; precocial: grouse, quail, ducks, waders, and metapodes.

yolk size and the amount of parental care after hatching (Oppenheimer 1970). The relative size of the yolk also varies within species (Goulden, Henry, and Berrigan 1987) so that, in some cases, egg volume is not consistently related to total energy content (McEdward and Coulter 1987).

4.6   Egg Size and Parental Care

In a wide variety of oviparous ectotherms, egg size increases in species where adults care for their eggs (see Shine 1978; Thresher 1984; Grahame and Branch 1985; Nussbaum 1985; Sargent, Taylor, and Gross 1987). For example, in at least four families of oviparous fish (Cichlidae, Percidae, Salmonidae, and Centrarchidae), egg size increases in relation to the quality or duration of parental care (see Figure 4.9; see also Barlow 1981; Thresher 1984). In

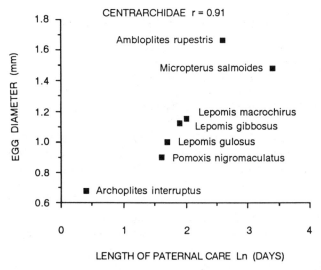

Figure 4.9   Egg size plotted on the duration of parental care in sun fish (Centrarchidae) (from Sargent, Taylor, and Gross 1987).

amphibia, too, egg size is smallest in species that distribute their eggs widely in water, larger in species that deposit eggs in a restricted area beneath stones in running water, and largest in those that lay on land, most of which guard their eggs (Salthe 1969).

The "safe harbor" hypothesis has commonly been invoked to explain this trend (Shine 1978; Sargent, Taylor, and Gross 1987): parental care of eggs presumably reduces their mortality relative to that of juveniles, lowering the costs of the longer developmental periods associated with large eggs. However, the causal sequence can be reversed: instead of egg care permitting the development of large egg size, large egg size may select for the evolution of egg care to alleviate the costs arising from the longer developmental period of big eggs (Nussbaum 1985, 1987). Except in cases where it is possible to determine whether parental care preceded large egg size or vice versa, it is difficult to distinguish between these two pathways. In practice, both are likely to operate, and parental care and egg size can be expected to coevolve regardless of which one is changed initially (Nussbaum and Schultz 1989; Shine 1989a).

In addition, associations between egg size and parental care could also arise because some ecological factor independently favors egg care and increased egg size (see Ito 1980; Ito and Iwasa 1981). For example, the introduction of a predator that reduces egg survival (favoring egg care) but could not eat larger juveniles (favoring increased egg size) could generate positive correlations between parental care and egg size (Shine 1989a). At the moment, firm inter- pretations of this trend are premature.

4.7 Is There a Single Optimal Propagule Size or an Optimal Range of Variation?

Several recent papers have suggested that in heterogeneous or unpredictable environments, selection will favor genotypes that produce a range of propagule sizes either because different-sized propagules are at an advantage in different situations or because variation reduces the chances that the genotype will disappear from the population altogether, resulting in zero fitness in subsequent generations (Janzen 1977a; Capinera 1979; Kaplan 1980a; Crump 1981; Kaplan and Cooper 1984, 1988; Philipi and Seger 1989). One way in which genotypes might achieve this would be to allow some level of random phenotypic expression. This could be achieved in different ways but is essentially analogous to deciding the course of development by tossing a coin, and so has been called "adaptive coin flipping" (Kaplan and Cooper 1984). Another option would be to produce eggs of different sizes in some fixed sequence (Philipi and Seger 1989).

Whether a measure of random gene expression is likely to be favored depends on the assumptions made by different models (McGinley, Temme, and Geber 1987) and their predictions consequently vary. For example, if large offspring are at an advantage in poor patches and small offspring in good ones (see Figure 4.10A) and parents are able to control the dispersal of young to patches of appropriate quality, they may increase their fitness by varying the size of young within broods (see Capinera 1979; Kaplan and Cooper 1984). However, where large offspring always outperform small ones but do so to a greater extent in poor patches than in good ones (Figure 4.10C) and the parent cannot control the dispersal of its offspring, a single optimal size of offspring will usually be favored (McGinley, Temme, and Geber 1987).

As McGinley points out, the latter situation is probably commoner than the former, but cases where the relative payoffs of large and small propagules vary with environmental quality or frequency in the population may exist. For example, in the gypsy moth, *Lymantria dispar*, individuals hatching from large eggs are more active dispersers, while individuals hatching from small eggs

KAPLAN & COOPER 1984

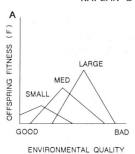
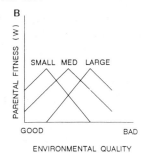

McGINLEY ET AL 1987

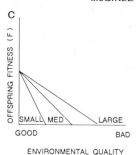
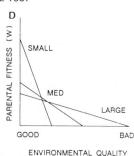

Figure 4.10 Schematic diagrams illustrating the effects of contrasting assumptions concerning the effects of propagule/offspring size on fitness (from McGinley, Temme, and Geber 1987). In (A) and (B), large offspring are at an advantage in bad environments while small offspring are at an advantage in good ones (see Kaplan and Cooper 1984). In (C) and (D), large offspring have substantially higher fitness than small ones in bad environments, but this difference disappears in good environments, though small offspring never have higher fitness than large ones.

may be more fecund (Capinera and Barbosa 1977). It would not be surprising if the relative benefits of fecundity and dispersal varied with habitat quality. Similarly, in the colonial western tent caterpillar, *Malacosoma californicum pluviale*, large eggs produce more active individuals that produce silk trails, while small eggs produce more sluggish caterpillars that follow silk trails and produce better silk tents than the leaders. Although caterpillars hatching from large eggs are more vigorous, they cannot survive in the absence of the more sluggish but good tent-making individuals that hatch from smaller eggs (Wellington 1957, 1965).

What evidence is there that selection has favored adaptive variability in propagule size? In plants, it is not uncommon for individuals to produce two distinct types of seed that are adapted to different ecological niches (Harper,

Lovell, and Moore 1970), but similar differences have not yet been found in animals. The most convincing evidence in animals is based on comparisons of variation in size among eggs *produced by the same females*. For example, Crump (1981) has shown that in tree frog (Hyla) species that breed in permanent water bodies, individual females produce more variable eggs than in species that live in temporary water bodies (see also Wilbur 1977b; Kaplan 1980a), though this trend can be explained in other ways.

4.8   Summary

The evolution of propagule size is more complex than early models suggest. Optimal egg size may be affected by the influence of egg size on the instantaneous rate of juvenile survival and on the duration of subsequent developmental stages. Both trade-offs probably vary widely in form. In addition, selection pressures on clutch size may modify those operating directly on egg size, while constraints on egg size may affect the absolute and the relative size of eggs. In some circumstances, selection may favor an optimal range of variation in egg size, but such situations are probably not common.

Despite these complexities, egg size shows consistent correlations with a variety of ecological and biological factors. Both within and across species, it is commonly related to maternal body size as well as to the harshness of the environments faced by young juveniles, the intensity of competition within and between species, and the duration of parental care. Several different hypotheses can account for each of these trends. To discriminate between them, we need improved measures of the costs and benefits of variation in egg size and of the extent to which these vary with ecological conditions and interact with clutch size—as well as a better understanding of anatomical and physiological constraints on egg size.

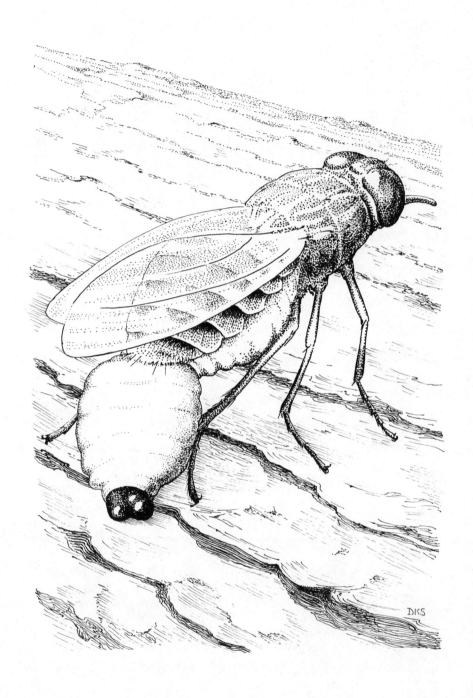

5   Viviparity

5.1   Introduction

Egg retention, ovoviviparity, and viviparity provide an effective way of increasing the survival of progeny and, in some cases, of accelerating their growth (Shine 1985, 1989b). However, most forms of egg bearing reduce the parent's mobility (Thibault and Schultz 1978; Shine 1980; Bauwens and Thoen 1981; Brodie 1989) and may also constrain clutch size (Crespi 1989). Viviparity might consequently be expected to develop where the survival or growth of viviparous offspring are substantially greater than under strict oviparity, or where the costs of retaining eggs are unusually low (Shine 1989b). These conditions have been specified formally by Blackburn and Evans (Table 5.1).

What environmental circumstances are likely to favor egg retention or viviparity? Variation in a wide variety of physical and biotic factors (including extremes of temperature, anoxia, and osmotic stress as well as high rates of predation or parasitism) could increase the relative survival of eggs retained *in utero* (Shine 1989b). In addition, in animals that maintain body temperatures above ambient temperature, viviparity may increase the rate of embryo development (Shine 1989b).

Three problems affect attempts to investigate the selection pressures favoring egg retention and viviparity and to account for their distribution. First, it is seldom possible to manipulate egg retention in order to investigate its effects, though experimental investigation of the costs of viviparity (for example, by adding weights to nongravid females) would not be impossible.

Second, intraspecific variation is limited and there are consequently relatively few cases where it is possible to investigate the ecological correlates within species. One exception is a thrips, *Elaphrothrips tuberculatus*, where females are facultatively viviparous, producing male offspring by viviparity and females by oviparity (Crespi 1989). As theories of the evolution of viviparity suggest, viviparous females produce half as many young as oviparous ones, but the survivorship of their offspring to the pupal stage is 1.4 times higher in spring and 2.1 times higher in summer (see Table 5.2). Unfortunately, the comparison of survivorship of oviparous versus viviparous offspring is confounded with a sex difference (see Chapter 13). However, there are other thrips species where males can either be produced by oviparity or

Table 5.1

Conditions favoring the evolution of egg retention (from Blackburn and Evans 1986).

Egg retention for some time period t_r will be favoured over strict oviparity (no egg retention) where

$$\frac{b'}{b} > \frac{E(1-p')}{E'^{t_r/t_d}\, E^{(t_d-t_r)/t_d}\,(1-p)}$$

where t_d is the total time to hatching, including the duration of egg retention. For strict oviparity, b is clutch size, E is egg survival to hatching, and p is the mother's survival from the beginning of one reproductive attempt to the next. For egg retention, b' is clutch size, E' is egg survival to hatching if eggs are retained over the entire incubation period, and p' is maternal survival from the beginning of one reproductive attempt to the beginning of the next, when eggs are retained for a period t_r. It is assumed that $E' > E$, $p > p'$ and t_d is independent of the mode of reproduction. High values of E' relative to E will encourage the evolution of viviparity, while low values of p' relative to p or b' relative to b will inhibit it.

Table 5.2

Fecundity of oviparous and viviparous females in *Elaphrothrips tuberculatus* and the relative survivorship of their offspring (from Crespi, 1989).

Season	Pupal Sex Ratio	Mode of Reproduction	% Females in Mode	Average Fecundity (No. of Eggs/ Embryos)	Relative Survivorship to Pupal Stage
Spring	0.19	Viviparity	0.25	21.5	1.4
		Oviparity	0.75	43.7	1.0
Summer	0.58	Viviparity	0.55	13.5	2.1
		Oviparity	0.45	26.1	1.0

NOTE: Relative survivorship is a multiple of the survivorship of offspring produced by oviparity, which is set arbitrarily at 1.0.

viviparity which, in future, will allow comparisons to be made with the same sex (Crespi 1989).

Third, viviparity has evolved a limited number of times in most animal groups. Since the ecological factors favoring the evolution of viviparity from strict oviparity require conditions that favor the intermediate stages of egg retention, they may be quite different from those affecting the radiation of viviparous taxa (Packard 1966; Shine and Bull 1979). For this reason, it is important to investigate the ecological correlates of the independent evolution of viviparity within particular taxonomic groups. In most groups, the number of times viviparity has evolved independently is low. Among vertebrates, estimates suggest that it has evolved only once or twice in mammals (Lillegraven 1975; Tyndale-Biscoe, and Renfree 1987), about six times in elasmobranchs (Wourms 1977), thirteen in teleost fish (Wourms 1981), and five in amphibians (Wake 1977). In contrast, it has probably evolved at least a hundred times in squammate reptiles (Shine 1985, 1989b).

As a result of these difficulties, explanations of the distribution of egg retention and viviparity are necessarily speculative. The following sections examine their distribution among invertebrates, fish, amphibia, reptiles, birds, and mammals.

5.2   Viviparity among Invertebrates

Though egg retention and viviparity have evolved many times among invertebrates (Hagan 1951; Hogarth 1976), the number of times they have evolved independently has not yet been estimated, nor have their ecological correlates been investigated systematically. Egg retention or viviparity occurs in coelenterates, such as the common beadlet anemone, *Actinia equina*, as well as in a number of molluscs including some species of periwinkle, *Littorina* (Hogarth 1976). In the majority of live-bearing molluscs, young do not derive nutrition from the parent, but, in some, like in the prosobranch *Veloplacenta* and the freshwater river snail *Viviparus*, young are nourished directly from the mother's soma, in some cases by a placental analog. Similarly, the annelids include viviparous or ovoviviparous representatives in at least five families.

Among the arthropods, at least thirteen orders of insects have viviparous or ovoviviparous members, and viviparity is relatively common in crustacea and arachnids. Orders that include viviparous species include the cockroaches, Dictyoptera; the true flies, Diptera; the endoparasitic Strepsiptera; the aphids, Aphidoidea; the book lice, Psocoptera; the true bugs, Hemiptera; and the plant bugs, Heteroptera. In addition, eggs are carried externally in some species, including giant water bugs, Belostomatidae, and sea spiders, Pycnogonidae.

The extent to which mothers feed their young before birth varies widely from species where eggs are retained but no feeding occurs, as in some cockroaches, to the nourishment of developing young with specialized secretions, as in the tsetse fly *Glossina* (see Chapter 2). Among terrestrial arthropods, embryos are nourished in external brood sacs in at least three orders, eleven show ovoviviparity or larviparity, and eight show true viviparity (Zeh and Smith 1985). Egg guarding occurs in a larger number of orders (20) than ovoviviparity or viviparity (14).

One of the most elaborate cases of viviparity concerns a member of the Strepsiptera, *Stylops*, which is parasitic on solitary bees (Hogarth 1976). Adult male *Stylops* are winged but the female is larval in form and lives within the hemocoel of the host. A mature female pushes her cephalothorax out through the thin cuticle between two adjacent segments of the host's abdomen. The male mates by injecting sperm directly through the female's cuticle. Embryos develop in the mother's hemocoel and subsequently in an external brood sac before being born from an orifice behind the mother's mouth. The larvae then emerge onto the surface of the host and transfer onto other hosts, returning to their nests and parasitizing their larvae. Larval development of *Stylops* is synchronized with that of its host so that both sets of larvae mature together.

In a number of animals, egg retention or viviparity is associated with harsh or unpredictable environments, while in other cases viviparity may help to buffer seasonal changes in food availability (Hogarth 1976). In the chrysomelid beetles, ovoviviparity is found most commonly in species living at high altitude (Crespi, pers. comm.). In sea cucumbers (Holothuria) and in some other echinoderms, viviparity is commoner in the Antarctic, where the seasonable availability of phytoplankton varies widely throughout the year, than in other areas, including the Arctic (Hogarth 1976). In other cases, viviparity may permit colonization of new habitats. For example, the oviparous ragworm, *Nereis diversicolor*, is restricted to breeding in salt and brackish water since its eggs and larvae die in freshwater, but its viviparous California counterpart, *N. limnicola*, can breed in freshwater as well as salt (Hogarth 1976). In some terrestrial land snails, developing young are retained within the mother's body in the dry season (Owiny 1974).

Many viviparous arthropods are parasitic (Hogarth 1976). For example, several viviparous beetles live inside ant or termite colonies where eggs might be recognized as alien. In other cases, such as in *Hemimerus talpoides* (an aberrant earwig that lives as an ectoparasite on giant rats), in the sheep keds *Melophagus*, and in some hemipterans that parasitize bats, viviparity may facilitate dispersal between hosts.

5.3   Viviparity in Fish

In the chondrichthyan fish (sharks, skates, rays, and chimeras) all recent species show internal fertilization. This, in conjunction with the relatively large size of eggs and their comparatively low fecundity, may have predisposed them to the evolution of egg retention and viviparity (see Wourms 1977; Shine 1989b). Forty families, 99 genera, and 420 species of sharks and rays out of an estimated 600–800 species show viviparity or ovoviviparity, compared to 13 families, 122 genera, and 510 species out of an estimated 422 families and 18,000 species of bony (teleost) fish (Wourms 1981). Among chondrichthyans, oviparous families include the horn sharks Heterodontidae, skates Rajidae, and three families of chimeras (see Wourms 1977, 1981).

In the sharks, oviparity is largely confined to small, benthic, littoral species. Among the rays, oviparous species occur chiefly in temperate and polar regions, while the tropical and subtropical rays, sawfish, and guitarfish are ovoviviparous or viviparous (Breder and Rosen 1966; Wourms 1977). Viviparity or ovoviviparity probably offer a variety of advantages to tropical species, including increased protection from predators and other hazards, more effective regulation of environmental factors, and the possibility of producing larger propagules that have a higher rate of survival (Amoroso 1968; Wourms 1977; Balon 1978; Baylis 1981).

Among bony fish, ovoviviparity and viviparity are uncommon in species breeding in the open sea (Breder and Rosen 1966; Wourms 1981; see Figure 5.1). A similar trend occurs within some families. For example, in the half-

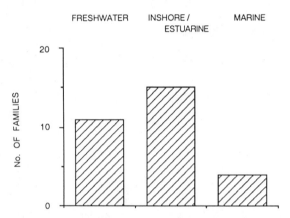

Figure 5.1   Numbers of families of bearing fish (where species are ovoviviparous or viviparous or carry eggs externally) found in three different habitats (from Baylis 1981).

beaks Hemirhamphidae, marine species typically produce pelagic eggs, while species breeding in fresh or brackish water are commonly viviparous (Breder and Rosen 1966). Among inshore and freshwater breeders, egg retention, ovoviviparity and viviparity occur in a minority of families and their incidence does not appear to differ between coastal and freshwater breeders (Breder and Rosen 1966).

In most cases, viviparity involves females, but cases of male viviparity are known. In the sea horses and pipefish of the family Syngnathidae, males of some species brood their offspring on their body or in a specialized brood pouch (Gronell 1984; Berglund, Rosenqvist, and Svensson 1986; Vincent 1990). In species that carry eggs externally, fertilization, too, is external, while in species where males brood the eggs in a specialised pouch, fertilization occurs within the brood pouch (Fiedler 1954). In some pipefish and sea horses, males have evolved a placenta through which embryos growing within the pouch receive some nutriment.

A wide variety of other forms of egg bearing also occur in bony fish. In a number of freshwater fish and a few marine ones, parents transport embryos or fry in their mouths to new sites when local conditions change, which may have led to the evolution of mouth brooding (Oppenheimer 1970; Figure 5.2).

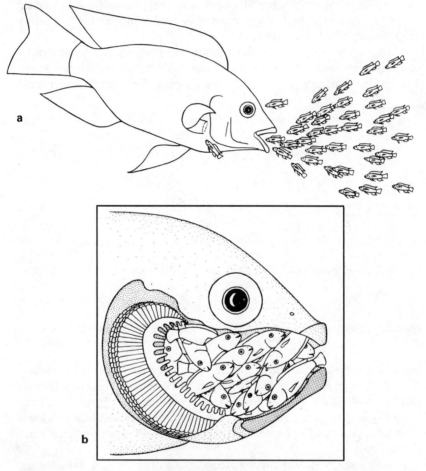

Figure 5.2 Female *Tilapia* (a) collecting its brood into its mouth, and (b) brooding well-developed young (from Fryer and Iles 1972).

Mouth brooding most commonly involves females but, in a few families (including the arapaimas Osteoglossidae, the sea catfish Ariidae, the cardinal fish Apogonidae, and a minority of cichlids) males are involved. In some species, males carry eggs on their head or lower lip (some Kurtidae and Loricariidae) or on their ventral surface (some pipefish of the subfamilies Nerophinae, Syngnathoidinae, and Solegnathidae).

Though several different ecological circumstances have probably favored the evolution of viviparity in teleosts (Turner 1947), external carrying of eggs and oral brooding of eggs or fry are commonly associated with intense competition, high predation rates, or environmental factors that make oviposition sites unsuitable for the subsequent development of eggs (Keenleyside 1978; Thibault and Schultz 1978; Blumer 1979). Among African cichlids, mouth brooding is commonest in species living in shallow, temporary water bodies (Fernald and Hirata 1977). This may either be because eggs are more likely to be predated in shallow water (Johannes 1978) or because water conditions show greater short-term variability, favoring the evolution of an efficient method of transporting them to other sites.

5.4   Viviparity in Amphibia

Viviparity occurs in all three groups of amphibia: the frogs Anura, the salamanders Urodeles, and caecilians Gymnophiona. Though it has been suggested that its usual function is to reduce the danger of desiccation to juveniles (Hogarth 1976; Jameson 1981), this cannot always be the case, for it is found in a number of aquatic species (Shine 1989b). As in fish, viviparity is associated with relatively unfavorable environments (Harms 1946; Salthe and Mecham 1974). For example, in the genus *Salamandra*, the only viviparous species is the Alpine *Salamandra atra*, while among the frogs the only truly viviparous species occur in the African montane genus, *Nectophrynoides* (Salthe and Mecham 1974). One advantage of viviparity is that it permits discontinuous growth of embryos: *Nectophrynoides* estivates in a pregnant condition during the dry season, but most growth takes place after feeding begins in the rainy season, while *S. atra* commonly carries embryos throughout at least three seasons, suspending their growth during unfavorable periods (Salthe and Mecham 1974).

The transport of eggs on a parent's body until hatching has evolved several times in anurans (Wells 1981; Inger, Voris, and Walker 1986). In all cases but one, females carry the eggs on their backs or in a dorsal brood pouch. The exception is the midwife toad, *Alytes obstetricans*, where the male carries the eggs entwined around his legs (Boulenger 1912). Female anurans rather than males may typically transport eggs rather than males because egg transport by males would interfere with territorial defense (Wells 1981). In addition, a single clutch is usually carried at a time, and egg transport would probably restrict the male's breeding opportunities. In the midwife toad, males can carry more than one clutch at a time, and mating opportunities are restricted to short pulses of breeding activity so that egg transport is unlikely to restrict the male's mating success (Wells 1981).

5.5 Viviparity in Reptiles

Among the reptiles, all turtles and crocodilians are oviparous. However, some 15% of snakes and lizards bear live young, and live birth has developed independently nearly a hundred times (Tinkle and Gibbons 1977; Shine 1985). Most of these species are ovoviviparous, but transfer of nutrients between the mother and the developing embryos occurs in a minority of groups (Vitt and Blackburn 1983).

Oviparity appears to be the ancestral state, followed by the evolution of egg retention in a proportion of species and by live-bearing in some of these (Neill 1964; Packard 1966; Packard, Tracy, and Roth 1977; Tinkle and Gibbons 1977; Shine and Bull 1979; Shine 1983a). The transition from oviparity to live birth may be facilitated by the fact that most oviparous species retain their eggs *in utero* for around half of their total period of development (Shine 1985). Within species, the period of egg retention is variable and may be partly under hormonal control (Shine and Guillette 1988). In at least ten species, some populations show viviparity while others are oviparous (Shine 1985).

Ovoviviparity or viviparity among species of snakes and lizards is usually associated with living in cold climates, and its incidence increases away from the equator (Figure 5.3; Tinkle and Gibbons 1977; Shine and Bull 1979; Shine 1981, 1983a,b, 1984, 1985, 1987a,b, 1989; Blackburn 1982). Viviparity is also commoner in genera where mothers brood their eggs after laying: for example, viviparity or ovoviviparity has evolved in 35% of genera of snakes where maternal care occurs in some species, compared to 6% of all genera (Shine 1985).

Especially among temperate reptiles, viviparity may help to increase the rate of development of the eggs (Mell 1929; Tinkle and Gibbons 1977; Shine and Bull 1979; Shine 1985, 1989b). By retaining eggs internally, females of some Australian lizards are able to increase their temperature by 7°C, shortening incubation periods by up to 40 days (Shine 1983a,b). Many reptiles are very sensitive to low temperatures: for example, eggs of the tropical iguana, *Iguana iguana*, which normally develop at 30°C, cannot tolerate even a week's exposure to 20°C (Licht and Moberly 1965). Rapid development and earlier hatching also reduce time spent in the soil when eggs are vulnerable to predators, increase the time during which offspring can feed before the onset of winter, and reduce the chance that they will hatch late when conditions are suboptimal (Shine 1985). Some northern snakes inhabit environments so cold that there is sometimes insufficient time to permit spring-laid eggs to hatch before winter:

NORTH AMERICA

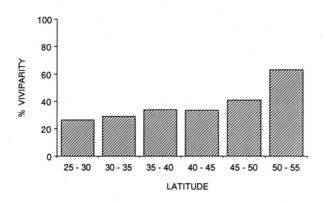

AUSTRALIA

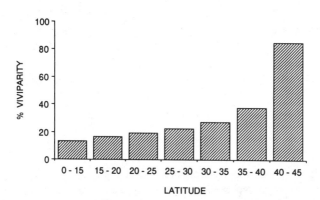

Figure 5.3   Percentage of different species of lizards and snakes found in different latitude bands in (a) North America and (b) Australia that show viviparity (data from Tinkle and
Gibbons 1977).

the grass snake *Natrix natrix* nests in compost heaps, the warmest microhabitat
available, but, despite this, eggs do not hatch if the summer is cold (M.A.
Smith 1973). By reducing the period of development, viviparity can permit the
occupation of habitats that would not otherwise be available (Shine 1987a).

5.6   On the Absence of Viviparity in Birds

With the exception of kiwis, *Apteryx*, egg retention is rare in birds and no bird
has evolved ovoviviparity or viviparity (Williams 1966b). This is surprising in

view of the prevalence of internal fertilization in birds and the high rate of egg loss. Since egg weight as a proportion of the mother's body weight declines with size (see Chapter 4), large monotocous birds such as some pelagic seabirds or some ratites would seem to be likely candidates (Anderson, Stoyan, and Ricklefs 1987).

One possible explanation of the absence of viviparity in birds is that selection seldom favors egg retention because of its effects on the mother's mobility and survival (Blackburn and Evans 1986). The large size of the cleidoic avian egg may limit simultaneous egg retention to very low clutch sizes (Blackburn and Evans 1986).

These conditions do not preclude the evolution of viviparity, nor do they explain why no birds have achieved viviparity via a drastic reduction of egg size, as have mammals. The absence of egg retention in birds suggests that some constraint may be imposed by the avian reproductive system, such as the relatively high temperatures of adults compared to developing embryos or the relatively low oxygen content of the oviduct (Anderson, Stoyan, and Rickleffs 1987). However, it is still difficult to see why these problems could not have been solved by metabolic adaptations. One possible explanation is that viviparity is not viable in birds without an extensive reduction in egg size (Dunbrack and Ramsay 1989), and that either the strong pressures maintaining egg size or the absence of an efficient form of supplying very altricial young with specialized nutrients (as in mammals) may have prevented reduction in egg size and blocked the evolution of egg retention and viviparity.

5.7   Viviparity in Mammals

All living mammals show viviparity with the exception of the two families of monotremes: the platypuses (Ornithorhynchidae), represented by a single species, and the spiny anteaters (Tachyglossidae), with two living species. In the platypus, *Ornithorhynchus anatinus*, females produce one or two eggs 11-12 days after copulation and brood them for 11-12 days in an underground burrow (Eisenberg 1981). The spiny anteater, *Tachyglossus aculeatus*, produces a single egg, which it retains in the oviduct for around 27 days (Broom 1895). The female subsequently transfers the egg to a pouchlike "incubatorium," curling herself to oppose the lips of her cloaca to the pouch opening. The egg is incubated within the pouch, hatching 10-12 days later (Griffiths 1968, 1978), and the young is brooded and nursed there for a further two months before being expelled (Eisenberg 1981). Lactation continues after expulsion until the juvenile is around five months old (Eisenberg 1981). In both species the eggs are small relative to adult size, and hatchlings are hardly more developed than fetuses (Griffiths 1968).

The breeding systems of therian mammals differ from those of reptiles in that mammals produce small eggs that hatch into immature, dependent offspring that receive prolonged and elaborate parental care (Hopson 1973). In addition, litter sizes are generally smaller than those of reptiles, though the small clutches of contemporary monotremes could indicate that the cynodont reptiles from which mammals evolved also produced relatively few eggs. The sequence of changes is still uncertain and is the subject of extensive debate (Long 1972; Hopson 1973; Pond 1977; Lillegraven, Thompson, et al. 1987; Dunbrack and Ramsay, 1989). In contrast to the evolution of viviparity in fish and reptiles, mammalian viviparity may have arisen primarily through a reduction in egg size (Tyndale-Biscoe and Renfree 1987). It seems likely that the evolution of endothermy preceded that of lactation and was already well advanced in the cynodonts (Crompton, Taylor, and Jagger 1978; Crompton and Jenkins 1979). Endothermy may have preadapted the first mammals for a nocturnal way of life. This may, in turn, have favored the evolution of fossorial habits, maternal brooding, and, eventually, lactation (Long 1972; Hopson 1973). Parental provisioning may have encouraged a reduction in egg size, allowing the development of egg retention and viviparity (see Pond 1977, 1983; Lillegraven, Thompson, et al. 1987; Dunbrack and Ramsay 1989).

Marsupials and eutherians subsequently developed alternative strategies of prenatal development. Contemporary marsupials possess a breeding system where prenatal expenditure and gestation length are minimized, and prenatal growth can be easily manipulated by delayed implantation or embryonic diapause (Low 1978). In contrast, eutherians show long and relatively inflexible gestation periods and produce large neonates that are weaned relatively quickly. For example, a 30 kg red kangaroo produces an offspring weighing less than one gram after a gestation period of around 33 days and weans it at around a year, while a white-tailed deer of similar size gives birth after around 200 days to an offspring weighing around 3,000 g that is finally weaned, at a similar weight, in less than six months (Low 1978; Sadleir 1969).

The eutherian mode of reproduction, associated with a higher and more flexible metabolic rate, may have some energetic advantages over the longer lactation periods and lower metabolic rates of marsupials, especially where food is seasonally abundant, temperatures are low, or body size is small (McNab 1978; Lillegraven 1974, 1975; Lillegraven, Thompson et al. 1987). In addition, the faster reproductive rates of eutherians may have advantages where adult mortality is high. In fluctuating and unpredictable environments, the marsupial system has the advantage that timing of parental expenditure can be easily manipulated as a consequence of delayed implantation and embryonic diapause, and investment can be easily terminated with little loss to the parent's fitness (Low 1978). When conditions improve, another embryo can be rapidly produced. However, only a small proportion of marsupials inhabit arid and highly unpredictable environments, while euthurians can (and frequently do) abort or resorb part or all of their litters (Russell 1982a,b; Morton, Richter, et al. 1982).

5.8 Summary

Egg retention, ovoviviparity, and viviparity are effective ways of increasing the survival of offspring and, in some cases, of accelerating their growth. Environmental conditions that might be expected to favor their evolution are similar to those favoring the evolution of large egg size: harsh conditions, high levels of predation or parasitism, or intense competition, especially between juveniles and adults. Where viviparity accelerates growth, it may be favored where breeding seasons are short.

The distribution of viviparity among invertebrates, fish, amphibia, and reptiles supports these generalizations. Among invertebrates, viviparity is commonly associated with harsh or unpredictable environments and is relatively common in parasitic species. In teleost fish, viviparity is most frequently found in species breeding in shallow coastal waters or freshwater, where water conditions are variable and predation rates are often high. Among the amphibia, it is again commonly associated with unfavorable environments, while in reptiles the proportion of viviparous species increases among genera living in cold climates. Among mammals, the evolution of endothermy and lactation probably preceded the evolution of viviparity. The evolution of egg retention and viviparity may have been facilitated by internal fertilization combined with a situation where mothers brooded eggs in underground burrows.

The absence of viviparity in birds is puzzling, for some large, monotocous species would appear to be good candidates for the development of egg retention and viviparity. Possible constraints include the relatively high temperatures of adults compared to young, the low oxygen tension of the oviduct, and the absence of lactation.

Incubation, Gestation, and Lactation

6.1   Introduction

This chapter reviews our understanding of the adaptive significance of variation in the length of incubation and gestation and of patterns of lactation in
endotherms. Though similar questions might usefully be asked about incubation and gestation periods in invertebrates, fish, reptiles, and amphibia, questions of this kind have been more frequently asked about birds and mammals
(see Lack 1960; Eisenberg 1981), and our understanding of the extent and
distribution of interspecific differences in all three parameters is more advanced in these than in other groups of animals.

The length of periods for which young are dependent on their parents varies
enormously, even among closely related species. For example, among primates, gestation periods range from 60 to 260 days and weaning ages from
slightly over one month to more than 100 months (Harvey and Clutton-Brock
1985). Much of this variation is related to differences in body size (Peters
1983; Calder 1984), but even among species of similar size, differences in the
length of developmental periods can be large. For example, among the seals,
weaning age varies from more than a year in some fur seals to an amazing 4
days in the hooded seal, *Cystophora cristata* (Bowen, Oftedal, and Boness
1985; Oftedal, Boness, and Tedman 1987). Across many different groups of
birds and mammals (and possibly across other animal groups, too) there is a
consistent tendency for some species to progress rapidly through the different
stages of development while others move more slowly, even when any effects
of body size are allowed for (Figure 6.1; Harvey and Clutton-Brock 1985;
Harvey and Zammuto 1985; Harvey, Read, and Promislow 1989; Harvey, Promislow, and Read 1989; Read and Harvey 1989; Promislow and Harvey
1990).

Despite an extensive descriptive literature (see Peters 1983; Calder 1984),
we currently lack satisfactory functional and causal explanations of variation
in development rate (Read and Harvey 1989; Harvey, Read, and Promislow
1989). Attempts to account for this variation differ. On the one hand, some
authors attribute most of the variation in development rates to constraints imposed by body size or by size-related variables, such as metabolic rate or brain
growth (see Blueweiss, Fox, et al. 1978; Western 1979; McNab 1980, 1988;

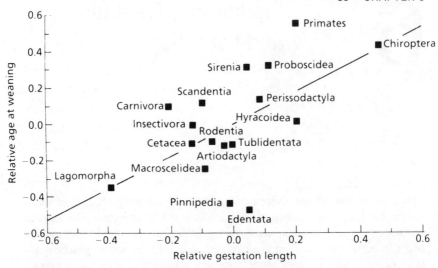

Figure 6.1 Comparisons between relative gestation period and relative age at weaning across different orders of placental mammals. Relative values refer to deviations from logarithmic regressions of gestation period and age at weaning on body weight, so that they are not correlated with body weight (from Harvey, Read and Promislow 1989).

Western and Ssemakula 1982; Hofman 1983; Martin 1984). On this view, selection is regarded as operating primarily on body size, which constrains rates of development and the duration of developmental stages. However, the presence of substantial differences in all life history parameters among animals of the same size and the similarity of development rates among closely related species of different sizes argues against this interpretation. Moreover, several of the size-related constraints that have been suggested (notably brain growth and metabolic rate) fail to account for much of the variation in development rate once size effects have been removed (Pagel and Harvey 1988; Read and Harvey 1989; Trevelyan, Harvey, and Pagel 1990; Harvey, Pagel, and Rees, in press).

An alternative view seeks to link variation in development rate to life history theory in general and to age-specific mortality schedules in particular (Harvey, Read, and Promislow 1989). The crucial variable here is the "intrinsic" rate of adult mortality—mortality arising for reasons not associated with the individual's age or reproductive effort. Where this is high relative to preadult mortality, selection will favor faster development and quicker reproduction than where it is low relative to preadult mortality (Williams 1966a,b; Pianka 1976; Charlesworth 1980; Boyce 1984). In practice, preadult mortality is usually high in stable populations of most mammals and species differences in the ratio of preadult to adult mortality are probably caused mainly by variation in adult mortality rates. It is also important to appreciate that these selection pressures may be modified by "safe harbor" trade-offs (see Section 4.3).

Table 6.1

Correlations (r) across families of placental mammals between life histories and mortality with the effects of body size removed (from Harvey, Read, and Promislow 1989).

	r	n	p
Gestation period	−0.44	21	0.048
Age at weaning	−0.50	18	0.040
Time from weaning to maturity	−0.52	18	0.026
Age at maturity	−0.61	20	0.004
Interlitter interval	−0.47	20	0.037
Maximum recorded lifespan	−0.03	17	0.240
Neonate weight	−0.39	20	0.120
Litter weight	−0.26	20	0.260
Number of offspring per litter	+0.59	21	0.005

NOTE: Data on age-specific mortality were obtained from life tables of 48 species belonging to 21 families from 10 orders. Family means (calculated by averaging generic means) were used rather than order means because of the few orders for which data were available. n = number of families for which data on life history variable and mortality were available. p = probability level.

As would be predicted, mammalian species that are characterized by low life expectancy in natural populations tend to progress rapidly (relative to their body size) through most or all of their developmental stages, while species showing high life expectancies move more slowly through the stages of development (Table 6.1: Harvey and Clutton-Brock 1985; Harvey and Zammuto 1985; Harvey and Read 1988; Read and Harvey 1989). Since large mammals are commonly characterized by low adult mortality and relatively high mortality of juveniles, while many small species show high mortality throughout the lifespan, part of the relationship between body size and reproductive rate may be caused by the correlation between body size and age-specific mortality schedules (Harvey, Read, and Promislow 1989).

This complicates the study of comparative rates of development where it would be convenient to distinguish between variation arising from differences in body size and variation arising from adaptive differences in rates of development. In particular, it raises the question of whether we should seek for ecological correlates of absolute durations of particular life history stages, accepting that part of the variation we observe is a consequence of variation in body size, or whether we should investigate relationships between ecology and the relative lengths of developmental stages, when the statistical effects of body size have been allowed for. The problem is an important one, for the rankings of different taxa on absolute and relative rates of reproduction commonly vary (Harvey, Promislow, and Read 1989). No general answer is yet available.

In this chapter, I adopt the second approach on the grounds that it is the more conservative. Section 6.2 describes correlations between body size and the duration of incubation, gestation, parental feeding, and lactation, while Sections 6.3 and 6.4 examine the extent to which deviations from these relationships can be related to ecological or behavioral variations. Sections 6.5 and 6.6 describe species differences in milk production and composition.

6.2 Body Size and the Duration of Development Periods

Among endotherms, the duration of most periods of development increase with size as approximately body weight$^{0.25}$ and incubation and gestation periods are no exception (see Peters 1983; Calder 1984; Boyce 1988). Across a wide variety of birds, incubation period (I, in days) increases with egg weight (EW, in grams) as $I = 12.03\,EW^{.217}$ (Rahn and Ar 1974). Thus increases in egg size are not associated with proportional increases in incubation time: for example, for a medium-sized bird, doubling egg size involves a 16% increase in incubation time. Since egg weight does not increase proportionately with body weight (Section 4.2), incubation time rises more slowly with maternal weight (MW) than with egg weight: for 100 assorted species, Blueweiss, Fox, et al. (1978) calculate a slope of $MW^{0.16}$.

Similar relationships exist between gestation time, neonatal weight and maternal weight in eutherian mammals (Blueweiss, Fox, et al. 1978; Western 1979; Western and Ssemakula 1982; Boyce 1988; Read and Harvey 1989). Across a diverse array of species, gestation period, G (in days), increases with maternal weight, MW (in grams), as $G = 11\,MW^{0.26}$ (Blueweiss, Fox, et al. 1978), though shallower slopes of around $MW^{0.13}$ and lower correlation coefficients are usually found within orders (Harvey and Clutton-Brock 1985; Gittleman 1986). Intercepts differ between orders: for example, primates tend to have longer gestation lengths for a given body weight than carnivores and artiodactyls.

The length of lactation periods in eutherians also increases as approximately maternal weight$^{0.15-0.28}$ (Mace 1979; Peters 1983; Reiss 1985), though, in primates, the slope is considerably steeper, lying around maternal weight$^{0.56}$ (Harvey and Clutton-Brock 1985). Why large primates should wean their offspring relatively later is not clear, but this is associated with a marked increase in longevity in the largest primates (see above).

One important consequence of the negative allometry of pre- and postnatal growth rates is that larger endotherms spend less of their daily energy budget on reproduction than small ones (see Chapter 4). Reproductive growth in mammals can be estimated by dividing total litter weight by gestation time in days, though this ignores the growth of the uterus. Defined in this way, daily reproductive growth increases across species as weight$^{0.5-0.6}$ (Blueweiss, Fox, et al. 1978; Peters 1983; Ross 1988). Since the rate of energy gain increases across species as around weight$^{0.75}$ (see Bennett and Harvey 1985a, 1987), reproductive expenditure relative to energy intake must decline as $W^{0.55}/W^{0.75} = W^{-0.2}$ (Peters 1983). A variety of other indices of relative reproductive expenditure are in use (Blueweiss, Fox, et al. 1978; Millar 1977, 1981; Peters 1983; Calder 1984), but all indicate that, across species, relative daily reproductive expenditure declines with increasing body size.

Females of smaller species may invest a higher proportion of their daily energy budgets in reproduction than larger ones for several reasons. Adult mortality is relatively high in most small animals, favoring early development and high rates of reproduction (see Section 6.1). In addition, the high surface area/weight ratio of the offspring of small species favors the evolution of rapid growth to minimize the costs of maintenance (see Lack 1968). Alternatively, the negative allometry of parental expenditure may arise in part from anatomical or energetic constraints operating in the mother (Brody 1945; Leutenegger 1976). In particular, Reiss (1985, 1987b) argues that because total energy intake, E_{in}, and energy required for everything except reproduction, E_{req}, both increase as around (maternal) body weight$^{0.5-0.9}$, the energy available for reproduction, E_{rep} (which is presumably proportional to the difference between E_{in} and E_{req}), must also show a similar slope.

6.3   Incubation

Both within and between major taxonomic groups, there are substantial differences in incubation periods that do not appear to be size related and are associated with ecological or social factors (see Bennett 1986; Harvey, Read, and Promislow 1989; Trevelyan, Harvey, and Pagel 1990). Birds that suffer particularly high predation rates on eggs and nests, such as the pheasants and the pigeons, appear to have short incubation periods relative to the size of their eggs (Lack 1968). In contrast, sea birds that feed offshore have longer incubation periods relative to their egg or body size than the inshore feeders. For example, fork-tailed storm petrels, *Oceanadroma furcata*, have incubation periods of around 71 days, around three times as long as would be expected for birds of their size (Boersma 1982).

Both among terns and the ducks, incubation periods are shorter in species breeding at high latitudes than in tropical or subtropical species, while in the latter they are also longer in hole-nesting species than in ground-nesting ones (Lack 1968). As Lack points out, it is not difficult to understand why shorter incubation periods are advantageous in species breeding in the Arctic or on the ground, but it is less easy to see what species breeding in holes or at lower latitudes gain from *longer* incubation periods.

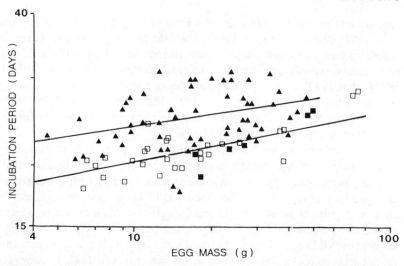

Figure 6.2 Incubation period plotted on egg mass for different species of shorebirds. Triangles and upper regression line: run-and-catch species; open squares and bottom regression line: pecking species; solid squares: begging species (from Nol 1986).

Among shorebirds, the length of incubation periods varies with the foraging mode of the young. Species where young feed by running and catching insects, like most plovers, have longer incubation periods, relative to egg weight, than species where the young seek for food or beg from their parents, as in snipe, woodcock, or oystercatcher (Nol 1986: see Figure 6.2). This relationship may occur because running and catching insects require greater maturation of neural function at hatching. If so, differences among the three groups might be expected to disappear if egg mass is replaced by chick brain weight.

6.4   Gestation and Age at Weaning

The length of gestation and the age at weaning relative to maternal size in mammals also vary widely. For example, at a maternal body weight of around 150 kg, lions (*Panthera leo*) produce an average of 2.6 offspring after a gestation of 106 days, pigs (*Sus scrofa*) a litter of 6 after a gestation of 120 days, and the dolphin (*Tursiops truncatus*) a single offspring after a gestation period of around a year (Harvey, Read, and Promislow 1989).

As described in Section 6.1, differences in the relative duration of different developmental stages are generally *positively* correlated with each other so that some mammals progress relatively quickly through all stages of development while others progress relatively slowly (Figure 6.1; Harvey and Clutton-Brock 1985; Read and Harvey 1989). This does not mean that trade-offs between the amount of time spent in consecutive stages do not occur—for example, com-

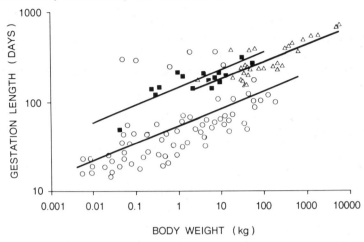

Figure 6.3   Log gestation length (days) plotted on log body weight (kg) for precocial and altricial mammals (from May and Rubenstein 1984). Circles: offspring kept in nest; solid squares: offspring carried; triangles: offspring independent.

pared to precocial species, altricial mammals have relatively shorter gestation lengths (Figure 6.3; R. D. Martin 1984; May and Rubenstein 1984; R. D. Martin and MacLarnon 1985).

As life-history theory predicts, interspecific differences in the relative duration of gestation and the relative age of offspring at weaning are inversely related to mortality rates (Table 6.1). Unfortunately, reliable measures of life expectancy in natural populations are too rare to permit quantitative comparisons to be made among closely related species. To check that these correlations do indeed arise from the adaptation of development rates to differences in age-specific mortality, more specific predictions need to be checked. Are differences in age-specific mortality among congeners similarly related to variation in development rates? Is gestation length unusually short in species where pregnant females are particularly likely to be killed by predators? Are weaning ages unusually low where the risk of juvenile mortality in the nest or the costs of lactation to the mother are particularly high?

The association between life expectancy and relative rates of development may help to explain some of the correlations between ecological variables and development rates that have been discovered. For example, both within and across families, forest-living primates appear to develop more slowly for their body size than savannah-dwelling species and show a lower maximum rate of reproduction (r_{max}) (Ross 1988; Harvey, Promislow, and Read 1989). Though data on life expectancies of wild primates are scarce, parallels with birds suggest that adult mortality is likely to be lower in forest-living species than those living in savannah (Lack 1968; O'Connor 1984). In other groups, relative rates of development are consistently related to diet, suggesting that energetic constraints may be involved. For example, relative to their body size, rodents that are food specialists (graminivores, frugivores, and insectivores) have longer gestation periods, are weaned later, reach sexual maturity later, and have longer lifespans than folivores (Mace 1979).

The correlations between life expectancy and relative rates of development discovered by Harvey and his colleagues pave the way for further investigations of correlations between relative development rates and ecological variables that are likely to be related to age-specific mortality schedules. Unlike previous attempts to relate variation in relative rates of development to ecology, these will have the advantage of a firm connection with life-history theory.

6.5 Lactation

The comparative energetics of lactation in mammals have been investigated more thoroughly and more extensively than those of parental provisioning in any other group of animals, but much is still unknown (Peaker 1977; Loudon and Racey 1987). Only recently have data been available which permit comparisons of the magnitude of parental expenditure among a wide range of different species (Oftedal 1981, 1984a,b), and these are still complicated by interspecific differences in the temporal patterning of milk yield and milk quality within the lactation period, especially in marsupials (Cockburn 1989).

Daily energy intake from milk, protein intake, and growth rate (g/day) of individual juveniles are closely correlated with (juvenile) body weight across mammalian species, all rising as around juvenile weight$^{0.83}$ (Oftedal 1981, 1984a), though it is unclear to what extent milk production is constrained by body size. Both daily milk yield and gross energy output at peak lactation rise with (maternal) weight$^{0.75}$. As a consequence of this allometry, daily milk production as a percentage of maternal weight declines with increasing maternal size, from 28% in pygmy shrews to 1.25% in elephants (Hanwell and Peaker 1977). However, because of the increased duration of lactation in large species, total expenditure on lactation per reproductive event or per lifespan increases as maternal body weight$^{0.8-1.14}$ (Gordon 1989; Reiss 1989).

When the effects of body weight are controlled, daily yields differ between species by a factor of four (Oftedal 1984a). For example, pigs produce nearly twice as much milk as predicted from their body weight while reindeer, humans, and baboons produce only one half as much (Oftedal 1981: see Figure 6.4). Females of multiparous species show consistently higher gross energy and protein yield than uniparous species, and among the latter, ungulates show higher yields than primates (Oftedal 1981, 1984a). As a result, there is a very close correlation between the mother's relative energy output in milk and the ratio of litter metabolic mass to mother's metabolic mass (Figure 6.5). A similar increase in gross energy production occurs within species: for example,

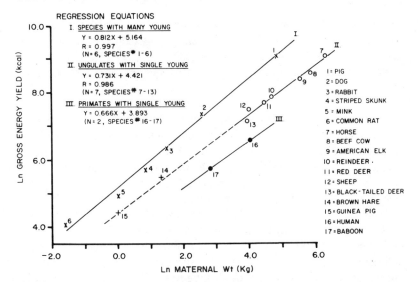

Figure 6.4 Log gross energy yield from milk plotted on log (female) body weight for seventeen mammals (from Oftedal 1981, 1984a).

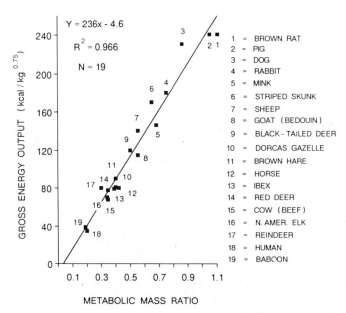

Figure 6.5 The relationship between relative energy output on milk and the ratio of litter metabolic mass (litter size × littermate weight$^{.83}$) to mother's metabolic mass (maternal weight$^{.75}$) in mammals (from Oftedal 1984a). Relative energy output on milk is calculated as energy output in kcals per day relative to maternal weight$^{.75}$.

sheep with twin lambs or triplets show substantially higher milk yields than those with singletons, and the proportion of their total energy turnover allocated to reproduction increases (Milne 1987).

Why do milk yields vary so widely? A variety of studies indicate that species living on food supplies that are rich or relatively abundant during the breeding season show high daily milk yields for their body weight, while species living on less abundant resources extend the duration of the lactation period and reduce daily expenditure. For example, detailed comparisons of hispid cotton rats (*Sigmodon hispidus*) and eastern wood rats (*Neotoma floridana*) show that the latter, which inhabit relatively poor ground, breed relatively slowly, extend their lactation period, and reduce their daily energy expenditure (McClure 1987). Within the carnivores, species with herbivorous or folivorous diets, like the black bear *Ursus americanus*, the giant panda *Ailuropoda melanoleuca*, and the red panda *Ailurus fulgens*, show relatively low daily energy outputs during lactation compared to more carnivorous species (Gittleman 1986). In contrast, species where several adults cooperate to rear the young (including coyotes, *Canis latrens*; red foxes, *Vulpes vulpes*; dholes, *Cuon alpinus*; and meerkats, *Suricata suricatta*) show relatively high total growth rates in relation to maternal body weights and probably achieve relatively high milk energy outputs for their weight (Case 1978; Gittleman and Oftedal 1987).

6.6   Milk Composition

Comparisons of milk composition between species are complicated by the presence of pronounced changes during the lactation period. For example, in macropods, there is an increase in the proportion of lipid in the milk at about 200 days, when young start to emerge from the pouch (see Figure 6.6). In addition, milk composition varies widely between species (Oftedal 1984a; Cockburn 1989). In eutherians, fat and protein contents are positively correlated and both are negatively correlated with sugar (Martin 1984). Fat and protein content (and hence the calorific value of milk) tend to decline across species with increasing maternal and neonatal weight, though exponents (−0.09 and −0.12) are small (Martin 1984). Once again, slopes calculated across groups of unrelated species mask important differences between families: if milk protein concentration is regressed on maternal weight, exponents for primates (−0.30) and ungulates (−0.44) exceed those for the overall plot, while carnivores show a smaller, positive slope (+0.16) (Martin 1984). Similar patterns are found if protein concentration is regressed on neonatal weight. The significance of these trends is not yet known, but Martin's analysis clearly demonstrates the dangers of placing any reliance on exponents derived from small samples of diverse species.

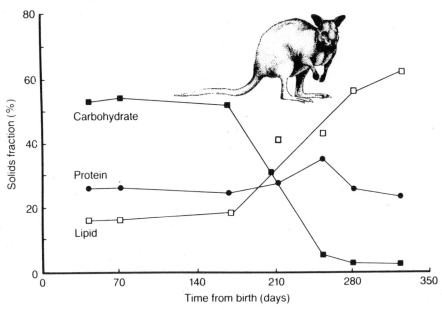

Figure 6.6 Changes in the relative concentrations of lipid, protein, and carbohydrate in the solid fraction of the milk of the wallaby, *Macropus eugenii* (from Cockburn 1989).

Why does milk content vary so widely among species? Though milk concentration apparently changes with body size, environmental factors, too, are clearly important. Mammals that suckle their young infrequently, including most altricial species, tend to have more concentrated milk than those that suckle at shorter intervals, including most precocial species (Ben Shaul 1962). In bats, flight places a premium on the reduction of weight, and milk is high in fat and dry matter (Jenness and Sloan 1970). Marsupials and primates have relatively dilute milk, while the milk of marine mammals is highly concentrated (Martin 1984; Green 1984; Bonner 1984).

The pinnipeds, in particular, provide a striking example of the adaptation of milk yield, milk composition, and lactation period to contrasting environments. They fall into two main groups: the otariids (fur seals and sea lions) and the phocids (true seals). The lactation periods of otariids are long, lasting for 4–12 months. Females interrupt the lactation period with foraging trips to sea, when they abandon their pups for 1–10 days. Across otariid species, the fat content of their milk is related to the number of days females spend at sea between suckling periods (see Figure 6.7). The length of lactation periods among otariid species are negatively correlated with latitude, possibly because breeding conditions at low latitudes are more prone to unpredictable environmental oscillations, such as El Niño effects, which delay maturation (Gentry and Kooyman 1986).

Figure 6.7   The relation between
(a) weaning age and the duration of the
mother's absence;
(b) weaning age and percentage fat content in milk; and
(c) duration of mother's absence and fat
content in milk in eight species of otariid
seals.
Numbers refer to species: (1) *Odobenus
rosmarus*; (2) *Zalophus c. californianus*;
(3) *Z. c. wollebaeki*; (4) *Euretopias juba-
tus*; (5) *Callorhinus ursinus*; (6) *Arctocephalus gazella*; (7) *A. galapagoensis*;
(8) *A. pusillus* (from Trillmich and Lechner
1986).

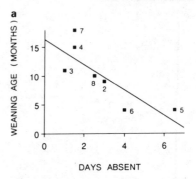

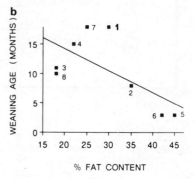

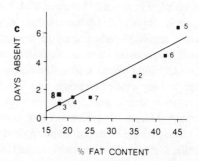

In contrast, female phocids remain close to their pups throughout the period
of lactation, and their young are provisioned primarily from maternal resources that have been stored in advance in the form of fat (see Bonner 1984;
Oftedal, Boness, and Tedman 1987; S. S. Anderson and Fedak 1987; Kovacs
and Lavigne 1986). This difference may have originated because the phocids
were originally adapted for feeding at depth, favoring the evolution of an insu-
lating blubber layer that was subsequently used as an energy store (S. S. Anderson and Fedak 1987). Partly as a result of continuous maternal attendance,
the duration of lactation is shorter in phocids than otariids, though it also varies

with breeding habitat. Species whelping on the pack ice show substantially shorter lactation periods than those breeding on land or fast ice (see Figure 6.8). For example, the hooded seal, *Cystophora cristata*, weans its pups in four

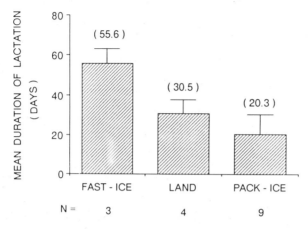

Figure 6.8   Mean duration of lactation in phocid seals breeding on land, fast ice, and pack ice. Extending lines show standard deviations (from Oftedal, Boness, and Tedman 1987). Removal of size effects influences individual values but not the relationship with breeding habitat (S. S. Anderson and Fedak 1987).

days (Bowen, Oftedal, and Boness 1985). Pack ice commonly breaks up in storms and this may favor early independence. Predation, too, may be involved: the northern pack ice breeders, like the hooded seal, are subject to land-based predation by polar bears, in contrast to southern pack ice breeders, which have longer lactation periods (Oftedal, Boness, and Tedman 1987).

These differences in lactation period are related to variation in milk content, milk yield, and energy output. All seals produce milk that is unusually high in energy content (4.5–5.9 kcal/g at mid- to late lactation), but the species with very short lactation periods have the highest fat contents. For example, the hooded seal produces milk that has an average fat content of over 60%, the highest known for any mammal. In contrast, all seals (and especially phocids) have relatively low crude protein contents in their milk compared to other mammals (4.9–11.6% compared to 20–35% for most terrestrial mammals, excluding primates).

Species with very short lactation periods and high milk energy content commonly have unusually large amounts of mammary tissue, very high daily milk yields (both in absolute terms and relative to the mother's energy budget), and relatively high weight gain per unit energy expended on pups. However, they show reduced total expenditure on lactation (see Figure 6.9). For example, the combined weight of mammary glands in female hooded seals exceeds 5 kg, daily milk production is around 8.5 kg/day or 4.8% of maternal body weight,

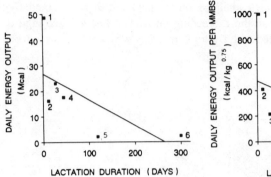

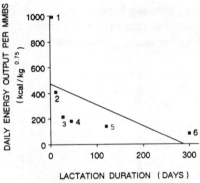

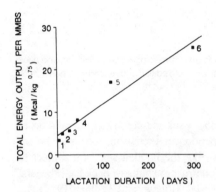

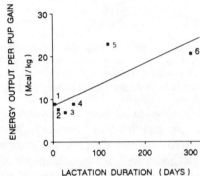

Figure 6.9 Daily energy output (Mcal), daily energy output per unit maternal metabolic weight (kcal/$kg_{0.75}$), total energy output per unit maternal metabolic weight cal/$kg_{0.75}$), and energy output per unit weight gain of pups (Mcal/kg) for six pinnipeds differing in lactation duration (data from Oftedal, Boness, and Tedman 1987): (1) hooded seal, *Cystophora cristata*; (2) harp seal, *Phoca groenlandica*; (3) Northern elephant seal, *Mirounga angustirostris*; (4) Weddell seal, *Leptonychotes weddellii*; (5) Northern fur seal, *Callorhinus ursinus*; (6) California sea lion, *Zalophus californianus*.

and pups gain an astonishing 5.7 kg/day during lactation—but the mother's total energy expenditure on lactation is less than a quarter of the expenditure of species with extended lactation periods (Oftedal, Boness, and Tedman 1987).

Compared to the phocids, the otariids show relatively high energy expenditure per unit weight gain in pups (Figure 6.9), though both families wean their pups at around a third of adult weight. This is partly because longer suckling periods are associated with a proportionally higher energy expenditure by the pup on maintenance versus growth, and partly because phocid pups have large fat reserves at weaning while otariid mothers provide energy for the energetically less efficient process of tissue gain (Oftedal, Boness, and Tedman 1987; S. S. Anderson and Fedak 1987).

6.7 Summary

Though rates of development in birds and mammals are allometrically related to body size, there is considerable variation in the length of most developmental stages among species of similar size, and the extent to which body size constrains the rate of juvenile development is not yet clear. Across species, differences in the relative duration of most periods of development are usually positively correlated with each other, so that some species move rapidly through all stages while others move slowly. In mammals, relative rates of development are inversely related to life expectancy, presumably because high age-specific mortality favors rapid development and reproduction.

Though positive correlations between the duration of consecutive development stages predominate, there is some evidence of trade-offs, too. Both in birds and mammals, precocial species have relatively long incubation/gestation periods but relatively short periods when young are dependent on their mothers for food.

Daily milk production, the duration of lactation, and milk quality are also allometrically related to body size. Total expenditure on milk is related to litter size and is generally higher in species living on rich or relatively abundant food supplies or where several adults provision a single breeding female. Among seals, pack-ice breeders have shorter lactation periods, higher daily milk yields, and richer milk but lower levels of total expenditure than land-breeding species.

7

Parental Care in Ectotherms

7.1 Introduction

The fitness costs to males and females of assuming parental care typically differ between ectotherms and endotherms, and for this reason it is convenient to deal with the evolution of care in these two groups in separate chapters. In most endotherms, the potential costs of parental care to males are likely to be very high. This is because the heavy involvement of females in parental care (associated with internal fertilization) and the usually small investment of males raises their potential reproductive rate so that any activity that constrains mating rate is likely to have heavy costs to males. Though it is theoretically possible to conceive of an endotherm where females produce large numbers of tiny propagules that are subsequently cared for by males, the selective forces favoring large propagule size in endotherms (see Chapter 3) probably militate against systems of this kind.

In ectotherms, sex differences in the costs of parental care can be smaller, though this is not always the case. Where males are able to care for large numbers of eggs or young and to maintain their mating rate while doing so, the fitness costs of assuming uniparental care may be relatively low (see Sections 7.2 and 7.5). Since the breeding success of females is commonly limited by egg production, which increases with body size, any effects of parental care on the females' ability to acquire resources or to increase body size may have substantial costs to their fitness (Gross and Sargent 1985).

Patterns of parental care (in the narrow sense of looking after eggs and young) among ectotherms range from the total absence of any form of care to the extended provisioning of young by both parents. The distribution of parental care raises three fundamental questions.

First, why do some species show elaborate and protracted care of offspring while other species (in some cases, close relatives) do not? Since most forms of parental care probably have substantial costs to the parents' survival or subsequent reproductive performance (Chapter 3), we might expect to find parental care in circumstances where it has appreciable benefits. Like large propagule size (Chapter 4), parental care might consequently be expected where environmental conditions are harsh, predation is heavy, or competition for resources is intense—or where the costs to the parent of providing care are reduced.

Second, why are both parents involved in care in some species while a single parent is involved in the majority? As soon as one parent cares for the young, the benefits of caring are likely to be reduced for the second parent (Maynard Smith 1977; Chase 1980). Consequently, biparental care might be expected only where the benefits of additional care are unusually high or the costs unusually low.

Third, why are males responsible for care in some species, and females in others? This is arguably the most important of the three questions because of its influence on the selection pressures operating on males and females (Chapter 1). While the benefits of parental care in terms of increased offspring survival will normally be similar for males and females, costs to the parent's survival and reproductive success may vary widely with ecological circumstances and may be influenced by antecedent conditions.

7.2   The Evolution of Parental Care by Males and Females

It is often suggested that females should usually be the care-giving sex because the costs of producing ova exceed those of producing sperm, a line of reasoning that is sometimes called the "anisogamy" argument. For example, Trivers (1972) argues that once the initial differentiation of gamete size occurred,

> sexual selection acting on spermatozoa favoured mobility at the expense of investment (in the form of cytoplasm). This meant that as long as the spermatozoa of different males competed directly to fertilize eggs (as in oysters) natural selection favouring increased parental investment could act only on the female. Since females were able to control which male fertilized their eggs, female choice or mortality selection on the young could act to favour some new form of male investment in addition to spermatozoa. But there exists a strong selection pressure against this. Since the female already invests more than the male, breeding failure for lack of an additional investment selects more strongly against her than against the male. In that sense, her very great investment commits her to additional investment more than the male's initial slight investment commits him. Furthermore, male-male competition will tend to operate against male parental investment, in that any male investment in one female's young should decrease the male's chance of inseminating other females. Sexual selection, then, is both controlled by the parental investment pattern and a force that tends to mould that pattern.

Trivers's argument requires some modification. First, breeding failure for lack of an additional investment selects strongly against both sexes. Deserted females are likely to continue to care for their young not because their "very great past investment commits them to additional investment" but because additional investment in offspring that are already partly developed is likely to yield a higher ratio of benefits to costs than beginning again on a new clutch (Dawkins and Carlisle 1976; Maynard Smith 1977). They are more likely to provide the additional investment than males because the costs of doing so will generally be lower.

Second, it is not the lower energetic costs of parental expenditure by males per se but their potentially higher rate of reproduction that leads to competition for females (see Section 1.2). The energetic costs of parental care are not the only factors affecting the relative reproductive rates of males and females. For example, in some pipefish where males brood eggs on or in their bodies, they appear to invest less energy per zygote in reproduction than females, yet reproductive rate is limited by the rate at which males can brood eggs rather than by the rate at which females can produce them (Berglund, Rosenqvist, and Svensson 1986).

Trivers's argument was extended by Baylis (1981), who suggested that it is the relative rate of gamete production that controls the evolution of parental care in fish. Baylis argued that the defense of oviposition sites (leading eventually to egg care) has evolved in male fish rather than in females because their rate of gamete production is higher and their breeding success is consequently limited by access to mating partners. His argument is correct where neither sex is involved in parental care. However, where either sex is already involved in caring for eggs or young, their effective rate of reproduction may be limited by the rate at which they can rear young rather than by their expenditure per gamete. For example, in the pipefish studied by Berglund, males retain their ability to produce gametes faster than females, and it is their involvement in parental care rather than any difference in gamete size that limits reproductive rate.

It would be possible to avoid this difficulty by arguing that the evolution of parental care is controlled by the relative rates of reproduction (as against gamete production) by the two sexes. However, this generalization, too, would be unsatisfactory, for the effects of reproductive rate on the evolution of parental care vary widely among organisms. For example, while their potentially high reproductive rate may favor the defense of oviposition sites and the evolution of paternal care in male fish because their mode of temperature regulation allows them to care for several clutches simultaneously, in mammals the potentially high reproductive rate of males favors the evolution of polygyny and the emancipation of males from care (Trivers 1972; Emlen and Oring 1977).

In practice, a wide variety of factors can affect the costs or benefits of parental care to the two sexes, including variation in paternity certainty and in the risk of cannibalism (see below). As a result, there need be no consistent relationship between sex differences in potential rates of reproduction and the evolution of parental care. The only reliable prediction that can be made is a very general one: that parental care is likely to evolve in whichever sex benefits most from caring for its young when the costs of care and the response of the other sex to desertion are taken into account.

Some confusion over explanations of parental care has in the past been caused by failure to distinguish clearly between (1) questions about the evolution of parental care where no-care by either sex is the antecedent state; (2) questions about the subsequent course of evolution of parental investment once either or both sexes care for the young; and (3) questions about the maintenance of sex differences in parental investment in contemporary populations. Sex differences in gamete size and in the rate of gamete production are most likely to influence the initial evolution of parental care. Once either or both sexes are involved in parental care, the nature of antecendent conditions is likely to be very important, and a wide variety of factors may affect the costs and benefits of additional care to the two sexes. Arguments about the maintenance of sex differences in parental care in contemporary populations commonly make more restrictive assumptions about biological constraints on the evolution of parental care than those concerning the past course of evolution.

7.3   The Maintenance of Parental Care by Males and Females

The selection pressures that maintain parental care by males and females have been specified by Maynard Smith (1977, 1978; see also Vehrencamp and Bradbury 1984). Where desertion by one parent affects investment by the other partner, game theory models are the appropriate way of specifying the conditions under which uniparental male and uniparental female care are likely to evolve. Maynard Smith's well-known model (1977) assumes that there are discrete breeding seasons; that a female's expenditure on egg laying and on parental care limits the number of young she can produce; and that uniparental care by males and females has the same effects on offspring survival.

Let P_0 be the probabilities of survival of eggs that are not cared for by either parent, P_1 the survival of eggs cared for by one parent and P_2 the survival of eggs cared for by both parents, such that $P_2 > P_1 > P_0$. Suppose that a male who deserts has a chance p of mating again while a female who deserts after egg laying produces W eggs, compared to w for a caring female. The payoff matrix for this game and the four ESS's that it generates are shown in Table 7.1. They are the following:

1. *When both sexes desert.* This requires that (a) the number of eggs laid by a noncaring female multiplied by their survival exceeds the number laid by a caring female multiplied by their survival, or the female will care; and (b) that

Table 7.1

Payoff matrix and four ESSs from Maynard Smith's (1977)
game theory model (see text).

Male		Female	
		Cares	*Deserts*
Cares	Female gets	wP_2	WP_1
	Male gets	wP_2	WP_1
Deserts	Female gets	wP_1	WP_0
	Male gets	$wP_1 (1+p)$	$WP_0 (1+p)$

ESS 1 : Female deserts and male deserts
Requires: $WP_0 > wP_1$ or female will care.
$\qquad$ $P_0 (1+p) > P_1$ or male will care.

ESS 2 ("Stickleback"): Female deserts and male cares
Requires: $WP_1 > wP_2$ or female will care.
$\qquad$ $P_1 > P_0 (1+p)$ or male will desert.

ESS 3 ("Duck"): Female cares and male deserts
Requires: $wP_1 > WP_0$ or female will desert.
$\qquad$ $P_1 (1+p) > P_2$ or male will care.

ESS 4: Female cares, male cares
Requires: $wP_2 > WP_1$ or female will desert.
$\qquad$ $P_2 > P_1 (1+p)$ or male will desert.

NOTE: This model assumes that $P_2 > P_1 > P_0$; that breeding seasons are
discrete; and that breeding success is limited by the costs of egg
laying and parental care.

the survival of eggs that are not cared for by either parent multiplied by the
number of matings achieved by a noncaring male $(1+p)$ cannot exceed the
survival of eggs under uniparental care, or the male will care.

2. *When the female deserts and the male cares ("Stickleback")*. This requires
that (a) the number of eggs laid by a noncaring female multiplied by egg
survival under uniparental care must exceed the number laid by a caring female multiplied by egg survival under biparental care, or the female will care;
and (b) that egg survival under uniparental care must exceed survival of un-
cared-for eggs multiplied by the number of matings that a noncaring male can
achieve, or the male will desert.

3. *Where the female cares and the male deserts ("Duck")*. This requires that
(a) the number of eggs laid by a caring female multiplied by egg survival under
uniparental care must exceed the number of eggs laid by a noncaring female
multiplied by the survival of eggs that are not cared for, or the female will

desert; and (b) that the mating success of a noncaring male multiplied by egg survival under uniparental care must exceed egg survival under biparental care, or the male will care.

4. *Where both partners care.* This requires that (a) the number of eggs laid by a caring female multiplied by egg survival under biparental care must exceed the number laid by a noncaring female multiplied by egg survival under uniparental care, or the female will desert; and (b) that the number of eggs that survive under biparental care must exceed egg survival under uniparental care multiplied by the mating success of the noncaring male, or the male will desert.

If breeding seasons are continuous so that each individual breeds many times and two parents are less than twice as effective as one at ensuring that their eggs survive, either uniparental male or uniparental female care can evolve from no-care (Maynard Smith 1977). Whether male or female care evolves depends upon biological factors affecting the relative costs and benefits of care to each sex as well as the antecedent condition. Where the initial condition is biparental care, and two parents are less than twice as efficient at egg care as one, males may be more likely to desert than females where the absence of males affects egg survival less than the absence of females (see Section 7.3). If two parents are *more* than twice as good as one at caring for their eggs, the only possible ESS's are desertion by both sexes or biparental care (Maynard Smith 1977).

Unfortunately, tests of Maynard Smith's model in contemporary populations may not tell us much about the initial evolution of parental care. This is because, after the initial evolution of parental care, the behavior of males and females and the development of eggs and juveniles are likely to become co-adapted. As a result, manipulation of contemporary breeding systems may provide little indication of the original payoffs of care to males and females. For example, after parental care has evolved, the development of young at hatching may change, and P_0 may consequently fall to zero. By removing both parents, we may be able to show that $P_1 \gg P_0$, but the original difference may have been far smaller. In species where one sex normally cares for eggs or young, the noncaring sex may gradually lose the capacity to provide effective care so that the experimental removal of one parent never leads to the assumption of care by the other and is usually followed by the death of the entire brood. Similarly, after the initial evolution of biparental provisioning, juvenile growth rates may increase, so that successful rearing requires biparental care and P_1 falls to zero.

The most relevant empirical approach to investigating the evolution of male and female care is probably to examine the costs and benefits of care to males and females in contemporary species where either sex may care for the young (e.g., Blumer 1986). However, even here the behavior of one sex may have become adapted to lower levels of investment by partners. As a result, explanations of the initial evolution and current distribution of male and female care rely on attempts to reconstruct its likely costs and benefits under particular antecedent conditions in the past and are necessarily speculative.

The four following sections review the distribution of parental care among invertebrates, fish, amphibia, and reptiles, attempting to account for the evolution of sex differences in parental care. The evolution of parental care in birds and mammals is described in Chapter 8.

7.4   Parental Care in Invertebrates

THE DISTRIBUTION OF CARE

Parental care of eggs or young (in the narrow sense) is uncommon among invertebrates, perhaps because parents are seldom able to protect their young effectively and selection favors the production of large numbers of eggs (Zeh and Smith 1985; Tallamy and Wood 1986). Among the terrestrial arthropods, eggs are covered by the female before being abandoned in at least twenty orders, egg care occurs in twenty orders, care of young in seventeen orders, and young are provisioned in seven orders (see Table 7.2). Orders showing the

Table 7.2
Parental care by terrestrial arthropods
(from Zeh and Smith 1985).

	Female Only	Female + Male	Male Only
EP	20	0	0
EC	20	3	3
YC	17	5	1
YP	7	3	0
YR	4	2	0
BS	4		
OVO	11		
VIV	8		

NOTES: The table shows the number of orders in which either or both parents cover the eggs before abandoning them (EP); remain with the eggs and care for them (EC); remain with the young and care for them (YC); provision the young (YP); or regurgitate food to the young (YR). In addition, the table shows the number of orders in which embryos are nourished in an external brood sac (BS); in which ovoviviparity or larviparity occur (OVO); and in which viviparity occurs (VIV).

most developed examples of parental care include the Hemiptera (true bugs), Thysanoptera (thrips), Embioptera (web spinners), Coleoptera (true beetles), as well as the Hymenoptera (ants, wasps, and bees) and Isoptera (termites) (Wilson 1971; A. C. Eickwort 1981; Tallamy and Wood 1986).

Parental care is confined to species where eggs or young are clumped in time and space and is commonly associated with physically harsh or biotically dangerous habitats (see Thorson 1950; Linsenmair and Linsenmair 1971; Wilson 1971; Ridley 1978; Hinton 1981; Zeh and Smith 1985; Tallamy and Wood 1986; Wyatt 1986). For example, in the intertidal beetle, *Bledius spectabilis*, adults are forced to remain with their young to prevent their suffocation by the incoming tide (Wyatt 1986), while, in some earwigs and crickets that nest in deeper soil and feed on moist vegetable matter, parental attendance is necessary to control invasions by fungi (West and Alexander 1963; Wilson 1971). Parental care of young is also common among insects whose young depend on rich but scattered and ephemeral resources, such as dung, dead wood, or carrion where direct competition for resources is intense (Wilson 1971). In many of these species, parents defend resources used by their growing offspring and are commonly equipped with horns or large mandibles used in direct competition with conspecifics (see Otte and Stayman 1979). Finally, parental care occurs in a number of terrestrial insects that depend on food supplies that nymphs or larvae could not easily exploit without parental assistance. For example, in some tree hoppers, Membracidae, where females guard eggs laid on twigs, the female moves down the twig shortly before the eggs hatch and makes a series of slits in the bark, which are later used by the developing nymphs to suck sap (Wood 1976, 1977). Similarly, in some tenebrionid beetles adults chew dead wood that is subsequently eaten by the young (A. C. Eickwort 1981).

Parental care among marine invertebrates is associated with small body size (Underwood 1979; Strathmann and Strathmann 1982; Grahame and Branch 1985), a trend that has parallels in reef fish (see Section 7.4). Several nonexclusive explanations have been proposed, including suggestions that (1) the capacity for egg production rises more rapidly with increasing body size than the capacity to brood eggs, favoring broadcast spawning in larger species (Heath 1977); (2) brooding ensures a minimum level of breeding success and is consequently favored in small species with relatively short breeding lifespans; and (3) broadcast spawning improves dispersal, which may have greater advantages to large species because suitable breeding sites are less abundant, established adults are longer lived, competition between larvae and adults of smaller species is more intense, or adult dispersal is less common (see Strathmann and Strathmann 1982). Though none of these explanations account for the observed trends in all groups, the existence of similar trends in groups varying widely in absolute size as well as in marine fish (see Section 7.4) suggests that allometries in egg production or brood size may well be involved.

THE DISTRIBUTION OF MALE VERSUS FEMALE CARE

In the comparatively small number of invertebrates that care for eggs or young, female care is commonest and biparental care and uniparental male care are unusual (Wood 1976, 1977; Ridley 1978; Smith 1980; Eickwort 1981; Tallamy 1984; Zeh and Smith 1985; Tallamy and Wood 1986: see Table 7.2). The prevalence of uniparental female care among insects compared to fish is probably related to the comparative scarcity of external fertilization. This favors the evolution of care by females rather than males because it is likely to be associated with the separation of the sexes before oviposition (see Section 7.5).

Uniparental male care occurs in at least two polychaetes, including a dioecious nereid, *Neanthes arenaceodentata*, where the male commonly eats the female after she has laid her eggs (Reish 1957). Among the terrestrial arthropods, it occurs in a minority of harvestmen, Opiliones (Rodriguez and Guerrero 1976), in the Polyzonida (Kaestner 1969) and in several Heteroptera, including members of the Coreidae, Dysodiidae, Reduviidae, and Belostomatidae. Among marine invertebrates, uniparental male care predominates among the sea spiders, Pycnogonidae, where males carry eggs attached to a pair of specialized ovigerous legs in seven out of ten families (Jarvis and King 1972, 1975; R. E. King 1973; Ridley 1978). At fertilization, which is believed to be external, the male extracts the eggs from the female's gonads with these legs.

Uniparental male care has probably developed by more than one evolutionary pathway. In some cases, the costs of egg care to males may be reduced because successive females lay their eggs in the same site so that the male can continue to mate while caring for eggs or young (see above). In the assassin bug, *Rhinocoris albopilosus*, where males brood egg masses, several females contribute clutches to the same egg mass and males compete to brood egg masses (Odhiambo 1959, 1960; see also Ralston 1971). *R. albopilosus* is of particular interest since in related species, including *R. carmelita*, females care for individual clutches, repelling conspecifics that attempt to add eggs to unattended broods, as well as predators (Edwards 1962). One possible explanation of these contrasting patterns of care could be that egg or nymph survival is positively related to brood size in *R. albopilosus* and negatively in *R. carmelita*, but whether this is actually the case is unknown. In the giant waterbug, *Abedus herberti* (Belostomatidae), several females may lay their eggs on the backs of a single male (R. L. Smith 1979a; A. C. Eickwort 1981; though see also Kruse 1990). Combined with heavy predation on eggs and a shortage of suitable oviposition sites, this may have facilitated the evolution of male care

(R. L. Smith 1979a). In some harvestmen and millipedes, too, males also defend oviposition sites where females lay their eggs (Rodriguez and Guerrero 1976; Kaestner 1969). Little is known of the breeding system of the two polychaetes where uniparental male care occurs but fertilization is generally external in this group, and it is possible that males guard eggs laid by more than one female.

In other cases, uniparental male care may develop where a male's opportunity for additional matings is low. For example, in sea spiders, prolonged courtship combined with male guarding may reduce the opportunity for remating and facilitate the evolution of male care. In *Pycnogonum littorale*, the female releases all her eggs in one spawning, mating pairs remain together for up to five weeks, and males carry the eggs of a single female in one mass for up to ten weeks (Jarvis and King 1972). However, in some other pycnogonids, males are polygynous, carrying several balls of eggs laid by different females (P. E. King and Jarvis 1970; P. E. King 1973).

BIPARENTAL CARE

Biparental care is rare among invertebrates. Where it does occur, it is often associated with nest building and the care of young as well as of eggs (A. C. Eickwort 1981). For example, it occurs in 50% of orders of terrestrial arthropods that regurgitate food to their young and 30% of orders that provision their young by any method, compared to 14% of orders that only show egg care. In species where parents feed their young, competition for larval food is usually intense and the benefits of parental care are likely to be depreciable. In particular, feeding competition appears to have been an important factor in the evolution of biparental care in the dung and burying beetles, such as *Nicrophorus* (Milne and Milne 1976; Heinrich and Bartholomew 1979; A. C. Eickwort 1981; Tallamy 1984; Scott 1990). In other cases, biparental care is associated with circumstances that restrict the male's opportunity to remate. For example, in the biparental desert isopod, *Hemilepistus reamuri*, where both sexes cooperate to build, clean, and alternately defend their burrow, males show high mortality (63–97%) during pair formation, and breeding seasons are brief so that a male's chances of remating are likely to be low (Linsenmair and Linsenmair 1971; Schachak, Chapman, and Steinberger 1976).

There is evidence that multiparental care can be more effective than uniparental care in a variety of communal and quasisocial insects (A. C. Eickwort 1981). For example, adult spider mites, Tetranychidae, defend communal nests and offspring against larvae of a phytoseiid predator, *Typhlodromus bambusae*, often killing the larvae. The death rate of larval predators released close to spider mite nests increases with the number of adult mites defending

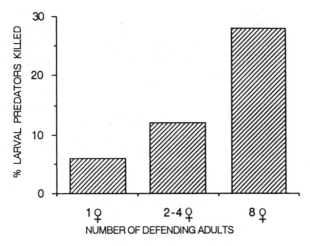

Figure 7.1   Percentage of larval predators killed by spider mites (*Shizotetranychus celarius*) in relation to number of adults defending the nest (from Saito 1986; Yamamura 1987).

the nest (see Figure 7.1; Yamamura 1987). There is apparently some degree of role separation: females locate and pursue predators and males kill them. As a result, the death rate of predators is higher when nests are guarded by both sexes than when they are guarded by females alone (Figure 7.1).

7.5   Parental Care in Fish

THE DISTRIBUTION OF PARENTAL CARE

Because of their great diversity of breeding systems, the evolution of parental care in fish poses some of the most testing questions. Parental care after egg laying is apparently absent among the chondrichthyan fish (sharks, skates, rays, and chimaeras) (Breder and Rosen 1966; Wourms 1977). Around 21% of the 422 families of bony fish show viviparity, ovoviviparity, or some form of parental care of eggs, while fewer than 6% also show guarding of newly hatched young (Breder and Rosen 1966; Keenleyside 1978; Blumer 1979; Baylis 1981; Gross and Shine 1981; Wourms 1981). In teleosts, parental guarding of eggs or young occurs in 57% of families breeding in freshwater compared to 16% of all fish (Baylis 1981). An association between parental care and breeding in fresh or shallow water also occurs within several groups.

Of the two families of catfish that include both marine and freshwater breeding species, one, the Plotosidae, includes freshwater forms that guard their eggs and marine forms that show no parental care, while the other, the Ariidae, includes marine forms that are external bearers and freshwater forms that guard eggs (Breder and Rosen 1966; Baylis 1981).

Egg guarding is the commonest form of parental care (see Table 7.3). As among birds, egg guarding has led to the development of brood parasitism in a number of fish including some North American minnows and shiners (Cyprinidae) that spawn in the nests of sunfish (Breder and Rosen 1966). In Lake Tanganyika, mouth-brooding cichlids are parasitized by a mochokid catfish *Synodontis multipunctatus* (Sato 1986). The catfish (presumably) lays its eggs at the same time as its hosts, and they are subsequently picked up and brooded orally. The eggs of the catfish hatch earlier than those of the host, and the catfish fry feed upon the fry of the host while still in its mouth.

Larval guarding is unusual, though it is highly developed in a few freshwater families (Breder and Rosen 1966). Among marine fish, it occurs definitely in only two species, both belonging to the family Pomacentridae, though it may also occur in single representatives of two other families (Baylis 1981). Excluding viviparous species, few fish feed their young. However, in some cichlids, parents will ingest food from the bottom and then expel it, allowing their fry to feed on it, while in others fry feed from mucus produced externally by their parents (see Section 2.7).

Several different factors probably contribute to the association between parental care and fresh or inshore waters. In the open sea, water conditions are relatively homogeneous compared to inshore or freshwater, egg predators are relatively scarce, and pelagic eggs probably have a reasonable chance of survival (Wourms 1977; Johannes 1978; Perrone and Zaret 1979; Baylis 1981). Production of pelagic eggs in freshwater is largely restricted to species that inhabit very large water bodies or migrate great distances upstream to spawn (Balon 1975; Baylis 1981). In inshore and freshwater, the spawning site exerts a strong influence on egg survival, and adhesive eggs laid on the bottom (demersal) are commonly favored (Breder and Rosen 1966; Johannes 1978; Baylis 1981). Favorable spawning sites are often limited, leading to local congregations of males. Defense of an optimal site may increase the breeding success of males by enabling them to attract and guard multiple mates (Baylis 1981). Once territorial defense by one or both sexes has evolved, the costs of improving the territory and of parental care to the guarding sex may be low: in most fish that show egg guarding, males defend sites where several females spawn in succession, and egg guarding by males is rare among species showing internal fertilization (Blumer 1979; Barlow 1981; Baylis 1981). Reduced egg predation or increased hatchability are probably the main benefits of guarding (Baylis 1981). In teleosts, egg guarding is associated with relatively large egg size and comparatively low fecundity, possibly because the long

Table 7.3

Occurrence of different forms of parental care in families of bony fishes
(from Blumer 1979).

Form of Parental Care	Number of Families		
	Either Sex	Males	Females
Guarding	63	57	27
Nest building, substrate cleaning, or both	39	37	18
Fanning	30	30	11
Internal gestation	14	0	14
Removal	12	12	5
Oral brooding	10	7	4
Retrieval	8	7	5
Cleaning eggs	6	6	4
External egg carrying	6	4	3
Egg burying	5	4	3
Moving	5	5	3
Coiling	4	3	2
Ectodermal feeding	3	3	3
Brood pouch	2	1	1
Splashing	2	2	0

NOTES: In the first column, each family is counted no more than once in any category
but could be counted in more than one category. Some families are allocated both
to male and female columns because they contain species in which males and females exhibit the same form of parental care.

developmental periods of larger eggs favor guarding to reduce mortality (see
Section 4.6).

As among marine invertebrates, parental care in marine fish is commonest
in small species with an adult length of less than 10 cm, which lay demersal
eggs (Barlow 1981; Thresher 1984). Large fish of 40 cm and above produce
mostly pelagic eggs, and those that produce demersal eggs generally either
scatter their eggs and do not care for them or prepare shallow nests in open,
sandy areas. One suggestion is that this association occurs because small fish
cannot produce sufficient eggs for planktonic dispersal to represent a viable
strategy, while large fish have difficulties finding caves and crevices big
enough to keep and guard eggs in (Baylis 1981; Thresher 1984; but see
Shapiro, Hensley, and Appeldoorn 1988).

THE DISTRIBUTION OF MALE VERSUS FEMALE CARE

In the majority of fish showing parental care of eggs or young, only one parent
is involved (Table 7.4). Male care is considerably commoner than female care:

of families showing ovoviviparity, viviparity or postovipositional care of eggs or young, uniparental male care occurs in 58%, uniparental female care (usually involving ovoviviparity or viviparity) in 36%, and biparental care in 25% (Blumer 1979; Gross and Shine 1981). In some species, patterns of parental care are variable, and uniparental male care, biparental care, or even uniparental female care can occur in the same population (Fryer and Iles 1972; Blumer 1985, 1986).

Comparisons of the distribution of parental care among groups of related species suggest that no-care is the usual ancestral state for fish with male-only care, male-only care is the usual ancestral state for biparental care, and biparental care is the usual ancestral state for uniparental female care (Barlow 1974; Loiselle 1978; Gittleman 1981; Gross and Sargent 1985). This hypothetical route of evolution is shown in Figure 7.2. Figures show percentages of externally fertilizing teleost families in each state, while the percentage of families including species in more than one state are shown in dashed circles (from Gross and Sargent 1985). Biparental care presumably evolved from uniparental male care followed, in a few species, by the evolution of uniparental female care.

Transport of eggs or fry and mouth brooding by males has probably usually developed from ancestors where males guarded eggs or, among cichlids, from biparental care (Fryer and Iles 1972; Blumer 1979). However, it is also known in one species of anglerfish, *Antennarius caudimaculatus*, which belong to a group that typically produce free-floating egg masses. In this case, external attachment of eggs may also serve to attract potential prey (Pietsch and Grobecker 1980).

There is a close association among bony fish between the pattern of care and

Table 7.4

Comparative incidence of parental care after egg laying and viviparity among teleost fish, amphibians, and reptiles (from Shine 1988b).

	Teleosts	Amphibians	Reptiles
Number of families for which data included	182	35	43
Number with male parental care (and proportion)	63	16	1
	(.35)	(.46)	(.02)
Number with female parental care (and proportion)	28	17	13
	(.15)	(.49)	(.30)
Number with parental care by either parent (and proportion)	68	23	13
	(.37)	(.66)	(.43)
Number with viviparity (and proportion)	11	6	18
	(.06)	(.17)	(.42)
Number with parental care and/or viviparity (and proportion)	77	25	24
	(.42)	(.71)	(.56)

NOTE: Data for teleosts and amphibians from Gross and Shine (1981) and references therein.

Figure 7.2 A hypothetical evolutionary phylogeny of parental care
in fish with external fertilization
(from Gross and Sargent 1985).
There are four possible states of
parental care: none, male, female,
or biparental. The likely direction
of evolution among states is
shown by arrows. The percentage
of externally fertilizing teleost
families in each state is shown,
and families including species in
more than one state are indicated
by dashed circles.

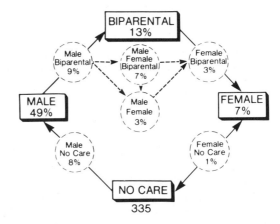

the mode of fertilization (see Figure 7.3): among species with external fertilization, approximately 76% of cases of parental care (including egg retention
and viviparity) involve males, which commonly defend territories used by
several females in succession (Williams 1975; Ridley 1978; Gross and Sargent
1985). In contrast, among internal fertilizers, 80% of cases involve females,
and care commonly takes the form of ovoviviparity or viviparity (Gross and
Shine 1981).

Figure 7.3 Occurrence of male
and female parental care (including biparental care) among 182
families of teleosts and 45 of
amphibia in relation to mode of
fertilization (from Gross and Shine
1981). Families may appear in
more than one category. Viviparity
and ovoviviparity are included as
forms of parental care, but not
nest preparation before mating.

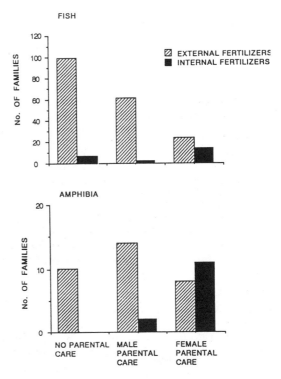

The distribution of parental care in fish raises at least five important questions:

1. WHY IS MALE CARE ASSOCIATED WITH EXTERNAL FERTILIZATION?

The association between male care and external fertilization is sufficiently close to form an obvious starting point of any attempt to account for sex differences in care among fish. Three explanations have been advanced. First, Trivers (1972) suggests that male care is associated with external fertilization because a male's confidence of paternity is higher when fertilization is external and females cannot store sperm (see also Blumer 1979; Perrone and Zaret 1979). However, as Grafen (1980) has shown, this depends on the male's ability to use the time or energy saved by deserting his previous offspring to increase his mating rate. If his only alternative use for resources saved by avoiding parental care in the first brood is to help later offspring of equally doubtful paternity, differences in the average level of paternity certainly will not affect the optimal allocation of parental care by males. Moreover, "sneaking," "streaking," and other alternative mating tactics are sufficiently common that it is not clear whether Trivers's basic assumption is correct (Baylis 1981; Gross 1984).

The second explanation is based on the assumption that females are the first to release their gametes in external fertilizers. This may favor desertion by females and enable them to put their partners in a "cruel bind" where they are forced to choose between caring for the clutch alone or deserting it (Trivers 1972; Dawkins and Carlisle 1976). Female care may predominate in internal fertilizers because the female is usually unable to desert the embryos before the male's departure. There are several objections to this explanation. In the majority of external fertilizers, gamete release is simultaneous in the two sexes (Gross and Shine 1981; Gross and Sargent 1985). Moreover, in some species, females release their gametes first and subsequently care for broods after males have fertilized them (Wickler 1965; Loiselle 1978). In other species, males spawn first but subsequently guard the eggs. For example, in some Callichthyidae (armored catfish), females draw spermatic fluid into their mouth from the male's vent, use this sperm to fertilize eggs held in their ventral fins, place the fertilized eggs in a foam nest built by the male, and then leave the male to care for the nest (Gross and Shine 1981).

The third explanation is based on the suggestion that while the benefits of parental care will be similar to males and females, the costs will often be higher for females, especially in external fertilizers (Williams 1975; Borgia 1979; Baylis 1981; Gross and Shine 1981; Gross and Sargent 1985). In many external fertilizers, males can guard large numbers of eggs, contributed by several females which spawn more than once per season (Blumer 1979). For example, in some darters, nests commonly contain 2000 or more eggs though females lay only around 150 eggs at a time (Gale and Deutsch 1985). In some cases, males which already have eggs in their nests are more likely to obtain further matings (Figures 7.4 and 7.5), either because females prefer to lay in nests that already contain eggs (Ridley and Rechten 1981; Unger and Sargent 1988) or because males that have recently spawned court more intensely (Jamieson and Colgan 1989). In internal fertilizers, by contrast, male defense of multiple clutches is only likely to evolve where oviposition occurs immediately after fertilization, so that mating partners are not separated.

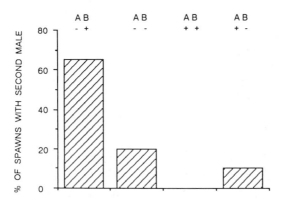

Figure 7.4 Proportion of female sticklebacks that spawned with or did not spawn with the second male (B) when presented in sequence with a male with (+) or without (–) eggs already in his nest (from Ridley and Rechten 1981). As the figure shows, females were more likely to spawn with males with eggs already in their nests.

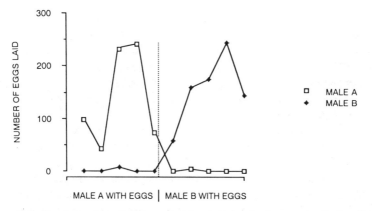

Figure 7.5 Numbers of eggs laid in nests of male fathead minnows that had or did not have eggs already in their nests (from Unger and Sargent 1988). Two nest sites, two males, and four females were placed in separate aquaria. After males had taken up defense of nests, eggs were randomly assigned to one of the two males (Male A) and females were allowed to spawn. Subsequently all eggs were removed from the original male in each aquarium and eggs were assigned to the other male (Male B) and females were allowed to spawn again. The figure shows the number of eggs received by A and B males in the first and second halves of the experiment. In the first half of the experiment, male A had eggs, while in the second half, male B had eggs and A had none.

The fitness costs of guarding may also be lower to males than females in external fertilizers because the energetic costs of spawning are generally lower to males (Loiselle and Barlow 1978; Perrone and Zaret 1979; Blumer 1986). In addition, body size may often have a weaker effect on mating success in males than on fecundity in females, so that any effects of guarding on growth have higher costs to females (Gross and Sargent 1985; but see De Martini 1976; Perrone 1978a; Shine 1988b). Female guarding in external fertilizers is often associated with situations where females breed no more than once a year and the costs of care to females are probably low (see below).

This third hypothesis provides a credible explanation of the association between male care and external fertilization as well as of the higher incidence of uniparental male care in fish than in insects, reptiles, birds, and mammals. It may also explain why in a number of monogamous fish, males are responsible for virtually all egg tending (Fricke 1974). It does, however, leave the four questions below unanswered:

2. WHY DOES UNIPARENTAL FEMALE CARE OCCUR IN SOME SPECIES WITH EXTERNAL FERTILIZATION?

Female care occurs in at least twenty-four families of external fertilizers (Gross and Shine 1981) and probably has more than one evolutionary cause. Uniparental female care in external fertilizers is often associated with short breeding seasons or with semelparity, both of which are likely to reduce any effects of care on the female's subsequent breeding opportunities (Qasim 1956; Perrone and Zaret 1979; Gross and Sargent 1985). For example, an association between female egg guarding and short breeding seasons is found among the blennies, where egg guarding is confined to northern species whose females usually lay all their eggs in a single clutch (Qasim 1956; Thresher 1984). In the Salmonidae, too, female care is associated with short breeding seasons, and care of eggs is most highly developed in Pacific salmon where the great majority of females are semelparous (van den Berghe 1984; Gross and Sargent 1985; van den Berghe and Gross 1989).

In other cases, female guarding in external fertilizers is associated with breeding systems where several females simultaneously defend small territories within the larger territory of a single male. Here, competition between females for limited nesting sites or persistent egg cannibalism could favor the evolution of uniparental female care either from a nonguarding ancestor or from uniparental male care. For example, in some cichlids, female care is associated with very restricted oviposition sites where males are able to defend territories large enough for several females to breed simultaneously. In two of the smallest genera of egg-guarding cichlids, *Apistogramma* and *Lamprologus*, males defend oviposition sites in cavities among rocks and tree roots (Fryer and Iles 1972). Several females lay their eggs within a male's territory, and females guard and tend their broods while the male defends the entire territory (Wickler 1965). Suitable oviposition sites may be unusually restricted in these species because guarding adults are particularly susceptible to predators on account of their size. In some marine fishes, female guarding is associated both with synchronized breeding and with the use of male territories by several females simultaneously. For example, in the triggerfish, *Odonus niger*, up to a dozen or more females defend small, clustered territories 70–100 cm in diameter within the larger territory of a single male (Fricke 1980). In this species, the male helps his females to guard the area against potential predators and also intervenes in territorial disputes between females. The costs of parental care to the females are probably low, for eggs are only guarded for one day after spawning, and breeding is closely synchronized.

Where biparental care is the ancestral state, uniparental female care may be more likely to develop than uniparental male care when conditions permit a return to uniparental care because the male has more to gain from liberation. For example, among the cichlids, uniparental female mouth brooding is widespread among external fertilizers and has apparently evolved from a biparental, egg-guarding ancestor (Fryer and Iles 1972). Biparental egg guarding in cichlids is usually associated with monogamy, which can extend over several breeding seasons, and with a relatively equal division of labour between the sexes (Fryer and Iles 1972). Uniparental female mouth brooding is probably commoner than uniparental male mouth brooding because the potential benefits of desertion are higher to males than females (see Gross and Sargent 1985). It may also be more likely to evolve where changing environmental conditions favor the transport of eggs by females to oviposition sites outside the male's territory (see Fryer and Iles 1972). For example, in some offshore species of mouth-brooding *Tilapia*, females mate in shallow water, then migrate to reedy areas to brood and release their young before migrating back to deeper water (Fryer and Iles 1972).

Where the costs of uniparental female care are unusually high, females may be selected to desert, leading to a return to no-care. Gross and Sargent (1985) suggest that the patterns of care may be unstable, moving in sequence through the different states shown in Figure 7.1, but whether or not this is the case is unknown.

3. WHY IS FEMALE CARE ASSOCIATED WITH INTERNAL FERTILIZATION?

Once internal fertilization has evolved in females, the evolution of egg re-- tention by the female is likely to require minimal reorganization and probably evolves easily (Williams 1966b). Moreover, the costs of egg retention to females may be relatively low, while benefits to the survival of eggs and young are likely to be substantial (Williams 1975; Johannes 1978; Gross and Sargent 1985). Internal brooding of eggs by males may seldom develop from an external fertilizing nonguarding ancestor because it usually constricts the number of matings that a male can achieve (Blumer 1979). External attachment of eggs to males, mouth brooding and internal brooding of eggs and young by males (as in pipefish and sea horses) have probably usually developed from uniparental male care of demersal or adhesive eggs where selection favors additional protection or the movement of eggs from site to site (Blumer 1979; Vincent 1990).

This argument should focus attention on the evolution of parental care in internally fertilizing species where females deposit their eggs immediately after fertilization. If the protracted association between females and zygotes is responsible for the predominance of female care among internal fertilizers, the relative frequency of male and female care should not differ from externally fertilizing species.

4. WHY DO MALES GUARD IN SOME INTERNAL FERTILIZERS?

Excluding the sea horses and pipefish, male care of eggs is known to be associated with internal fertilization in only three families: the sculpins, Cottidae; the Pantodontidae (a relative of the arapaimas, living in freshwater); and the cardinal fish, Apogonidae. In all three groups, females are oviparous and do not appear to retain fertilized eggs (Breder and Rosen 1966).

The sculpins are shallow marine and freshwater fish with a circumpolar distribution. In a few species, males possess an intromittent organ and fertilization is internal (Breder and Rosen 1966). Males commonly mate with several females, whereas females often breed only once, depositing their entire mass of demersal eggs in the territory of a single male (Downhower and Brown 1981). Eggs and, in some cases, fry are then guarded by the male for around six weeks. The breeding system of the pantodonts is not well known, but in some species males again possess a specialized anal fin and fertilization is probably internal (Breder and Rosen 1966). Males may stay with the eggs, which float freely or are attached to floating vegetation. Finally, in the cardinal fish, a marine group that is primarily nocturnal, several species are male mouth brooders. In some species, including *Apogon imberbis* (Garnaud 1950, 1962) and *A. notatus* (Usaki 1977), fertilization appears to be internal. In *Apogon notatus*, pairs defend spawning sites on the seabed, where females deposit their eggs immediately after mating (Usaki 1977). Males subsequently inhale the eggs and brood them in their mouths. Females spawn several times per season, and males are thought to carry eggs produced by more than one female. Females are primarily responsible for defense of the spawning site and may guard brooding males.

No satisfactory explanation of the association between internal fertilization and male care in these cases is yet available. However, in all three groups, internal fertilization may be a secondary development following the evolution of uniparental male care. Where internal fertilization enhances the successful transfer of sperm to the female, there is no reason why it should not evolve in external fertilizers with male care, and no reason to predict that it should lead to a reversal of parental roles unless it is accompanied by egg retention in females. The benefits of egg retention (which might be expected to encourage the evolution of female care) may be small when parental guarding by the male is already well developed.

5. WHY IS BIPARENTAL CARE RELATIVELY UNCOMMON IN FISH?

Recent studies of normally biparental species show that biparental care of fry is often more effective than uniparental care (Keenleyside 1978). Experiments with a biparental cichlid, *Herotilapia multispinosa*, where both parents are equally involved in guarding and tending fry, show that the removal of either parent causes the survival rate of eggs and fry to fall to low levels when predators are present but has little effect when they are absent (see Figure 7.6). Removal of males has a slightly smaller effect than the removal of females, though this trend is reversed when parents are removed 6 days after spawning (Keenleyside 1978).

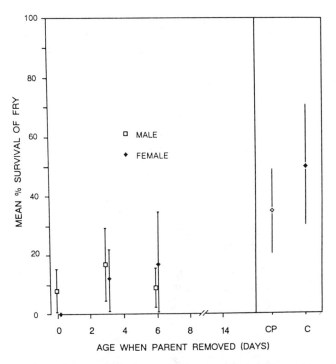

Figure 7.6 Effects of the removal of males (squares) or females (diamonds) on the survival of eggs and fry to 15 days post spawning in a biparental guarding cichlid, *Herotilapia multispinosa* (from Keenleyside 1978). Parents were removed at different times after spawning (Day 0). Eggs hatched approximately 48 hours post spawning and fry were free-swimming 72 hours after this. Predators were male *Cichlasoma nigrofasciatum*. CP: control with predators present; C: control with no predators. Extending lines show ±1 S.E.

A second benefit of biparental care to females may be that it permits them to assume brood care if their partner dies. For example, female brown bullheads, *Ictalurus nebulosus*, commonly remain in the vicinity of their brood, though males are primarily responsible for guarding (Blumer 1986). However, if the male disappears, the female takes over the male's role. In this species, females breed once per season, with the result that the costs of attendance are likely to be low (Blumer 1986). Finally, biparental guarding permits parents to feed alternately, and, in some monogamous cichlids, males and females alternate in parental duties (Ward and Samarakoon 1981).

The relative rarity of biparental care in fish compared to birds is probably connected both with differences in thermoregulation and with the fact that few ectotherms feed their young. As a result, most forms of egg care may either be nondepreciable or only partly depreciable, and the benefits of care by a second parent are likely to be reduced (Perrone and Zaret 1979; Gross and Sargent 1985; but see Keenleyside 1980). Among freshwater fish, there is close association between biparental care and the care of fry (Barlow 1974): where caretaking is limited to eggs, it is usually confined to one sex alone while biparental care typically involves fry as well as eggs (Blumer 1979). Unless biparental care is substantially more effective than uniparental care, it is only likely to be stable where the opportunities of both sexes to breed again are relatively low (see Section 7.2). The distribution of biparental care is closely associated with the distribution of monogamy among freshwater and temperate coastal marine fish (see Figure 7.7) which is, in turn, associated with low numbers of breeding attempts per season by females (Barlow 1984).

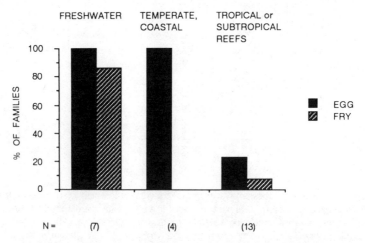

Figure 7.7 Percentage of families of teleost fish allocated to three ecological categories (freshwater, temperate coastal, and tropical or subtropical reefs) where monogamy is associated with biparental care of eggs and fry (from Barlow 1984).

Among mouth-brooding cichlids, biparental care would be expected to have few benefits unless the size of the brood is too large to fit in the mouth of a single parent (Perrone and Zaret 1979). As expected, biparental care is not common among mouth brooders. Where it does occur, clutch sizes are unusually large and all the fry cannot fit into the mouth of one parent (Myrberg 1965, 1966; Perrone and Zaret 1979). There is even one African cichlid where eggs are guarded by a single parent while both parents cooperate to brood the fry (Fryer and Iles 1972).

7.6   Parental Care in Amphibia

PARENTAL CARE IN FROGS

Among amphibia, a relatively high proportion of families contain species showing parental care by one or both sexes (Table 7.4) Parental care of eggs or young after oviposition is usually associated with terrestrial reproduction, large egg size, and small clutch size (Salthe and Duellman 1973; Salthe and Mecham, 1974; McDiarmid, 1978). Some form of parental care (including ovoviviparity and viviparity) occurs in around 71% of amphibian families and biparental care in around 20% (Gross and Shine 1981). Uniparental male and female guarding of eggs or young occurs with approximately equal frequency. However, evidence of egg guarding is usually based only on observations of attendance. In some species, "guarding" parents provide little protection for eggs or young (Woodruff 1977), and parents (especially males) may remain with eggs in order to increase their access to mates.

Among the frogs and toads, parental care of eggs or young is commonest among terrestrial breeders in the humid tropics (Wells 1981). Terrestrial oviposition may represent an adaptation to the high frequency of predation on eggs by conspecifics, as well as other predators, in permanent water bodies (Wells 1981). Eggs need to be kept moist and this may favor parental care in terrestrial species (Wells 1981).

Egg guarding, often by the male, is the commonest form of care. In some species, males guard more than one clutch laid in the same nest while, in others, either the same female lays successive clutches in the male's territory or several females lay in his territory and the male guards different clutches in turn (Drewry 1970, 1974; McDiarmid 1978; Wells 1981; Weygoldt 1987). In most species, the number of different females that lay in a single nest or territory is unknown.

Care of young occurs in a minority of species. In some groups, including the poison arrow frogs, Dendrobatidae, eggs are laid terrestrially and the tadpoles are transported by either or both parents to water, where they develop into adults (Weygoldt 1987). Transport by males appears to be rare where males compete intensely for mating territories, unless the time needed for tadpole transport is very short (Wells 1981; Weygoldt 1987). If transport occurs under these circumstances, it typically involves females, perhaps both because tadpoles may be damaged in fights between males and because the costs to males of leaving territories are high. Where males are nonterritorial or competition for mating territories is less intense, either sex may carry the tadpoles.

In some families, either the male (Scheel 1970) or the female (Vaz-Ferreira and Gehrau 1975) guard tadpoles after hatching. However, this is relatively uncommon, perhaps because tadpoles are aquatic while the adults in most care-providing frogs are terrestrial.

Young are provisioned in a small minority of species. For example, in some dendrobatids that breed in small bodies of water, females feed the developing tadpoles with nutritive eggs (Weygoldt 1987). In *Dendrobates quinque-vittatus* and *D. reticulatus*, both sexes maintain territories against members of the same sex. Small clutches of 2–3 eggs are laid on bromeliad leaf axils where they are guarded by the male, though the female may also join him. In some species where males transport tadpoles between bromeliads, they also attempt to induce additional females to lay eggs in the bromeliad they are guarding, which the developing tadpole will feed on. In other species, females feed tadpoles with unfertilized eggs without inducement by the male (Weygoldt 1980, 1984).

Virtually all frogs are external fertilizers (Ridley 1978). Whether the male or female attends the eggs appears to depend primarily on the costs of egg attendance to the male's subsequent mating success (see McDiarmid 1978; Wells 1981; Weygoldt 1987). Where females lay in limited oviposition sites, such as tree holes, rock cavities, and specially constructed nests, males usually guard the eggs and continue to call from these sites (Wells 1981). In contrast, where eggs are laid in leaf litter, in burrows constructed by the female, or in foam nests formed by pairs in amplexus, females rather than males typically guard the eggs (Wells 1981). This association may occur because the defense of limiting nesting sites allows males to continue to attract females, thus reducing the costs of parental care (McDiarmid 1978; Wells 1981). However, it is not yet clear whether male care is usually associated with situations where several females lay in the same nest.

PARENTAL CARE IN SALAMANDERS

In salamanders, parental care in the narrow sense is not as developed as in anurans and is limited to egg attendance. As in frogs, care is associated with terrestrial breeding or is found in species breeding in streams, while species breeding in ponds or lakes typically abandon their eggs in open water or attach them to small clumps of stones or to submerged vegetation (Salthe 1969; Nussbaum 1985). This difference is associated with differences in egg size and larval food supplies. In streams, hatchlings feed on aquatic insects and other benthic organisms, and this may have selected for relatively large egg and hatchling size. The long development periods of large eggs may, in turn, have favored the evolution of parental care (Nussbaum 1985). In contrast, the hatchlings of pond-dwelling species feed mostly on zooplankton, and large size may not have been so important.

Either sex may care for the eggs, but there is no firm evidence of biparental care (Nussbaum 1985). As in fish, the sex of the care-giving parent is closely associated with the mode of fertilization (Wells 1977a,b; Ridley 1978; Gross and Shine 1981): in all five salamanders showing parental care and external fertilization, males care for the eggs, while in all twelve showing internal fertilization and parental care, females do so.

Here, too, the distribution of parental care appears to be associated with the defense of limited oviposition sites that attract successive females (Nussbaum 1985). Internal fertilization involving a spermatophore appears to occur where there are advantages to oviposition away from courtship sites, possibly to avoid egg cannibalism, and may have led to the disassociation of males and zygotes and the evolution of maternal care (Nussbaum 1985). One difficulty with this argument is that it removes the hypothetical advantage to females of choosing particular mating sites in the first place.

7.7   Parental Care in Reptiles

Postlaying care of eggs is not highly developed in reptiles, many of which bury their eggs and do not guard the site (Brattstrom 1974; Tinkle and Gibbons 1977; Shine 1988c), though viviparity is relatively common (see Chapter 5). Egg guarding by females occurs in around 3% of oviparous snakes and 1% of oviparous lizards as well as in all living crocodilians (Oliver 1956; Greer 1971; Tinkle and Gibbons 1977; Shine 1988a,c). In the latter, it has been shown to improve the survival of eggs in at least one species (Metzen 1977).

No consistent differences in the proportion of species showing parental care are apparent between tropical and temperate reptiles. Among snakes, egg guarding is most commonly found in large, powerful species, such as the pythons, or in venomous snakes like the African and Asian cobras—presumably because they can defend their eggs effectively (Neil 1964; Shine 1988a,c).

Parental care is best documented in squamates, where females have been reported to remain with their eggs after oviposition in over a hundred species (Shine 1988a,c). In most species, care is restricted to the defense of nest sites. However, in a few, including the pythons, females coil around their eggs, defending them against predators and warming them by shivering thermogenesis and, in some, they aid and defend newly hatched young (Shine 1988a,c). Though the belief that female snakes swallow their young to protect them from predators has been widespread from Egyptian times, there is no firm evidence that this occurs (Shine 1988a,c). Biparental care occurs in some crocodiles and possibly in a few snakes, too (Oliver 1956; Shine 1988a,c), while uniparental male care has not been documented.

The rarity of male parental care in reptiles may be related both to the prevalence of internal fertilization and to the nesting habits of females (Perrone and Zaret 1979; Shine 1988a,c). Suitable oviposition sites are less likely to be limited among reptiles, which do not need to keep their eggs as moist during development as do amphibia. Moreover, females commonly bury their eggs immediately after egg laying, and this, too, may have constrained the evolution of multiple laying by females in the same nest. Finally, the relatively high frequency of viviparity (see Table 7.4) may be involved: both internal fertilization and behavioral thermoregulation may tend to favor viviparity over oviparity combined with egg care (see Chapter 5).

7.8 Summary

Parental care of eggs and young among invertebrates, fish, amphibia, and reptiles is associated with environmental conditions likely to maximize the benefits of additional parental investment: harsh or unpredictable conditions, high levels of predation or parasitism, and intense inter- or intraspecific competition.

Which sex should show parental care depends on the relative costs and benefits of care to the fitness of male and female parents. These are commonly influenced by the extent to which parental care restricts additional breeding opportunities in the two sexes, but a variety of other factors can affect the relative costs or benefits of care.

The frequency of uniparental care by males and females differs among groups of ectotherms. Females are more commonly involved among terrestrial invertebrates and reptiles, males and females are involved with approximately equal frequency among amphibia, while male care predominates among fish. Both within and across taxonomic groups, uniparental male care is associated with external fertilization and female care with internal fertilization, though there are numerous exceptions to this rule, especially among the bony fish.

Several explanations have been proposed to account for the association between male care and external fertilization, including arguments that male care should predominate where paternity certainty is high and that care is likely to evolve in whichever sex produces its gametes last. However, there is no close association between the order of gamete production and the frequency of male and female care (Section 7.4), and there is no firm evidence that paternity certainty is higher under external fertilization.

Instead, the only satisfactory generalization that can be made about the distribution of male versus female care is that it is related to variation in the costs and, less commonly, in the benefits of parental care to the survival and breeding success of males and females. In external fertilizers, uniparental male care is commonly associated either with circumstances where several females mate successively or simultaneously in a male's territory or with low female densities or short breeding seasons that constrain the male's ability to mate with several partners. The first of these two conditions is common among externally fertilizing fish and amphibia. In contrast, it is comparatively uncommon among terrestrial invertebrates and reptiles, where oviposition sites are not as closely constrained by the need to keep eggs moist and parents commonly rely on crypsis to avoid egg predation.

Uniparental female care among external fertilizers is often associated either with conditions where breeding seasons are short, females breed once per year, and the costs of parental care to the female's fitness are likely to be low (as in a number of temperate fish); with simultaneous polygyny by males, where females may need to guard their eggs against female conspecifics breeding in the same male's territory (as in some cichlid fishes and, possibly, in some dendrobatid frogs); or with oviposition or brooding in cryptic sites separated from the mating territories of males (as in some frogs, salamanders and cichlid fish).

Female care may predominate among internal fertilizers because the evolutionary pathway leading from internal fertilization to egg retention is a simple one that facilitates the disassociation of males and their progeny. In the small number of species that combine internal fertilization with uniparental male care, there are closely related species that are external fertilizers, and the evolution of male care probably preceded the evolution of internal fertilization (see Section 7.3). Under these circumstances, there is little reason to expect a change to maternal care.

Biparental care is rare among ectotherms, probably because two parents are usually little more effective than one at caring for eggs. In species where selection favors guarding or provisioning offspring, parental care is more likely to be depreciable and biparental care is relatively common. In at least some species, there is clear evidence that the presence of both parents increases the survival of the young. Biparental care may also occur where breeding seasons are short and females do not have the opportunity to produce several clutches of eggs in the course of a single season.

These arguments emphasize the importance of antecedent conditions to the evolution of male and female care. For example, where internal fertilization precedes the evolution of parental care, we might expect females to care for the young; but, where the evolution of parental care precedes the evolution of internal fertilization, there is no reason to predict a reversal in the pattern of care (see above). Similarly, among externally fertilizing fish where no care is the ancestral state, male care may predominate because it is less likely to reduce the potential reproductive rates of males than females. Conversely, where biparental care is the ancestral state and females are territorial, males may have more to gain from desertion than females, and uniparental female care may be most likely to evolve, as in the cichlids.

In a few species, including some fish and amphibia, either the male or the female guards the young or the larvae. Studies of the factors affecting the probability of male and female care in these species may, in the future, have a particularly important role to play in clarifying the costs and benefits of male and female care.

8

Parental Care in Birds and Mammals

8.1 Parental Care and Breeding Systems in Birds and Mammals

Virtually all endotherms face the dual problems of keeping their young warm and feeding them. Both in birds and mammals, parents may often be close to their maximum metabolic expenditure when feeding young (Chapter 3), and their ability to find and collect food (or, in mammals, to process it for supply in the form of milk) commonly constrains the number of young they can rear.

These constraints exert a profound influence on the costs and benefits of parental care to males and females. In contrast to ectotherms, males are usually unable to care for the clutches of several females at the same time so that paternal care is likely to constrict a male's breeding rate. In addition, all birds and mammals show internal fertilization, increasing the likelihood that males will be disassociated from their progeny (see Chapter 7) and (possibly) reducing paternity certainty.

Parental care of eggs occurs in all birds except for brood parasites and the megapodes, which lay successive eggs in mounds from which the chicks emerge well feathered and fully independent (Kendeigh 1952; Frith 1959, 1962; Lack 1968). However, the extent to which parents care for their young after hatching varies widely among birds. With the same two exceptions, all show some form of protection, but in many nidifugous birds parents do not feed their young after hatching (Kendeigh 1952). Most of these species are cursorial or aquatic, nesting on the ground or on freshwater (Kendeigh 1952; Lack 1968; Ar and Yom-Tov 1978; Silver, Andrews, and Ball 1985).

The extent of post fledging care also varies widely. In some species, recently fledged young continue to be fed by their parents, while in others they are guarded but not fed (Lack 1968). In some communal breeders, young of one or both sexes remain in their parents' territory for an extended period after they are nutritionally independent, especially when the surrounding habitat is saturated by conspecifics (Lack 1968; Brown 1987). Fledged young also associate with their parents in many monogamous species (Lack 1968). In these species, family groups are usually maintained within flocks for the duration of the nonbreeding season and split up before the next breeding season. Among waterfowl, these associations appear to be linked to the growth rate of the young: in most ducks, immatures reach adult weight in their first autumn and parental bonds are broken shortly after fledging, whereas in some geese and swans immatures do not reach adult weight until their third or fourth year and offspring commonly associate with their parents during their first two winters, gaining priority of access to feeding sites from the proximity of their parents (Scott 1980, 1984).

Four main forms of parental care occur among birds and mammals:

(1) BIPARENTAL CARE

Birds are unique among vertebrates in that biparental care is the norm, occurring in over 90% of the 9,000+ living species (Lack 1968). Biparental care, usually involving shared incubation of eggs and feeding of young, is found in almost all monogamous birds with altricial young (see Figure 8.1). However, care is not always shared equally in biparental species, and females are commonly more heavily involved than males (Kendeigh 1952; Lack 1968). Males and females may also adopt different roles in caring for the young. For example, in many passerines, females incubate more than males but males spend more time in territory defense (Kendeigh 1952). In addition, males and females may differ in their feeding behavior: in budgerigars and pied flycatchers, males do not distribute food to nestlings in relation to the needs of young but tend to give it to the nearest chick, while females feed younger or smaller chicks before older or larger ones (Stamps, Clark, et al. 1985; Gottlander 1987).

Biparental care in birds is associated with at least three different types of breeding systems: where both sexes are monogamous, as in many passerines (Kendeigh 1952); where males are polygynous and females are monogamous, as some of the marsh-nesting icterids (Orians 1980; Muldal, Moffatt, and Robertson 1986); and in a small number of species where females are polyandrous and several males cooperate with a breeding female to rear their joint young. Examples include the Tasmanian native hen, *Tribonyx mortierii* (Ridpath 1964; Maynard Smith and Ridpath 1972), and the Galapagos hawk (Faaborg, de Vries, et al. 1980). A number of passerines, too, occasionally breed in this way (Jenni 1974; Davies and Houston 1986).

Although females care for their young in all mammals, direct care by males occurs in fewer than 5% of species. Some form of care by males is relatively common in three mammalian orders—the primates, carnivores, and perissodactyls—where it has been reported in 30–40% of genera (G.D. Mitchell 1968; Kleiman and Malcolm 1981). Male care occurs in fewer than 10% of genera in insectivores, bats, lagomorphs, rodents, cetaceans, and artiodactyls, though in a substantial number of these species it is rare or irregular. Care by males has so far not been observed in edentates or pinnipeds (but see Barlow 1972).

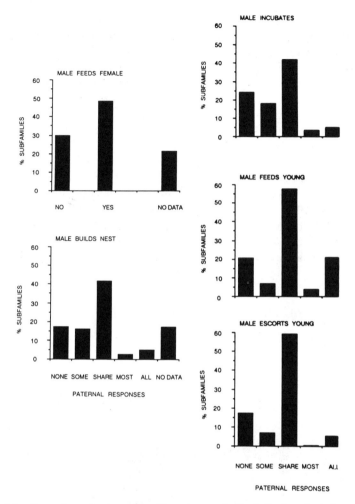

Figure 8.1 Percentage of avian subfamilies where male feeds the female during court-ship and on the nest, assists in nestbuilding, incubates, feeds young, and escorts young (from Silver, Andrews, and Ball 1985).

However, it is not always easy to identify male care in mammals. For exam-ple, in some polygynous species, male territoriality may help to preserve a food supply for a male's mates and progeny or to protect offspring against infanticide by unrelated males (van Schaik and Dunbar, in press). Conversely, male care of infants may sometimes serve primarily to increase the male's survival or mating success. For example, in some macaques and baboons, males carry or associate with infants, apparently in order to deflect aggression from other members of the same group ("agonistic buffering"), to enlist support if they are attacked, or to increase their access to the infant's mother (Deag and Crook 1971; Deag 1980; Packer 1980; Busse and Hamilton 1981; Collins 1986). It is not yet clear whether "care" of this kind is to the infant's benefit or not (Strum 1983). Though it may sometimes help to protect infants against predators or immigrant males or increase their access to resources in competitive situations (Packer 1980; Busse and Hamilton 1981), it may also increase the risk that they will be injured in fights between males and can even lead to starvation (e.g., Kuester and Paul 1986).

There is a close association among mammals between highly developed and habitual male care and obligate monogamy, and some form of male care is commoner in social species than in solitary ones (Clutton-Brock and Harvey 1976; Ralls 1977; Kleiman and Malcolm 1981; Rood 1986). Like monogamy, male care is commonest in the carnivores and the primates (Kleiman 1977, 1985; Malcolm 1985; Clutton-Brock 1989a). Within orders, male care is more common in monogamous mammals than in polygynous ones (see Figure 8.2),

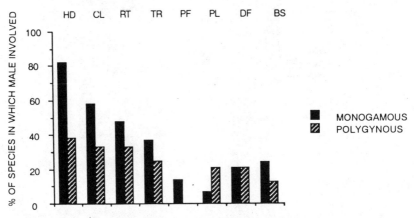

Figure 8.2   Proportion of species where males engage in eight different aspects of direct parental care in twenty-nine monogamous mammals, and twenty-four polygynous species showing some form of male care (from Kleiman and Malcolm 1981, Table 1). Carnivores are excluded, as are polygynous species living in multimale groups where infants are often used in agonistic buffering. HD: huddle with young; CL: groom and clean young; RT: retrieve young; TR: carry and transport young; PF: provide food; PL: play with young; DF: actively defend young; BS: babysitting.

and the most highly developed examples of male care are almost all found in monogamous species. For example, in the monogamous prairie vole, *Microtus ochrogaster*, males and females contribute equally to all aspects of care apart from lactation, and paternal activities include nest and runway construction, food caching, grooming, brooding, and retrieving young (Getz and Carter 1980; Getz, Carter, and Gavish 1981; see also Gubernick and Alberts 1989).

In cases where the female produces a litter of above average size, males may construct a second nest and the litter is subdivided, with the male caring for one of the nests (Thomas and Birney 1979). Similarly, in the monogamous Japanese raccoon dog, *Nyctereutes procyonoides*, males attend the female during parturition and take part in all forms of care shown by females except for lactation (Yamomoto 1987).

In a small number of mammals, litters are cared for cooperatively by a female and more than one breeding male. For example, in the African wild dog, *Lycaon pictus*, where groups consist of a single breeding female, 1–10 adult males, yearlings, and pups (Frame, Malcolm, et al. 1979), more than one male may mate with the female, and lactating mothers and their dependent pups rely almost entirely on food brought to the den by other pack members (Malcolm and Marten 1982; Moehlmann 1988). Similarly, in some species of tamarins, several males associate and mate with a single breeding female and help to carry and care for her young (Goldizen 1987; Sussman and Garber 1987). These breeding systems appear to be analogous to those of some communal breeding birds (see below).

(2) Uniparental Female Care

Uniparental female care is found in at least eighty-five species of birds belonging to eleven families, including the manakins, cotingas, hummingbirds, flycatchers, pheasants, grouse, turkeys, bustards, lyre birds, bower birds, and birds of paradise (Kendeigh 1952; Welty 1982). Many of these groups are polygynous and have precocial young, but uniparental female care also occurs in some monogamous species with precocial young, including some ducks and grouse, and in a few monogamous species with altricial young, including some hummingbirds (Kendeigh 1952; Kear 1970; Wiley 1974). In contrast, uniparental female care is the predominant form of parental care among mammals, though in some cases it is difficult to tell to what extent the male contributes to parental care (see above).

(3) Uniparental Male Care

Females care for their young in all mammals, but in a small minority of birds, eggs are incubated and chicks cared for predominantly or solely by the male. All of these species have precocial young, and most are cursorial or wading birds that nest on the ground (Lack 1968). The evolution of paternal care is of particular interest in these species for in some cases it is associated with role reversal and competition among females for males.

Uniparental male care is associated with at least four different breeding systems, as described below:

MONOGAMY COMBINED WITH THE PRODUCTION OF
ONE OR TWO BROODS

In a few monogamous birds, males are responsible for most or all care of eggs and young. Examples include the kiwis, some cassowaries, some of the tinamous, at least one partridge, and a number of shorebirds. For example, in the killdeer plover, *Charadrius vociferus*, males are responsible for the bulk of incubation and brood care though females may assist to some extent (Lack 1968; Demaree 1975). Pairs apparently breed monogamously and may persist through several seasons (Lennington and Mace 1975). In moorhens (*Gallinula chloropus*), males are usually responsible for around three-quarters of incubation, and females appear to select males on their body condition, which reflects their ability to incubate with little or no assistance (Petrie 1983a,b). In two species of tinamou, *Nothoprocta ornata* and *Crypturellus variegatus*, pairs appear to be monogamous, clutch sizes are small, females may be role-reversed, and males are responsible for all parental care (Beebe 1925; Pearson and Pearson 1955). At least some emus (*Dromiceius novae-hollandiae*) breed in this fashion (S. Davies 1976), though multiple laying by several females in the same nest is also reported (Kendeigh 1952).

In many of the birds belonging to this group, polyandry occurs in closely related species and, in several, females sometimes breed polyandrously (see Ridley 1978). It is likely that its frequency is often underestimated.

DOUBLE CLUTCHING COMBINED WITH MONOGAMY OR
SEQUENTIAL POLYANDRY IN FEMALES

In some species, females typically lay one clutch, which is incubated by their mate, then a second, which they incubate themselves. Examples include a number of wading birds, including sanderling, *Calidris alba*; Temminck's stint, *Calidris temmincki*; little stint, *C. minuta*; and mountain plover, *Charadrius montanus* (Ridley 1978; Erckmann 1983). Consecutive clutches may either be fathered by the same male or by different males, as in Temminck's stint (Hildén 1975).

SEQUENTIAL AND SIMULTANEOUS POLYANDRY

Polyandry combined with uniparental care and monogamous breeding in males occurs in fewer than 1% of all bird species and is largely confined to shorebirds belonging to five families: the jacanas, the painted snipe, the plovers, the sandpipers, and the phalaropes (see Table 8.1). It has also been recorded in the mesites, cassowaries, and in the button quail. Many of these species show an increase in female size relative to male size (Jenni 1974) and a reduction in relative egg size (H. Ross 1979).

In some polyandrous species, including dotterel (*Eudromias morinellus*) and red-necked phalarope (*Phalaropus lobatus*) and some cassowaries, a pro-

Table 8.1
Distribution of breeding systems where females lay more than one clutch
in shorebirds (from Erckmann 1983).

Species	Mating System	Breeding	
		Latitude	Habitat
Scolopacidae			
Wilson's phalarope	FAP	Temperate	Small ponds
Northern phalarope	FAP	Arctic-Subarctic	Tundra ponds
Red phalarope	FAP	Arctic	Tundra ponds
Spotted sandpiper	RDP	Temperate	Ponds, rivers
Sanderling	DC	High Arctic	Tundra
Temminck's stint	DC	Subarctic	Tundra
Little stint	DC	Subarctic	Tundra
Charadriidae			
Dotterel	FAP	Subarctic-Arctic	Montane tundra
Mountain plover	DC	Temperate	Short-grass prairie
Rostratulidae			
Painted snipe (old-world)	RDP(?)	Tropics-Subtropics	Swamps, marshes
Jacanidae			
Northern jacana	RDP	Tropics-Subtropics	Swamps, marshes
Wattled jacana	RDP	Tropics-Subtropics	Swamps, marshes
African jacana	RDP	Tropics-Subtropics	Swamps, marshes
Bronze-winged jacana	RDP	Tropics-Subtropics	Swamps, marshes
Pheasant-tailed jacana	RDP	Tropics-Subtropics	Swamps, marshes
Pedionomidae			
Plains wanderer	?	Subtropics	Open plains

NOTES: FAP, female access polyandry; RDP, resource defense polyandry; DC, double-clutching.

portion of females pair successfully with more than one male in succession in
the course of a single breeding season, while males usually breed with a single
female per season (Erckmann 1983; Handford and Mares 1985; Whitfield,
pers. comm.). The proportion of polyandrous females is often low: for example, of fifty nine female red-necked phalarope in one Canadian study population, only 8% were polyandrous (Reynolds 1987). Though male care usually
predominates, female involvement varies. In the red-necked phalarope only
males incubate the eggs, while in dotterel females usually share incubation
(Pulliainen 1971). Females commonly defend their mating partners against
other females in these species.

In other species, some females are simultaneously polyandrous while males
are monogamous within seasons (Graul, Derrickson, and Mock 1977; Faaborg
and Patterson 1981). In at least three species, females actively defend resource-
based territories within which one or more males incubate different clutches of
eggs. Examples include the American jacana, *Jacana spinosa* (Jenni and Col-

lier 1972); the painted snipe, *Rostratula benghalensis*; and the spotted sand-piper, *Arctitis macularia* (Oring and Knudson 1972; Oring and Maxson 1978). Females may not assist in brood care at all, as in jacanas, or may assist their final mating partner, as in spotted sandpipers (Hays 1972).

SIMULTANEOUS POLYGYNY COMBINED WITH
SEQUENTIAL POLYANDRY

In some cursorial birds, males compete for harems of females who lay in a single nest, which the male then incubates (see Handford and Mares 1985)—a breeding system that resembles those of externally fertilizing fish (see Chapter 7). Numbers of eggs in nests can be large: for example, in the greater rhea, *Rhea americana*, males can acquire harems of up to fifteen females and clutches can include 20–50 eggs (Bruning 1974a,b), though in the tinamous clutches of 1–10 eggs are usual (see Table 8.2).

Birds breeding in this fashion include the ostrich, *Struthio camelus* (Bertram 1978; Hurxthal 1979) emu, *Dromiceius novae-hollandiae* (Kendeigh 1952); the brushland tinamou, *Nothoprocta cinerascans* (Lancaster 1964a); the Bou-card tinamou, *Crypturellus boucardi* (Lancaster 1964b); and the highland tina-mou, *Nothocercus bonapartei* (Schafer 1954). In some species, including os-triches, one or more females may assist with incubation or brood care, while in others only the male incubates and cares for the brood. Females are thought to lay in the nests of more than one male in turn in some cases, but in others, like the highland tinamou, females remain with males and replace the clutch if it is lost.

There is an interesting contrast between the tinamous, cassowaries, kiwis, and the emu, where males typically care for clutches of fewer than ten eggs, and the ostrich and the two rheas, where males commonly incubate more than twenty eggs at a time (Handford and Mares 1985). Female territoriality has been recorded in several species of the first group (Handford and Mares 1985). In addition, females are larger than males in kiwis and emus, and both larger and brighter in tinamous and cassowaries. In contrast, in ostriches and the two rheas, clutch size is large, males compete for females and are larger, and, in ostriches, males are more strikingly colored. A possible explanation is that because of the large size of clutches, males can "process" eggs faster than females can lay them in ostriches and rheas, while the situation is reversed in tinamous, cassowaries, and kiwis (see Chapter 1).

(4) COMMUNAL BREEDING

In a small minority of birds, several adults of either or both sexes may assist in rearing young (Emlen and Vehrencamp 1983; Brown 1987). Communal breeders include species where several monogamous pairs breed coopera-tively, as in the groove-billed ani, *Crotophaga sulcirostris*; where one male and several females or one female and several males breed cooperatively, as in

Table 8.2

Breeding systems of ratites and tinamous (from Handford and Mares 1985)

	Habitat	Total Clutch Size and Incubation Period	Social Organization in Winter	Sexual Dimorphism	Mating System	Parental Care
Tinamous	Various	1–10 eggs 16–22 days	Probably solitary in most spp. Perhaps gregarious in *Eudromia*	F slightly larger and brighter	Varies according to species. Monogamy simultaneous polygyny, sequential polyandry, promiscuity probable.	M
Cassowaries	Mainly forest	3–8 eggs 49–56 days	Solitary: family units?	F larger and brighter	Appears predominantly monogamous; sequential polyandry apparent in *C. casuarius*	M
Emu	Grassland and desert scrub	7–10 eggs 36–60 days	Mostly in pairs. Some solitary or in groups	F slightly larger	Appears largely monogamous; simultaneous polygyny and sequential polyandry possible, known in captivity	M: perhaps rarely both
Kiwis	Mainly forest	1–2 eggs 74–84 days	Largely solitary	F 20% heavier	Monogamous	M: female may feed male
Ostrich	Desert to savannah	16–35 eggs occasionally more 42 days	Complex; large herds based on family units plus some single males	M slightly larger, black and white, nuptial reddening of neck, legs, F brownish	Labile, monogamy, simultaneous polygyny, sequential polyandry, promiscuity; harems usually small: 3–5 females, one major hen; brood fostering	M + major F
Greater rhea	Desert, thorn scrub, grassland	25–35 eggs occasionally more 36–37 days	Large herds plus some solitary males	M slightly larger	Promiscuity. Mean harem size 6–7: some harems of up to 10–12; brood fostering	M
Darwin's rhea	Desert, steppe, puna	10–40 eggs? ? days	Large herds plus some solitary males	M slightly larger	Simultaneous polygyny, probably promiscuity	M

NOTES: M = male; F = female. Total clutch size shows the average number of eggs laid by all females in a single nest.

the Tasmanian native hen, *Tribonyx mortierii*, the Harris hawk, *Parabuteo unicinctus*, and the dunnock, *Prunella modularis*; and where groups are made up of several individuals of both sexes and mating is promiscuous, as (apparently) in the noisy miner, *Manorina melanocephala*, and the pukeko, *Porphyrio porphyrio* (Faaborg and Patterson 1981; Davies 1990). Where several males are involved in parental care, they are usually also involved in copulation.

8.2   Models of Parental Care

How can we account for these diverse patterns of parental care? Theoretical treatments have concentrated mostly on the benefits of deserting versus assisting one's mate. Selection can be expected to favor desertion by one parent when the chance of breeding again is high, the current brood requires little extra attention to reach independence, and the parent's contribution to biparental care is small (Trivers 1972). However, predicting the evolution of parental care is complicated by the fact that both males and females are subject to selection pressures to desert, and the payoffs to the deserting sex will be affected by the response of the sex that is deserted (see Section 7.2).

Under these circumstances, game theory models that seek to identify the evolutionary stable strategy for each sex, given the response of its partner to being deserted, provide the appropriate theoretical framework (Maynard Smith 1977, 1982a,b; Grafen and Sibly 1978; Bradbury and Vehrencamp and Bradbury 1984; Lazarus 1989). However, as noted in Section 7.3, the predictions of these models are difficult to test without better measures of the costs and benefits of parental care to each sex than are currently available. At present, explanations of the evolution and distribution of parental care are necessarily based on educated guesses at the likely costs and benefits of care to males and females.

The next five sections discuss five fundamental questions about the distribution of male care in birds and mammals:

8.3   Why Do Males Help in Biparental Species?

Males would only be expected to assist their mates in brood care where the number of eggs or young that survive under biparental care exceeds survival under uniparental care multiplied by the mating success of a noncaring male (Section 7.3). Is it likely to do so?

So far, little attempt has been made to measure the consequences of male care in mammals, and evidence for birds is mixed. There is no consistent tendency for species showing biparental care to show higher fledging success than

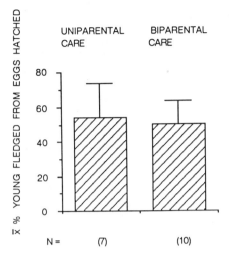

Figure 8.3 Comparative fledging success of shorebirds with uniparental and biparental brood care (from Erckmann 1983). Histograms show means calculated across species.

species with uniparental care (see Figure 8.3), but this could occur because biparental care has evolved in circumstances where brood survival is otherwise low. The disappearance of one partner in some altricial species is sometimes associated with a reduction in the number of young fledged, but in others the success of the current brood is usually unaffected (Beissinger and Snyder 1987)—though it may be necessary to monitor the survival of chicks after fledging and their eventual breeding success to be sure that this is the case (Clutton-Brock 1988).

Natural "experiments" on the effects of male care are provided by intra-specific variation in breeding behavior in facultatively polygynous, altricial birds. Secondary females often get little or no assistance from their mating partner, while primary females may get as much help as in monogamous species. Secondary females commonly have lower reproductive success than primary females (e.g., Patterson 1979; Alatalo, Carlson, et al. 1981; Hannon 1984; Richter 1984; Dhondt 1987). For example, in red-winged blackbirds (*Agelaius phoeniceus*), male assistance enhances the fledging success of those secondary females that receive it (Muldal, Moffatt, and Robertson 1986). Unfortunately, differences in the amount of male care received by primary and secondary females are often confounded with other factors that may affect the relative breeding success of primary and secondary females, including variation in female age, nest site quality, and timing of breeding.

Firmer evidence of the effects of male care is provided by experimental removal of males in biparental species (Mock and Fujioka 1990). In many altricial birds, the removal of males reduces the growth rate of chicks, the survival of nestlings or fledglings, or the survival of juveniles (see Table 8.3). In a number of species, male care has little effect on fledging success but an important influence on the subsequent survival or reproductive success of off-

Table 8.3
Summary of results of male removal studies on passerines
(from Wolf, Ketterson, and Nolan 1988).

Species	Mass of Young	Tarsus Length of Young	Survival to Nest-Leaving	Survival to Inde-pendence	Source
I. *Removal of male affects brood size or growth*					
Tree swallow	+		+		Robertson and Stuchbury (unpubl. data)
Tree swallow	+		+		Quinney (unpubl. data)
Blue tit	+		+		Sasvari (1986)
Great tit	+		+		Sasvari (1986)
Great tit	+	+	+		Bjorklund and Westman (1986)
Eastern bluebird	+		+		Robertson (pers. comm.)
Pied flycatcher	+	+	+		Alatalo, Lundberg, and Stahlbrandt (1982)
Savannah sparrow	+		+		Weatherhead (1979a)
Yellow-headed blackbird			+		Rutberg and Rohwer (1980)
House sparrow	+				Schifferli (1976)
Snow bunting	+		+		B. Lyon, Montgomerie, and Hamilton (1987)
II. *Removal of male has little or no effect prior to nest-leaving*					
Eastern bluebird	=		=		Gowaty (1983)
Northern cardinal	=		=		Richmond (1978)
Seaside sparrow	=		=	+	Greenlaw and Post (1985)
Dark-eyed junco	+	=	=/+	+	Wolf et al. (1988)
Song sparrow	+		=	=	Smith, Yom-Tov, and Moses (1982)
White-throated sparrow	+		=		Whillans-Browning and Falls (unpubl. data)

NOTES: The absence of an entry indicates no relevant data. Symbols: + indicates value significantly greater when male present, = indicates no significant difference when male present or absent, =/+ indicates male made measurable contribution in some years but not in others. The absence of an entry indicates no relevant data.

spring. In dark-eyed juncos (*Junco hyemalis*), for example, experimental removal of males within 2 days of broods hatching had little effect on the survival of chicks to fledging, but had a stronger influence after fledging (Wolf, Ketterson, and Nolan 1988, 1989: see Figure 8.4). In some species, results suggest that male assistance is more important early in the nestling period

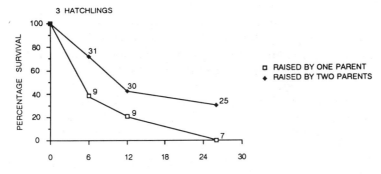

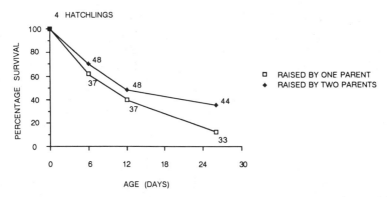

Figure 8.4 Survivorship of dark-eyed junco chicks in three and four hatchling broods raised by one (squares) and two (diamonds) parents from hatching to independence, 1983–86 pooled (from Wolf, Ketterson, and Nolan 1988). At hatching, 100% were alive. Values were calculated as the percentage of each brood surviving from hatching to independence, retaining 0 values for broods in which all young died; the mean of these values constitutes each point. Sample size is shown at each point. Broods whose fates were unknown following fledging were excluded from the sample at independence.

when chicks need brooding and females cannot easily increase their foraging time to compensate for male removal (Sasvari 1986; Burley 1988). Reduced food availability, too, may increase the effects of male care on juvenile survival (e.g. Bart and Tornes 1989). Passerine species where male removal has the greatest effect mostly either breed at high latitudes or are cavity nesters. This may be because nutritional demands are high in cold climates or where clutch size is relatively large (Wolf, Ketterson, and Nolan 1988), while the effects of male removal are typically small or absent in species where females are responsible for most of the feeding (Cowaty 1983).

The effects of male removal are also variable in birds with precocial young. In western sandpipers (*Calidris mauri*), the removal of either parent during incubation leads to immediate desertion by the other parent, presumably because a single parent cannot care effectively for the young (Erckmann 1983). In contrast, in some populations of willow ptarmigan (*Lagopus lagopus*), where both sexes again normally care for their offspring, the removal of males at incubation or hatching has little or no effect on the survival of chicks to 15 days (see Figure 8.5) and does not affect the number of young surviving to the

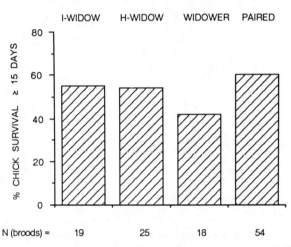

Figure 8.5 Survival of willow ptarmigan chicks reared by one or two parents (from K. Martin and Cooke 1987). I, widow: male mate removed at onset of incubation; H, widow: male mate removed at hatching; widower: female mate removed at hatching.

next season or the survival of breeding females in some populations (Martin and Cooke 1987). However, in other populations, male removal affects the survival of chicks as well as that of the mother (Hannon 1984). In most populations, the opportunity for polygyny is low and selection may favor males that remain with their mates, initially to guard their paternity and subsequently as an insurance against brood loss or mate loss (when the male commonly assumes care of the brood). Once males remain with their mates, the costs of parental care may be low and biparental care may be likely to evolve. In at least some facultatively polygynous species, males may minimize the costs of parental care by delaying their involvement until their probability of obtaining further matings is low. For example, in red-winged blackbirds, the presence of new females settling in their territories causes males to defer parental care to later broods (Muldal, Moffatt, and Robertson 1986).

This leaves the question of why biparental care is so common among birds and so rare in mammals. In birds, the relative ease with which young can be located by predators may favor rapid rates of development and biparental feeding. This may be enhanced by the sparse nature of many avian food supplies: most birds that show uniparental female care live on plant foods or fruit rather than animal matter (see below). The relative scarcity of biparental care in mammals is almost certainly associated with the fact that, except in carnivores, most forms of care to which males can contribute are nondepreciable or only mildly depreciable, so that the benefits of additional care are probably seldom large enough to offset costs to a male's mating success.

It is relevant to ask why lactation has not evolved in the males of monogamous mammals, for the physiological barriers to male lactation do not appear to be insurmountable (Maynard Smith 1977; Daly 1979). Daly (1979) suggests that it may be absent because the reproductive success of monogamous pairs is limited not by the female's lactational capacity but by food availability in the pair's territory during the most arduous season of the year. However, it is difficult to believe that assistance in lactation would not increase a female's reproductive capacity and the growth and survival of her progeny. An alternative possibility is that monogamy in mammals may represent a mate-guarding strategy by the male, and the energetic costs of effective defense of his territory or mate preclude the evolution of male lactation.

8.4 Why Don't Males Always Assist with Parental Care?

In some birds, males do not contribute directly to parental care at all. First, most subfamilies of birds where the male provides little or no care have polygynous mating systems, and the costs of care to male mating success would probably be high (Lack 1968; Pitelka, Holmes, and MacLean 1974; Mock 1983; van Rhijn 1984; Silver, Andrews, and Ball 1985; Miller 1986). Second, the majority of species also have precocial, young which are fed little by their parents or not at all. And third, many of them are terrestrial breeders, and the emancipation of males may make it easier to hide nests and may consequently reduce predation (Skutch 1949; Lill 1986). In all three cases, the net benefits of male care may be low.

Male care is also restricted or absent in a few polygynous species with altricial young, including some manakins, cotingas, flycatchers, birds of paradise, bower birds, and hummingbirds (Kendeigh 1952; Lack 1968). Here, polygyny combined with the restriction or absence of male care is thought to be associated with diets that are highly nutritious or relatively easy to collect and/or with small clutch size and slow developmental rates among nestlings (D. W. Snow 1962, 1971, 1980; Lill 1974, 1986; Snow and Snow 1979; Beehler 1983). For example, both among the weaverbirds and among the manakins and cotingas, polygynous species with restricted male care eat proportionately fewer insects and more fruit or seeds than monogamous species with biparental care (B. K. Snow 1961; D. W. Snow 1961, 1962; Crook 1962, 1964; Lack 1968). Among birds of paradise, both polygyny and monogamy occur among fruit eaters, but polygynous species tend to be more catholic in their choice of food while monogamous species feed on a narrower range of species (Beehler 1983). In some lyre birds, polygyny and uniparental female care is associated with the production of a single egg and with unusually low levels of energy expenditure by the female on incubation and feeding of young (Lill 1986).

However, polygyny is not a necessary condition for the absence of male care, for a number of monogamous birds show predominant or uniparental female care (see above). For example, while males assist their mates in protecting young in most swans and geese, they desert shortly after the beginning of incubation in many species of ducks (Kear 1970). Males play the smallest part in parental care in genera where ducklings are best adapted to water, probably because the benefits of assistance by a second parent are low in these species. The contrast between ducks and geese may arise either because ducks are too small to defend their broods effectively against predators and geese are not, or because the potential for successive matings is generally greater in ducks, which can fledge their young more quickly and generally do not pair for life.

8.5 Why Do Males Agree to Being the Only Care-giver in Some Monogamous Birds?

The evolution of predominant *male* care among monogamous birds is not well understood, and firm evidence of the costs and benefits of care to both sexes is not available for any species.

Several different pathways may lead to the evolution of male care in monogamous species. In the kiwis, unusually high costs of egg production (see Section 4.2) may constrain the female's ability to incubate or to care for the young, though it is also possible that the evolution of male care may permit subsequent enlargement of egg size. In other cases, the high rates of brood loss may favor rapid clutch replacement with the result that selection on both sexes favors male care, though, as yet, there is no convincing evidence to support this explanation (see Erckmann 1983). Finally, many of these species belong to groups where polyandry occurs, and occasional polyandry has been reported in a number of them. Opportunities to breed polyandrously are likely to raise the costs of parental care to females and may consequently favor the evolution of paternal care. Since sequential polyandry is often difficult to detect in the field, at least some apparently monogamous species may often be polyandrous.

In truly monogamous species, desertion by females should, in theory, be as common as desertion by males. However, in monogamous species where one sex sometimes deserts, it is generally the male. In addition, in most animal groups, uniparental female care is more likely to evolve from biparental care than uniparental male care (Kendeigh 1952; Lack 1968). Maynard Smith (1977) attributes this to the longer refractory period of females following egg laying, which may reduce their chances of finding another mate. Perhaps a more likely explanation is that there are relatively few species where biparental care is shared exactly equally between the sexes. For example, in many monogamous species, males are more heavily involved in territory defense than females, while females spend more time caring for the young (Breitwisch 1988). This may be why the removal of females commonly depresses chick survival more than the removal of males (see Figure 8.5).

Double-clutching may have evolved either from a situation where pairs produced two clutches in succession and the male took over care of the first brood before they were independent, freeing the female to start a second clutch (double-brooding), or from mate desertion and attempted polygamy by both sexes (Erckmann 1983). The ecological conditions leading to double-clutching are unclear: in shorebirds, it does not appear to be associated with unusually high rates of nest predation or unusually low food availability (Erckmann 1983). It is commonest in arctic-breeding species with short breeding seasons, which may indicate that it increases the fitness of both sexes where breeding seasons are too short to permit two consecutive broods to be reared.

8.6   Why Do Males Agree to Polyandry?

The evolution of uniparental male care in association with polyandry raises the question why males agree to mating polyandrously. Four general hypotheses have been suggested:

1. *Polyandry and male care evolve where females are nutritionally stressed by egg laying and either cannot afford to contribute to incubation or are unable to take over full responsibility if deserted. Where females mate with the same males in successive years, both sexes may benefit from female desertion* (Graul, Derrickson, and Mock 1977; Maynard Smith 1977; Ashkenazie and Safriel 1979; Oring and Lank 1984). In fact, there is no firm evidence that females are unusually heavily stressed by egg laying in polyandrous species. Among shorebirds, females of polyandrous species produce relatively small clutches and do not appear to deplete body resources by egg laying (Erckmann 1981, 1983; Reynolds 1987). In spotted sandpipers, for example, females can produce four-egg clutches at eight-day intervals at least three times (Lank, Oring, and Maxson 1985). Moreover, food shortage might, in general, be expected to favor the evolution of biparental care rather than uniparental male care (Erckmann 1983).

2. *Polyandry and male care have evolved where unusually high rates of brood loss favor the emancipation of females from care so that they can quickly produce a new clutch* (Jenni 1974; Oring 1982). Only among shorebirds has a systematic attempt been made to test this hypothesis (Erckmann 1983). Here, the frequency of nest failure declines at high latitudes and is no higher for polyandrous species than for other shorebirds (Figure 8.6). More-

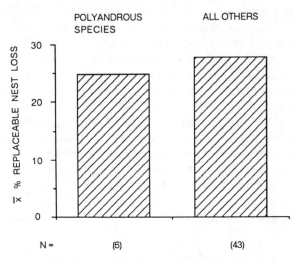

Figure 8.6 Replaceable nesting losses in polyandrous shorebirds versus all other species (from Erckmann 1983). Replaceable nesting failure was the proportion of nests lost to predators or to weather-related causes. Histograms show means calculated across species. Other losses due to human disturbance, infertility, or abandonment for unknown reasons were excluded.

over, renesting is not common among arctic-breeding shorebirds, including polyandrous species, and is largely confined to cases where clutches are lost very early in the nesting cycle (Erckmann 1983). For example, less than a quarter of male northern phalaropes and dotterels that lose clutches renest (Schamel and Tracy 1977; Michelson 1979). Data for shorebirds provide no indication that renesting intervals are shorter for polyandrous species than for monogamous ones (Erckmann 1983). Finally, evidence that female spotted sandpipers will initially assist in incubation but desert as soon as the opportunity for polyandry arises (Maxson and Oring 1980) suggests that females are quite capable of assistance and that their emancipation is unlikely to increase their partner's fitness.

3. *Polyandry and male care allow females to take advantage of fluctuating food availability by rapid production of multiple clutches at times when food is abundant* (Parmelee and Payne 1973; Graul 1974). However, there is no evidence either that the food supplies of polyandrous species are particularly

variable or that polyandrous species show particularly high variability in the number of clutches produced per year (Erckmann 1983). In most polyandrous species, polyandry occurs in good seasons as well as in poor ones and appears to depend on the availability of potential mates rather than on food availability.

4. *Once double clutching has evolved, the evolution of polyandry combined with uniparental male care is likely to follow since females will increase their fitness by laying clutches for several males and their mates are unable to prevent this* (Jenni 1974; Pitelka, Holmes, and MacLean 1974; Pienkowski and Greenwood 1979). This is sometimes called the "stepping-stone" hypothesis because it is based on the premise that monogamy with biparental care is likely to facilitate the evolution of double-clutching, which, in turn, facilitates the evolution of polyandry. However, comparative evidence suggests that polyandry probably usually develops from monogamy combined with biparental care rather than from double-clutching. Most polyandrous shorebirds belong to the subfamily Tringinae, which includes no double-clutching species (Erckmann 1983). In addition, females of several polyandrous species sometimes share incubation with one of their mates—a pattern unreported in any double-clutching species (Erckmann 1983).

Thus none of these four explanations accounts satisfactorily for the distribution of polyandry, nor does there seem to be any simple association between polyandry and other ecological variables. Polyandry has developed in species that currently show a wide range of laying seasons, ranging from less than a month in the Arctic sandpipers, phalaropes, and dotterel to year-round breeding in the tropical jacanas (Erckmann 1983). Male-biased sex ratios may sometimes increase competition for mates between males and favor polyandry (Hildén and Vuolanto 1972; Maynard Smith and Ridpath 1972; Schamel and Tracy 1977; Ridley 1978, but a wide range of adult sex ratios are found in polyandrous species (Erckmann 1983).

The evolution and distribution of polyandry and uniparental male care remain a puzzle, but several characteristics of shorebirds may reduce the costs of care to males and increase the benefits of desertion to females (Erckmann 1983). These characteristics include the following:

1. The ability of single parents to rear as many young as two parents when foraging conditions are good. This may be particularly likely in relatively small species where parents are ineffective at defending nests against predators.

2. Males that are already extensively involved in care in the likely antecedent state (monogamy combined with biparental care).

3. Small clutch sizes of less than four eggs combined with the potential for rapid egg production. Clutch size may be constrained in shorebirds by the need to keep relatively large eggs warm and resources for egg production may be plentiful. Under these conditions, females can benefit from emancipation by

laying clutches for several males during a short period of time. Small clutch size is also found in some polyandrous cassowaries (Crome 1976; Handford and Mares 1985) as well as in button quail, which are unusual among galliforms in producing clutches of only 2–4 eggs and in breeding polyandrously (Hoesch 1959, 1960; Lack 1968; Trollope 1970; Wintle 1975). As in the scolopacids, their young are precocial, and biparental care of chicks may be ineffective. The large clutch sizes of most other galliforms, ducks, and geese may help to explain why polyandry is rare or absent.

4. Females that are commonly larger than and dominant to males, making it difficult for males to control their partners.

Once male care and (occasional) sequential polyandry have evolved, simultaneous polyandry combined with female defense of resource-based territories may arise in cases in which shorebirds colonize temperate (spotted sandpiper) or tropical (jacanas) habitats, where breeding seasons are relatively long and parents can produce several clutches per year and resource distribution is clumped (Erckmann 1983; Davies 1990).

8.7 Why Are Males Responsible for All Care in Some Polygynous Birds?

Uniparental male care combined with polygyny is apparently confined to species with precocial young where more than one female lays in the same nest and clutch sizes are large (see Table 8.2). In the tinamous that breed in this fashion, the male takes sole charge of the nest, while in ostriches they may be helped by the dominant female (Lack 1968). These breeding systems have parallels among invertebrates and fish (see Chapter 7, Sections 7.4, 7.5).

Breeding systems where several females lay in the same nest could have evolved either from egg dumping by unrelated females (which is not uncommon among ground-nesting birds with large clutch sizes) or from cooperative breeding by related or unrelated females (Wilson 1975). Unfortunately, the degree of relatedness among females laying in the same nest is unknown in any of these species. The initial benefits to females of habitual laying in the same nest could include multiparental care, any positive effects of brood size on egg or clutch survival, and, in some cases, a reduction in the period between laying and the start of incubation (Lack 1968; Handford and Mares 1985). One useful approach to understanding the evolution of use of the same nest by several females would be to examine the fitness consequences of brood size on developing young in species where parents do versus do not welcome extra eggs (see Section 7.4).

Once several females lay in a single nest, male care may be favored because each clutch represents a larger contribution to the male's fitness than to that of any single female, and the benefit/cost ratio of caring may consequently be highest for the male. In addition, by caring for the eggs himself, the male may reduce the risk that females will reduce clutch size by ejecting each other's eggs from the nest (see Bertram 1979; Vehrencamp, Koford, and Bowen 1988). The advantages of caring to females may be reduced in these systems as a result of uncertainty over how many of their eggs are still present in the clutch (see Handford and Mares 1985). Where males are assisted by the dominant female, this is probably because dominant mothers are likely to contribute a larger number of eggs to the clutch than subordinates (see Sauer and Sauer 1966; Bertram 1979).

Finally, we need to ask why simultaneous polygyny and sequential polyandry, combined with male care, are relatively common among ratites, while polygyny combined with uniparental female care is the usual form of polygamy in galliforms and other birds with precocial young. Assuming that monogamy combined with biparental care is the ancestral state in both groups, the large clutch sizes laid by females in most galliforms may increase the potential costs of desertion to females, reduce selection for sequential polyandry, and increase the potential benefits of additional matings to males. If so, they are likely to facilitate the evolution of polygyny combined with uniparental female care. In ratites, use of a single nest by several females may be favored if the size of clutches laid by individual females is constrained and there is a positive relationship between brood size and egg or chick survival, or if this helps to minimize the period between laying and the onset of incubation (see above). Detailed studies of the breeding ecology of these species are now badly needed.

8.8 Summary

Theoretical treatments of the evolution of parental care by males or females have concentrated primarily on the evolution of uniparental care from biparental care. Under these circumstances, it is necessary to consider the response of the remaining partner to desertion as well as the benefits of desertion to the first partner and its effects on the survival of young. Where the benefits of biparental care are low, both parents are likely to gain from deserting. The outcome of these conflicts between the sexes is likely to depend on ecological and social factors affecting the relative costs and benefits of desertion to each sex as well as on their responses to desertion.

In most endotherms, female fecundity and the ability of males to care for multiple clutches of eggs or young at the same time are constrained by the dual problems of keeping the young warm and feeding them. As a result, parental care usually has higher costs to males and may have lower costs to females than among ectotherms. In conjunction with the prevalence of internal fertilization, this probably accounts for the relative rarity of predominant or uniparental male care in endotherms.

Both in birds and in mammals, biparental care is usually associated with monogamy. Evidence that male care improves the survival of young in the nest is mixed, but several recent studies of altricial birds show that biparental care can have an important influence on the proportion of fledglings that survive.

Uniparental female care is the commonest breeding system in mammals and is found in a minority of birds. Most birds with uniparental female care are polygynous and have precocial young. In these species, the benefits of male care are probably relatively low, while costs to the male's mating success may be substantial.

Uniparental male care does not occur in mammals but is found in some birds, including species showing monogamy, double-clutching, and polyandry. The reasons for the evolution of male care in these species are not well understood and probably vary between species. With the exception of cases where one female and several males cooperate to rear their joint young, polyandry is associated with small clutch size and is found in families where males are usually heavily involved in parental care and young are precocial. Here the previous evolution of male care may reduce the costs of polyandry to males, while small clutch size may increase the benefits of emancipation to females. Precocial young may usually be necessary for uniparental care to evolve from biparental care.

Finally, in some cursorial birds, several females lay in each nest and eggs are incubated by males who subsequently care for the young. Male care may develop once several females lay their eggs in a single nest because the combined clutch represents a large portion of the male's fitness. Conditions favoring this situation are unknown but may include selection for minimizing the interval between egg laying and the start of incubation and any benefits of increasing brood size to chick survival.

9

Parental Tactics 1: Variation in Care in Relation to Benefits

9.1 Introduction

Previous chapters have been primarily concerned with interspecific differences in the extent and form of parental care. In contrast, the next five chapters focus on adaptive variation in parental care within species.

Parents might be expected to adjust their expenditure on parental care in relation to variation in its benefits to their offspring and in its costs to themselves so as to maximize their fitness (Winkler 1987). At any point in an animal's lifespan, its expectation of future reproductive success can be divided into two components: (1) the survival and future reproduction of the brood that is currently being produced; and (2) the parent's expected future reproduction in subsequent breeding attempts. Models of parental care (Williams 1966a,b; Schaffer 1974a,b; Pianka 1976; Carlisle 1982; Sargent and Gross 1985, 1986; Winkler 1987; Curio 1988; Montgomerie and Weatherhead 1988) assume that parental expenditure in the current brood will increase the brood's reproductive value but lower the parent's subsequent reproductive success, measured after investment in the current brood is complete. In contrast, conservation of resources for future reproduction will lower the reproductive value of the present brood but increase the parent's subsequent reproductive success. Parents would be expected to maximize the sum of reproductive success derived from their current and future broods, weighing the value of their current brood relative to the value of their own subsequent reproduction and assessing the likely effects of current expenditure on future reproduction (see Carlisle 1982; Winkler 1987).

It is generally assumed that relationships between current reproductive expenditure and future reproductive success are curvilinear. Concave relationships between current and future reproductive success will favor "big bang" reproduction, while convex ones will favor an intermediate level of current expenditure (see Figure 9.1). In the latter case, many different factors may affect the position of the optimum, including the resources available to the parent, the effects of parental expenditure on the parent's survival, the probable relatedness of parents to current and future offspring, and the parent's reproductive value (see Carlisle 1982; Winkler 1987).

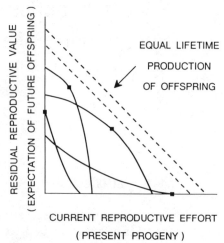

Figure 9.1 Trade-offs between current reproductive expenditure and the parent's subsequent reproductive value (from Pianka 1976). Four hypothetical curves relate costs to the parent's residual reproductive value (measured after expenditure on the current breeding attempt) to benefits in terms of the fitness of present progeny. The point of intersection of each curve with the line of equal lifetime production marks the optimal balance of expenditure (solid point). In many animals, this point is likely to change with parental age (see Figure 10.2).

One of the principal problems of predicting patterns of parental care is that the form of cost and benefit functions is unknown (see Chapters 2 and 3). In addition to being curvilinear, functions may sometimes be linear or even stepped (see Carlisle 1982; Winkler 1987). As a result, it is seldom possible to test particular models.

Most current models have at least two further limitations (Winkler 1987). First, they are static in nature. When the benefits and costs of parental expenditure are time dependent, as is often likely to be the case, dynamic optimization models provide a better theoretical approach but have not yet been applied to problems of parental care (Mangel and Clark 1988; McNamara and Houston 1986). In addition, where several individuals provide parental care, the behavior of one care-giver is likely to affect the optimal expenditure of others (Chase 1980).

Second, it is important to remember that all predictions derived from existing models rely on *ceteris paribus* assumptions (Winkler 1987) and may apply to a very limited range of situations. For example, Winkler (1987) predicts that parents should invest less heavily in late-born young because time to the next breeding season is reduced and parents consequently have an increased chance of breeding again. However, in this case, all other things will seldom be equal. Late-born young have notoriously low rates of survival, and increased parental expenditure may be unable to reverse this trend, reinforcing Winkler's prediction for a different reason. Conversely, parental expenditure in late-born young may sometimes have increased effects on their fitness. The specific situation envisaged by Winkler, where reproductive effort depends solely on time to the next breeding season, is likely to be very unusual.

Let me give a practical example of the problem of making realistic predictions about variation in parental investment. A considerable number of studies have attempted to make predictions about how parental investment should change in relation to the age of offspring. However, as offspring get older, at least three different changes commonly occur. First, older juveniles have greater energy requirements for growth than smaller ones, and the relative ability of the parent versus the juvenile to provide the necessary energy is likely to change (see Williams 1966a,b).

Second, the probability that older juveniles will survive to breeding age is generally higher than for younger offspring, both because older juveniles are closer to maturation and because the instantaneous rate of juvenile mortality usually declines with increasing age. They are consequently of greater value to parents, who might be prepared to take greater risks to ensure their survival (see Andersson, Wiklund, and Rundgren 1980; Winkler 1987).

Third, the effects of parental expenditure on the fitness of offspring are likely to change (Emlen 1970; West-Eberhardt 1975; Rubenstein 1982; Sargent and Gross 1986). In general, they are likely to increase initially and then decline as juveniles approach nutritional independence (see Figure 9.4).

Since increases in offspring value with age will favor *increased* expenditure in older offspring while the declining effects of parental expenditure on offspring development may favor *reduced* expenditure, predictions concerning changes in parental care depend on the relative importance that different theorists attach to these two effects. At the moment, they have little to go on (see Chapter 2).

The rest of this chapter reviews the evidence that parents adjust their investment in relation to its effects on the fitness of their offspring and thus on their own fitness. Sections 9.2–9.5 examine effects of paternity certainty, brood size, offspring age, and offspring quality on the magnitude of parental investment, while Sections 9.6 and 9.7 examine the evidence for adaptive brood reduction and desertion.

9.2 Parental Care and Parent-Offspring Relatedness

Parental expenditure might be expected to increase in relation to the certainty of relatedness (Winkler 1987; Montgomerie and Weatherhead 1988). Though both brood parasitism (Andersson 1984; Gowaty, Plissner, and Williams 1989) and adoption of unrelated individuals do occur (McKaye 1981; Andersson 1984; Constanz 1985; Thresher 1985; Thierry and Anderson 1986;

Mrowka 1987a), there is now abundant evidence that parental discrimination between offspring and unrelated juveniles is common where there is a substantial risk of adoption and the benefits of care are depreciable (Riedmann and Le Boeuf 1982; Pierotti and Murphy 1987).

When parental care is nondepreciable, a parent's own young may benefit from the presence of other juveniles because this reduces mortality from predation (see Hamilton 1961; Lazarus and Inglis 1978). This is probably why some fish actively solicit the addition of unrelated fry to their broods. For example, in some cichlids, parents will actively kidnap fry from other broods (McKaye and McKaye 1977), while in fathead minnows (*Pimephales promelas*), males that take over established nests will continue to care for eggs fathered by the previous owner (Sargent 1989). In the latter case, males possibly benefit by adopting the previous occupant's eggs, for females prefer to lay in nests that already contain eggs (Unger and Sargent 1988). However, males evidently discriminate between their own eggs and the previous occupant's, and their own show consistently higher survival (Sargent 1989).

The best evidence that parental care is related to variation in the probability of relatedness comes from studies of birds in which several mature adults cooperate to rear young. In several species, females attempt to bury or evict eggs laid by other females in the communal nest (Vehrencamp 1977; Bertram 1979; Mumme Koenig, and Pitelka 1983). In tropical house wrens (*Troglodytes aedon*) males that replace previous territory holders sometimes attack their broods (Freed 1987), while in polyandrous groups of dunnocks (*Prunella modularis*) subordinate males will only assist in rearing if they are permitted breeding access to the female and, if not, may destroy the eggs (Davies 1985; Houston and Davies 1985; Burke, Davies et al. 1989). Similarly, in groove-billed anis (*Crotophaga sulcirostris*) females that have contributed few or no eggs to a breeding attempt may try to force the group to desert and nest again (Vehrencamp 1977; Vehrencamp, Koford, and Bowen 1988).

Firm evidence that parental care varies in relation to the probability of relatedness is also provided by studies of infanticide in mammals. Among a variety of rodent species, the probability that males will kill litters of pups is closely related to the probability that they are the father of the litter (see Figure 9.2; Mallory and Brooks 1978; Brooks 1984).

9.3 Parental Care and Brood Size

Where the benefits of parental care are depreciable, total parental expenditure should usually increase with brood size (Lazarus and Inglis 1986; Winkler 1987; Montgomerie and Weatherhead 1988). Both in birds and mammals there is extensive evidence that it generally does so, though parental expenditure per

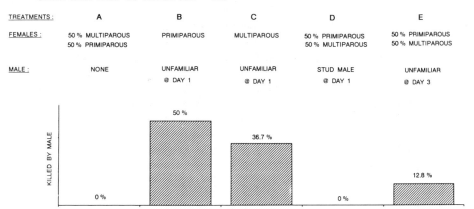

TREATMENTS:	A	B	C	D	E
FEMALES:	50 % MULTIPAROUS 50 % PRIMIPAROUS	PRIMIPAROUS	MULTIPAROUS	50 % PRIMIPAROUS 50 % MULTIPAROUS	50 % PRIMIPAROUS 50 % MULTIPAROUS
MALE:	NONE	UNFAMILIAR @ DAY 1	UNFAMILIAR @ DAY 1	STUD MALE @ DAY 1	UNFAMILIAR @ DAY 3

Figure 9.2 Infanticide in collared lemmings, *Dicrostonyx groenlandicus* (from Mallory and Brooks 1978). Female lemmings received one of four treatments at parturition: (A) they were maintained in isolation with their young; (B) and (C) they were exposed to an unfamiliar male for 24 hours on day 1 post partum; (D) they were exposed to the stud male (who had fathered the litter) for 24 hours at 1 day post partum; and (E) they were exposed to an unfamiliar male for 24 hours on day 3 post partum. Only young caged with males unfamiliar to the mother were killed.

head of offspring often declines (Lack 1954; Pearson 1968; Pinkowski 1978; Walsberg 1978; Robertson and Biermann 1979; Biermann and Sealy 1982; Johnson and Best, 1982). For example, in house mice, mothers respond to larger brood sizes by increasing both the amount and the quality of milk, but pups from larger litters show reduced growth (König, Riesler, and Markl 1988). In polygynous birds, male care, too, often increases with brood size. For example, in red-winged blackbirds, males are more likely to help females with larger broods, whether they are primary or secondary mates, especially if they are unable to feed the brood adequately on their own (Whittingham 1989). However, in other cases, increasing brood size has little effect on the amount of food provided by parents, possibly because their feeding rate is constrained by the availability of economical food items or by the amount of time they can spend foraging (see Smith, Kallander, et al. 1988).

Where the benefits of parental care are nondepreciable, a positive relationship between total parental expenditure and brood size would again be expected if care influences the survival or reproductive success of the entire brood so that the benefits of investment to the parent rise in direct relation to brood size (Lazarus and Inglis 1986). Several studies of fish and birds have shown positive relationships between brood size and measures of nest defense (e.g., Kramer 1973; Ricklefs 1977a; Gottfried 1979; Robertson and Biermann 1979; Greig-Smith 1980). However, experimental manipulation of brood size is necessary to demonstrate this effect convincingly, for correlations between parental defense and brood size may arise because qualitatively superior parents or those living in superior habitats are likely to produce more eggs and to expend more resources on caring for them (Lazarus and Inglis 1986; Curio 1987).

Recent experimental studies confirm that changing brood size commonly affects nest defense. In sticklebacks (*Gasterosteus aculeatus*), males guarding three clutches of eggs show more frequent defense of their broods at all stages of development than males guarding a single clutch (Sargent 1981; see Figure 9.3). Similarly, in the Central American cichlid, *Aequidens coeruleopunctatus*,

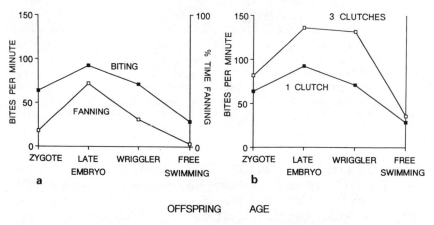

Figure 9.3 Parental behavior versus offspring age in sticklebacks (from Sargent 1981). (a) Both fanning and brood defense initially increase with offspring age and subsequently decline. (b) Males allowed to defend three clutches show more frequent brood defense than males defending a single clutch.

experimental reduction of brood size causes guarding females to leave their broods more readily and for longer when scared (Carlisle 1985), while in the biparental convict cichlid, *Cichlasoma nigrofasciatum*, time spent fanning eggs by both parents increases with brood size (Townshend and Wootton 1985; see also Coleman, Gross, and Sargent 1985; Sargent 1988). In redwinged blackbirds and great tits, experimental manipulation of clutch size affects the intensity with which parents harass predators (Robertson and Biermann 1979; Windt and Curio 1986), and experimental enlargement of broods of American goldfinches (*Carduelis tristis*) raises the rate at which parents give alarm calls (Knight and Temple 1986c).

When the risks incurred by parents who increase care are very high, selection will not necessarily favor a relationship between brood size and parental care, since the chance that the parent itself will fail to survive (and will consequently be unable to rear the brood) has to be set against the immediate benefits of care to the offspring (Lazarus and Inglis 1986). Some evidence supports this prediction. For example, in three-spined sticklebacks, experience of predation tends to reduce the relationship between egg number and measures of defense, and more dangerous activities, such as physical attacks on predators, are less likely to be influenced by brood size (Pressley 1981). However, other studies suggest that relationships between brood size and nest defense are *stronger* in circumstances where risks are higher. For example, in three out of the four bird species studied by Gottfried (1979), nest defense against a model snake increased with clutch size, while against a model blue jay (*Cyanocitta cristata*), it was independent of clutch size in all species. Here again, interpretation is hampered by the absence of reliable estimates of risk and of the costs of different levels of parental care.

Where predators are likely to take individual young rather than the entire brood, the optimal level of parental expenditure should be independent of size as long as it is determined by the parent rather than the offspring (Lazarus and Inglis 1986). At least two studies support this prediction; in pink-footed geese (*Anser brachyrhynchus*) (Lazarus and Inglis 1978) and Bewick's swans (*Cygnus columbianus*) (Scott 1980), parents spend more time watching for predators than nonbreeding adults, but vigilance shows no consistent increase with brood size. In a study of plovers, parental vigilance during one of two developmental states increased with brood size in one species but was independent of brood size in another (Walters 1982).

9.4   Parental Care and Offspring Age

As Section 9.1 argues, it is difficult to make general predictions about the way in which parental care should change with offspring age, since the increasing reproductive value of offspring is likely to favor an increase in parental care, while changes in the effects of care on offspring fitness are likely to favor a reduction (See figure 9.4).

Some of the clearest evidence that care is adjusted both to the reproductive value of offspring and to its effects on offspring fitness comes from studies of parental care in relation to offspring age in fish. For example, three-spined sticklebacks fan eggs in the later stages of development more frequently and defend them more intensely than recently fertilized eggs (see Figure 9.3). The increase in fanning rate may be a response to the greater oxygen requirements of the developing eggs, but the increased intensity of defense presumably occurs because eggs in the later stages of development are worth more to the parent than they are in earlier stages (Sargent and Gross 1986). After the fry hatch, both fanning and attacks on intruders decline in frequency, presumably because larger juveniles are more likely to survive without parental attention. Similar changes in guarding behavior are found in other fish (see Colgan and Gross 1977) as well as in some birds (Andersson, Wiklund, and Rundgren

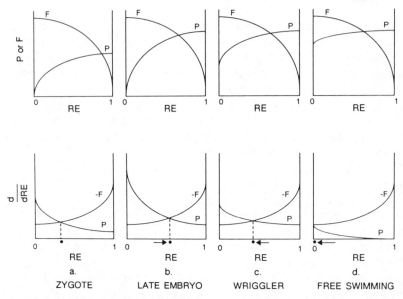

Figure 9.4 Hypothetical parental tactics for four different stages of offspring development in fish (from Sargent and Gross 1986). Figures in the top row show the parent's expectation of present (P) and future reproductive success (F) in relation to parental expenditure (RE). Figures in the second row show rates of return on expenditure into present and future reproduction: optimal reproductive expenditure (•) increases between stages a–b and then decreases between stages b–c and c–d. This is because (1) at spawning, offspring survivorship is zero if the parent offers no care but improves slightly with maximum expenditure (RE = 1); (2) just before hatching, young are still completely dependent on the parent but the parent can now achieve higher offspring survival with maximum parental care (RE = 1) simply because its offspring are closer to maturity, leading to an increase in optimal reproductive expenditure (•); (3) when the offspring are at the wriggler stage, they have a finite chance of surviving without parental care and the parent enjoys a lower dP/dRE than for hatching eggs and so should lower its optimal RE; (4) when the fry are free-swimming, the effect of RE on their survival is small, and because the parent will produce more young in the future if RE is low, optimal expenditure declines further.

1980; Blancher and Robertson, 1982; Walters 1982; Curio, Regelmann, and Zimmermann 1984; Reid and Montgomerie 1985; Montgomerie and Weatherhead 1988; Redondo and Carranza, 1989; Westmoreland 1989). Parental care may also be adjusted to relatively short-term changes in the likely effects on the fitness of offspring. For example, female vervet monkeys with dependent offspring are more responsive to alarm calls that indicate the presence of predators just before times of year when infant mortality is highest (Hauser 1988).

The influence of changes in the effects of parental expenditure is often difficult to distinguish from those of changes in the value of offspring. For example, in several species of baboons and macaques where the dominance and reproductive success of daughters depends on their mother's rank, successive

Figure 9.5 (a) Reproductive value for female rhesus macaques on Cayo Santiago Island. (b) Theoretical curve to illustrate differences in the reproductive value of two sisters, aged one and four, respectively (B, A), and of the same two animals four years later (from Schulman and Chapais 1980). When B is aged one, its reproductive value is lower than that of A, but four years later the position is reversed (from Schulman and Chapais 1980).

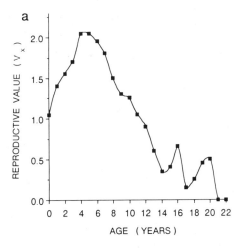

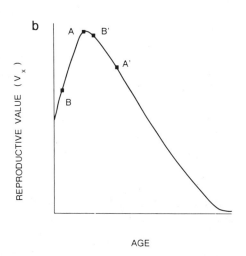

daughters rank immediately after their mother, so that the rank of sisters within matrilines is inversely correlated with their age (Kawai 1958; Sade 1972). Schulman and Chapais (1980) interpret this as a response to changes in the reproductive value of offspring, arguing that mothers prefer to assist their younger offspring over their mature sisters because younger females who have not yet started to breed have higher reproductive value (see Figure 9.5). However, mothers preferentially support their younger daughters *before* their reproductive value exceeds that of their older sisters (Horrocks and Hunte 1983b). An alternative explanation is that mothers prefer to support younger daughters because assistance has a greater *effect* on their survival or breeding success. A third possibility is that by ensuring that younger daughters outrank their older sisters, mothers reduce the potential benefits that younger daughters might derive from forming coalitions with their older sisters to outrank their mothers (Horrocks and Hunte 1983b).

9.5   Parental Care and Offspring Quality

Depending on circumstances, selection might favor either *increased* investment in superior broods because their reproductive value is higher or *reduced* investment where the effects of parental expenditure on offspring fitness are smaller for superior offspring. Testing theoretical predictions is further complicated by the fact that parents might often be expected to adjust expenditure on their broods in relation to the average quality of their own broods rather than in relation to the population average (Montgomerie and Weatherhead 1988).

So far there have been relatively few attempts to investigate how parental care changes in response to brood quality. In tawny owls, *Strix aluco*, Wallin (1987) found that the intensity of nest defense declined during the breeding season, irrespective of brood size, matching a decline in the survival of broods. However, in great tits, Curio and Regelmann (1987) found no relationship between the average weight of nestlings (which predicts their probability of survival) and measures of parental defense against a captive owl.

Parents might also be expected to discriminate between offspring within broods, either favoring large nestlings because they are more likely to survive or favoring small ones because the fitness return on expenditure is greater. For example, some plants abort offspring (fruit) that are poorly developed and are unlikely to germinate successfully (Stephenson and Winsor 1986).

Few studies have yet been able to distinguish satisfactorily between parental discrimination and variation in the ability of offspring to monopolize access to parents by varying their begging rate or their position in the nest. The existing evidence for birds suggests that parents commonly feed the closest chick that is begging actively (see Ryden and Bengtsson 1980; Mock and Parker 1986; Göttlander 1987). Where they discriminate consistently between chicks, active discrimination in favor of larger young (e.g., Drummond, Gonzalez, and Osorno 1986) is apparently less common than discrimination in favor of younger or smaller offspring. For example, in budgerigars and pied flycatchers, females typically feed younger or smaller offspring before older or larger ones (Stamps, Clark, et al. 1985; Gottlander 1987). In budgerigars, these differences are not correlated with variation in begging rates by the young (though male parents respond to begging rates rather than offspring weight or age). In pied flycatchers, active begging increases the chance that a chick will be fed, and females tended to favor light nestlings more than males (Gottlander

1987). In fieldfares, too, parental feeding is negatively related to nestling weight, while in several other passerines, including great tits and blackbirds, food is relatively evenly distributed among members of the brood (Ryden and Bengtsson 1980; Bengtsson and Ryden 1981, 1983). In contrast, in blue-footed boobies (*Sula nebouxii*) parents appear to feed the larger of the two chicks in preference to the smaller one, which commonly dies when food is in short supply (Drummond, Gonzalez, and Osorno 1986).

Parents may have a variety of techniques for equalizing the distribution of food. In some species, they vary the position on the nest that they feed from while in others they refuse to feed the same nestling more than once in succession (Best 1977; Ryden and Bengtsson 1980). For example, in coots (*Fulica atra*), parents usually feed the chick closest to them, apparently regulating proximity by attacking chicks that persistently follow them. On those occasions when they do not feed the closest chick, they usually feed a chick smaller than the one closest to them (Horsfall 1984). In some species, including cattle egrets, older chicks have priority of access to food brought to the nest and younger chicks commonly starve (Mock 1984a,b; Fujioka 1985a,b; Mock and Parker 1986), but where parents intervene it is usually to help the younger or smaller chick. For example, in South Polar skuas (*Catharacta maccormicki*), where siblicide is common, parents sometimes intervene to prevent chicks from killing each other (Spellerberg 1971).

Little attempt has so far been made to investigate the extent to which female mammals control the supply of milk to different littermates. However, during the later stages of lactation, ewes with twins are reluctant to allow lambs to suck unless they do so simultaneously, which may serve to distribute resources equally between the pair (Alexander 1960; Ewbank 1964). In addition, in goats and rhesus macaques, mothers with twins are reported to provide more care for the weaker member of the pair (Spencer-Booth 1969; Klopfer and Klopfer 1977).

9.6 Brood Reduction

Where the reproductive value of young falls to very low levels, parents might increase their fitness by premature termination of parental care of some of their brood if this enhances either the fitness of survivors or their own capacity for future expenditure (Howe 1976; O'Connor 1978; Tait 1980; Gosling 1986; Lloyd 1987; Parker and Mock 1987). Specifically, parents should favor brood reduction as soon as their fitness, if they produce a full brood, is lower than their fitness if they reduce brood size by one or more offspring (O'Connor 1978; see also Lloyd 1987). Brood reduction is most likely to be favored early in the breeding attempt since its potential benefits to all parties are likely to fall as fledging or weaning are approached. The reduction in average daily mortality rate necessary to favor brood reduction is lower where the nesting period is long or the brood size is large (see Figure 9.6). It is also lower for surviving sibs than for the parent, especially where multiple paternity of broods is common.

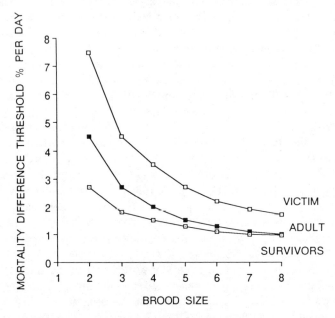

Figure 9.6 Circumstances favoring brood reduction in the parent (adult), sibs (survivors), and the victim (from O'Connor 1978). The vertical axis shows the difference in average daily mortality of offspring at different brood sizes necessary to favor brood reduction in each category of animals. Survivors are assumed to be related to each other and to the victim by 0.5. As brood size increases, the potential benefits of brood reduction to parents, survivors, and victims are likely to increase, and the difference in daily mortality necessary to favor reduction will decline.

O'Connor's model makes two assumptions that are unlikely to be realistic. First, it ignores any effects of brood size on the adult's own survival or on the survival or breeding success of young after fledging. The latter effect may well be important, for early development can have a substantial influence on first winter survival in many birds, but the inclusion of both effects would be likely to strengthen O'Connor's predictions. Second, it assumes that the negative effects of large brood size are inflicted equally on all brood members. However, large brood size may have the greatest effect on the fitness of the weakest members. If so, this is unlikely to reduce the average benefits of brood reduction, for the elimination of the smallest or most subordinate members of the brood may have little effect on the fitness of their larger and more dominant sibs.

Several empirical studies have produced results that coincide with predictions of brood reduction models (O'Connor 1978; Tait 1980; Gosling 1986). Starvation of nestlings has been recorded in sixteen of seventeen bird species in which empirical measures of the relationship between brood size and fledging rate indicate that brood reduction would be beneficial to parents or survivors, but in only one of six species where it was not expected to occur (O'Connor 1978). There is also some evidence that starvation mortality is commoner in young nestlings and in large broods while sibling rivalry and mortality of the entire brood is commoner in species with small broods (O'Connor 1978).

However, many of these trends may be consequences of ecological constraints on parental expenditure or of the effects of food shortage on sibling competition rather than of evolved parental behavior. For example, in blue-footed boobies, mortality of junior chicks is associated with reduced food availability because this increases the frequency of attacks by senior chicks (Drummond, Gonzalez, and Osorno 1986; Anderson 1989a,b; Drummond and Chavelas 1989). It is unsurprising that starvation of nestlings is most commonly reported in species where increasing brood size exerts a strong effect on nestling survival, for both effects are likely to be found where food supplies are limiting during nestling growth. Nestling starvation may be more likely to occur in large broods because of the greater energetic strain imposed on the parent or because large broods are commonly associated with reduced nestling size. The greater susceptibility of small animals to starvation could also account for the higher incidence of nestling mortality soon after hatching. It is interesting to note that one of the most direct predictions of O'Connor's model was not fulfilled: contrary to expectation, starvation mortality was *more* prevalent in species with fast growth rates and short fledging periods than in species with slow ones.

A clear indication of whether brood reduction is a consequence of ecological constraints or of active parental manipulation is only likely to be achieved by detailed studies of the way in which parents treat individual offspring. In birds, there is, as yet, more evidence that parents attempt to distribute resources equally between chicks or that they favor smaller young (see above) than of active discrimination favoring larger chicks (Stamps, Clark et al. 1985). In his extensive review of avian infanticide, Mock (1984a) does not quote a single example of active parental discrimination in favor of larger chicks. Where parents feed the most accessible young or provide food in such a way that larger chicks can monopolize resources, their failure to intervene or to distribute resources may either be an evolved parental strategy or may arise because parents cannot regulate competition among siblings effectively.

In this respect, there appears to be a difference between birds and multiparous mammals, where there is clear evidence that parents do sometimes kill their own offspring. In some rodents, mothers sometimes kill a proportion of their young if food is short, brood size is unusually large, or the chance of infanticide by males is high (Day and Galef 1977; Gandelman and Simon 1978; Fuchs 1982; Huck 1984; Bronson and Marsteller 1985; Perrigo 1987; Mendl 1988). The responses of females vary among species. For example, female house mice respond to increasing energetic burdens by killing a proportion of their pups soon after birth, with rapid bites to the head, thereby helping to maintain the growth rates of survivors (Perrigo 1987). In contrast, female deer mice that are energetically stressed sometimes eliminate their entire litter but rarely kill individual pups (Perrigo 1987). One possible explanation of the contrast between birds and mammals is that it is difficult for female mammals to restrict nursing to a proportion of their young while birds can do so more easily by feeding the closest chick. This explanation could also account for the apparent absence of maternal infanticide in uniparous mammals apart from man (Daly and Wilson 1984) since in uniparous species mothers can more easily control the supply of resources to the young.

Even if birds seldom intervene actively to reduce the size of their broods, they may achieve the same effect by ensuring that their offspring hatch asynchronously (Lack 1947; Hahn 1981; Fujioka 1985a,b; Hébert and Barclay 1986; Mock and Ploger 1987), or by varying the relative size of eggs within clutches (Slagsvold, Sandvik, et al. 1984; Stokland and Amundsen 1988). Hatching asynchrony varies widely between species and commonly leads to differences in survival between early and late hatching chicks, especially when food availability is low (Parsons 1975; but see also Stamps, Clark, et al. 1985, 1989; Husby 1986; Slagsvold 1986a,b; Anderson 1989a,b; Lessells and Avery 1989; Bryant and Tatner 1990). For example, asynchronously hatching broods of blackbirds (*Turdus merula*) produce more young than synchronous broods when food availability, is low but the provision of extra food removes this effect (see Figure 9.7). As Lack originally suggested, hatching asynchrony apparently increases chick survival because it is associated with a relatively early reduction of brood size and, consequently, with an increase in food available to survivors (Gibbons 1987; Anderson 1989a,b; Magrath 1989). Hatching asynchrony may also have other benefits, including a reduction in the period necessary for incubating or fledging the entire brood and a spreading of the peak demand for food by chicks (Hussell 1972; Bryant 1978; Clark and Wilson 1981).

Parents may also reduce brood size selectively by varying the size of eggs, thus generating differences in the relative size of chicks. However, it is not clear to what extent consistent differences in egg size within clutches represent an evolved strategy leading to selective brood reduction and to what extent they are nonadaptive consequences of variation in resource availability throughout the laying period. At least some evidence suggests that they can represent an adaptive parental strategy. For example, in herring gulls (*Larus argentatus*), experimental removal of the second egg causes parents to increase

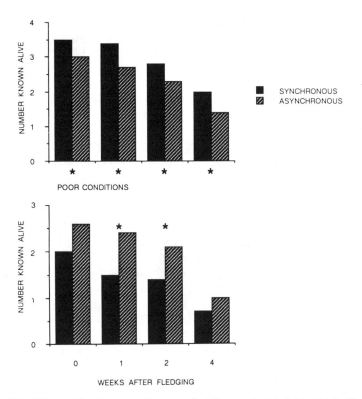

Figure 9.7 Effects of hatching asynchrony on fledgling production in blackbirds (*Turdus merula*) under good and poor conditions (from Magrath 1989). Eggs were manipulated so as to hatch synchronously or asynchronously. Histograms show the mean number of nestlings known to have fledged, from synchronous (plain bars) and asynchronous (hatched bars) broods, and to have survived to 1, 2, and 4 weeks from fledging, in broods from which at least one nestling was known to have fledged. An asterisk below a bar indicates a statistically significant difference between good and poor conditions; one above columns shows a significant difference between synchronous and asynchronous broods. Lack of an asterisk indicates that the difference was not significant between synchronous and asynchronous broods. The most accurate measures of brood productivity were at 1 and 2 weeks, when territories were searched for dependent young. Weight at 8 days influenced survival until 4 weeks after fledging, but breeding birds were a random sample of those alive at 4 weeks.

the relative size of their third egg (Parsons 1976). The size of last eggs relative to the clutch mean also varies consistently between bird species: as predicted by the brood-reduction hypothesis, last eggs are relatively smaller in altricial than precocial birds; in large species (which typically have long incubation periods) compared to small ones; and in open-nesting passerines compared to hole-nesting species (Slagsvold, Sandvik, et al. 1984). However, several anomalies remain. For example, in penguins of the genus *Eudyptes*, the second (final) egg is much larger than the first, and the nestling hatching from the second egg is usually the only one to survive (Warham 1975; Williams 1989). Why *Eudyptes* has adopted this bizarre arrangement is unknown.

9.7   Brood Desertion

When food availability falls to exceptionally low levels, disturbance by predators is particularly common or the chance that the brood will survive is low for other reasons, fish and birds may desert their entire brood (Mock 1984a), and mammals may respond by aborting or resorbing litters or by deserting or eating their young (Bradbury and Vehrencamp 1977; Gosling 1986). The net benefits of desertion will commonly depend on the parent's ability to raise another brood in the same season (Mock and Parker 1986), on the costs of raising broods to the parent's residual reproductive value and on the probability that young raised under adverse conditions will survive to breed successfully.

In some fish that normally show biparental care, removal of one parent can cause the remaining parent to eat the eggs (Keenleyside 1978). However, attacks on young may have other functions, too. In some coral reef fish, parents attack and may even kill some of their young 1–2 weeks after hatching (Thresher 1985). Though individual young have not been followed, results suggest that expelled young commonly join the broods of neighboring pairs and that parents increase the rate at which they can produce broods by prematurely terminating investment in this way, even though their behavior may reduce the survival of their young.

In birds, desertion usually occurs relatively early in the breeding attempt, parents are more likely to desert eggs than chicks (see Ricklefs 1969, 1977a) and relatively long-lived species like wood storks or flamingoes are more likely to desert than species that have a lower life expectancy (Mock 1984a). In some species, parents are more likely to desert very small broods early in the breeding season, possibly because it is more profitable to start again if there is sufficient time to rear a larger brood (Mock and Parker 1986). Some of the best evidence of adaptive brood desertion comes from studies of the effects of experimental reduction of clutch size in blue-winged teal, *Anas discors* (Armstrong and Robertson 1988). When clutch size was artificially reduced by removing eggs, the probability that the clutch would be deserted by the parents increased relative to control clutches that were disturbed but not reduced. Nests reduced to four eggs were more likely to be deserted than those reduced to seven (see Figure 9.8).

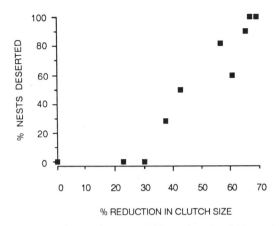

Figure 9.8   Percentage of nests deserted by blue-winged teal (*Anas discors*) following experimental reduction of clutch size by 0–70% (from Armstrong and Robertson 1988).

In mammals, a relationship between brood desertion and resource availability has been recognized for some time:

> Because the ground is chapt, for there was no rain in the earth, the plowmen were ashamed, they bowed their heads. Yea, the hind also calved in the field, and forsook it, because there was no grass. (Jer. 14:4–5)

Desertion also occurs in multiparous mammals. Following the death of one of their progeny, female brown bears sometimes abandon the surviving cub. Tait (1980) shows that mothers could increase their reproductive success by abandoning lone cubs if this allowed them to reconceive a larger litter directly instead of requiring them to wait until the end of the two-year lactation period. Similarly, mothers refuse to suckle very small litters in some marsupials (Lee and Cockburn 1985). In coypu (*Myocastor coypus*), young females that are in above average physical condition selectively abort small litters of predominantly female embryos two-thirds of the way through the nineteen-week gestation period, while large litters or small, predominantly male litters are retained (Gosling 1986). Females conceive soon after abortion, bearing litters that are significantly larger than those aborted (5.82 versus 4.17). Gosling suggests that small, female-biased litters are aborted because females in superior condition can express their reproductive potential more effectively by conceiving a larger litter, while small male-biased ones are retained because the increased investment in males that is possible in small litters continues to increment their fitness (see Chapter 11). While this may be the case, it is surprising that an average increase in litter size of 1.65 pups is sufficient to offset the costs of an extra thirteen weeks of gestation.

9.8 Summary

Since the quantitative effects of parental investment on offspring fitness are largely unknown (Chapter 2), attempts to investigate whether parents adjust their expenditure in an adaptive fashion have concentrated on testing qualitative predictions. Depreciable parental care is sensitive to the likely relatedness of parents and offspring and usually increases with brood size. In at least some cases, parents invest more heavily in older zygotes that have a higher chance of surviving than in younger ones, though parental care eventually declines as offspring approach independence.

No general prediction can be made about how parental investment should vary with offspring quality. Depending on circumstances, selection might favor either increased investment in superior offspring because their reproductive value is higher or reduced investment where the effects of parental expenditure on offspring fitness are smaller for superior offspring. In birds, parents more commonly bias investment toward weaker offspring, presumably in an attempt to maximize their survival, though both hatching asynchrony and variation in egg size may serve to reduce brood size when resources are inadequate to rear the entire brood. In multiparous rodents, mothers commonly kill a proportion of their young if food is short, brood size is unusually large, or the chance of infanticide by males is high. Where the fitness of young falls to very low levels, parents may increase their fitness by prematurely terminating investment in their entire brood. As would be expected, desertion is apparently more likely to occur if brood size is small.

The results described in this chapter emphasize three points relevant to future studies. First, they underline the need for better measures of the costs and benefits of parental expenditure. Second, they show the importance of field experiments designed to test the predictions of specific theoretical models. And, third, they illustrate the need to understand the behavioral and physiological processes responsible for changes in the fitness of parents and offspring arising from parental care—and the dangers of testing theoretical predictions with demographic data alone.

10

Parental Tactics 2: Variation in Care in Relation to Costs

10.1 Introduction

Though parents would be expected to adjust parental care in relation to its costs to their own fitness, our ability to predict variation in parental care is once again constrained by our inability to measure the costs of parental care precisely (see Chapter 3). This chapter examines three qualitative predictions based on the assumption that parental expenditure will decline as its costs to the parents' fitness increase: that parental care will be reduced where resource availability is low (Section 10.2); that where an individual's reproductive value declines with increasing age, old animals should invest more heavily than young ones (Sections 10.3 and 10.4); and that paternal care should decline as access to other mating partners increases (Section 10.5).

The last part of the chapter considers two related issues. In species with biparental care, a parent's best strategy may be to persuade its mate to invest as heavily as possible while minimizing its own investment (Chase 1980; Houston and Davies 1985). Since both partners are likely to be subject to this selection pressure, conflicts of interest between partners may be common. Section 10.6 describes the evidence that these conflicts influence parental behavior.

Since calculations of the costs and benefits of parental care are likely to be complex and involved, animals may sometimes base their level of investment on simple "rules of thumb," especially where environments are relatively stable. One such rule is to invest in relation to the level of past investment. In Section 10.7 we consider the evidence that animals behave in this way.

10.2 Parental Investment and Resource Availability

Where parents are better able to survive adverse conditions than their offspring, they would usually be expected to increase parental investment when resource availability is high and the fitness costs of expenditure are low (Wittenberger 1979b; Carlisle 1982). In addition, where offspring are better able to withstand resource limitation than parents, reduced food availability may favor increased reproductive effort by parents (e.g., Calow and Woolhead 1977), though this situation is probably rare in animals showing parental care.

The amount of energy transferred by parents to their offspring commonly declines as the availability of resources declines. For example, in the solitary bee, *Osmia lignia*, both energy intake per offspring and offspring growth decrease as food availability declines during the foraging season (Torchio and Tepedino 1980). Several studies of fish indicate that the intensity or duration of care declines with food availability: low food rations reduce time spent fanning eggs by males in sticklebacks (Stanley 1983), and by females in convict cichlids, *Cichlasoma nigrofasciatum* (Townshend and Wootton 1985). Feeding rate in birds and milk yield in mammals also decline with resource availability (Drent and Daan 1980; Loudon, Darroch, and Milne 1984; Loudon 1985; Prentice and Whitehead 1987) as does the intensity of nest defense (Montgomerie and Weatherhead 1988). Similarly, in white-tailed deer (*Odocoileus virginianus*) mothers that have recently given birth are more likely to attack intruders in years when food is abundant and body condition is high than in years when food availability and condition are low (W. P. Smith 1987).

However, since optimal levels of expenditure cannot be identified, it is usually difficult to distinguish adaptive changes in parental expenditure from direct, nonadaptive consequences of reduced food availability. Interest in the sensitivity of parental expenditure to changing costs has consequently focused on the effects of changes in the parent's reproductive value.

10.3   Reduced Reproductive Success in Young Breeders: Restraint or Constraint?

The fitness costs of reproduction to young animals may be higher than to old animals for at least three reasons. First, both the energetic costs of a given level of reproductive output and their influence on the parent's survival may be greater in novice or young breeders than in mature animals. For example, analysis of the time budgets of breeding rooks (*Corvus frugilegus*) indicates that younger animals expend more energy on breeding than older birds, although they lay smaller clutches and fledge fewer young (Røskaft, Espmark, and Järvi 1983), while young red deer are more likely to die after breeding successfully than older mothers, although younger mothers produce smaller calves (see Figure 10.1). Second, in animals where individuals continue to grow after reaching breeding age, reproduction may delay the attainment of maximum size, in some cases leading to permanent stunting and impairing reproductive output over the rest of the lifespan (see Chapter 3). And third, where survival declines with age, a given risk of dying as a result of breeding

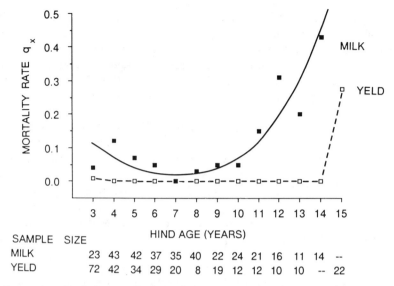

Figure 10.1   Survival costs of breeding in female red deer of different ages. Points show the overwinter survival of females of different ages who reared calves successfully (milk) versus those that either failed to produce calves or lost them shortly after birth (yeld), based on a sample of around 500 hinds (from Clutton-Brock 1984).

represents a larger cost to young animals than to old ones, especially in species that breed on a large number of occasions (Pianka, 1976; Curio 1988).

Because of the higher costs of breeding, young animals would often be expected to expend a lower proportion of the energy available to them on reproduction than mature individuals (Williams 1966a; Pianka 1976; Curio 1983, 1988). Extended iteroparity, low adult mortality, continuing adult growth, improvements in reproductive efficiency, low rates of population increase, and a high sensitivity of survival and growth to reproductive expenditure should all reinforce this trend (Charlesworth and Leon 1976; Pianka 1976). However, there may also be some circumstances where reproductive effort should *decline* with age (Fagen 1972). For example, if young animals are less efficient at acquiring resources but there are certain minimal costs of reproduction that must be paid by all animals attempting to breed, young breeders may need to expend a *higher* proportion of the energy available to them to have any chance of breeding successfully. In addition, where parental survival and reproductive capacity increase with age (as in some fish and reptiles), individuals might increase their fitness by expending a *lower* proportion of their available resources on reproduction as they get older.

Empirical evidence indicates that in the great majority of species, young animals show lower reproductive success than older individuals (Lack 1966;

Skutch 1976; Curio 1983; Clutton-Brock 1988; Ollason and Dunnet 1988; Partridge 1988), but there is little firm evidence that this is because young parents expend a lower proportion of their available resources. A variety of field studies suggest that feeding and reproductive skills increase with age in many iteroparous species (Lack 1966; De Steven 1978), and the costs of rearing offspring may often be greater for inexperienced adults than for older individuals (see above). Correlations between age and reproductive performance may also arise because poor-quality individuals show inferior reproductive performance and also die young (see Curio 1983; Clutton-Brock 1988): in some cases, age effects are reduced or even disappear altogether when the reproductive performance of the same individual is compared at different ages (Clutton-Brock 1988). Since it is difficult to manipulate maturation processes experimentally, studies of changes in parental investment at the end of the lifespan provide a more promising approach to testing whether parental investment varies adaptively with age.

10.4 Reproductive Value and Terminal Investment

In iteroparous species where survival or fecundity decline with increasing parental age, the proportion of available resources devoted to reproduction should increase with age, especially where the potential number of reproductive attempts is high (Williams 1966a,b: see Figure 10.2). A variety of theoretical papers have explored this prediction and come to similar conclusions (Gadgil and Bossert 1970; Charlesworth and Leon 1976; Pianka 1976; Stearns 1976; Michod 1979; Charlesworth 1980; Begon and Parker 1986; Curio 1988), though there may also be circumstances where reproductive effort should *decline* with age (Fagen 1972; Schaffer 1974a).

Although several papers have claimed to demonstrate adaptive increases in parental expenditure with age, few are convincing. A satisfactory demonstration requires evidence that older individuals devote a higher proportion of their available resources to reproduction or take greater risks in protecting their offspring and consequently show reduced survival or subsequent reproductive performance. Evidence of increased energetic expenditure or reproductive performance by itself does not constitute firm evidence of increased investment, since older animals may be able to afford higher levels of expenditure because of their increased size, improved feeding skills, or greater breeding experience (Pianka and Parker 1975). Conversely, evidence that older individuals show *lower* reproductive performance does not refute the prediction, for the ability to acquire, process, or store resources may decline with age (Clutton-Brock 1984).

The situation is complicated by problems of measuring the proportion of resources allocated to reproduction. Most attempts to estimate proportional

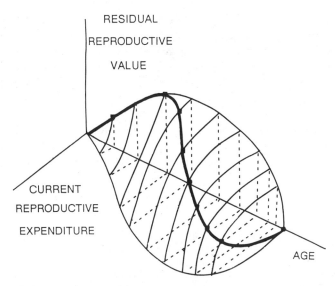

Figure 10.2 During the lifetime of an iteroparous organism, the trade-offs between current and future reproduction may change, favoring a progressive increase in reproductive effort with age. The light curves relate costs in future progeny to benefits to present progeny (and vice versa) at different ages (from Pianka 1976). The dark solid curve traces the optimal reproductive tactic that maximizes lifetime reproductive success throughout the lifespan. The shape of the three-dimensional surface will vary with immediate environmental conditions as well as with the reproductive strategy of the organism concerned.

expenditure are unsatisfactory. For example, studies of several lizards have interpreted increases in the ratio of clutch or litter weight to maternal weight as evidence that parental investment increases in older individuals (Tinkle and Hadley 1973, 1975; Pianka and Parker 1975). However, older individuals are also larger and may require less energy per unit body weight for maintenance and growth (see Chapter 6). If the rate at which individuals can acquire food increases with their size (for example, because the largest individuals monopolize the best feeding sites), larger individuals may be able to expend a higher proportion of their energy budgets on reproduction for a given fitness cost (Clutton-Brock 1984).

Finally, evidence that the fitness costs of reproduction increase with age does not, on its own, provide a reliable indication of the adaptive increase in reproductive effort. In red deer, the difference in overwinter survival between hinds that rear calves through the summer (milks) and those that do not (yelds) increase with age (Clutton-Brock 1984; see Figure 10.4). However, this increase could occur because the ability of hinds to acquire or process resources declines with age, while there are certain minimal costs of breeding—with the result that reproduction is more dangerous for older mothers. Firm evidence of an adaptive increase in investment with age would also require that older mothers produce better offspring than prime-aged females.

Despite these problems, there is suggestive evidence that reproductive effort or parental investment may increase with age in some long-lived iteroparous animals. In a number of teleost fish, increases in the proportion of energy devoted to reproduction with rising age appear to be too large to be caused by allometric increases in resource acquisition (see Woodhead 1979). For example, in whiting (*Merlangius merlangus*) daily percentage loss of weight due to spawning increases by a factor of four between small (young) fish and large (old) ones (see Figure 10.3). In California gulls (*Larus californicus*) old par-

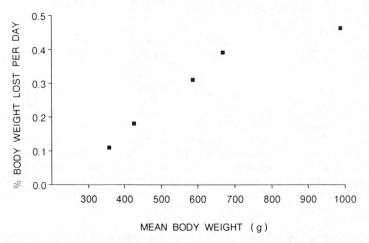

Figure 10.3 Daily percentage loss in body weight due to spawning in whiting (*Merlangius merlangus*) of different sizes (from Hislop 1975).

ents fledge more young than young or middle-aged parents, apparently because they feed their chicks longer and more frequently (Pugesek 1981; Pugesek and Diem 1983). In snow geese (*Anser caerulescens*) females more than 5 years old defend their nests more aggressively than 2–5-year-olds (Ratcliffe 1974), while in several short-lived passerines where parental survival does not decline with age, no changes in nest defense have been found (Montgomerie and Weatherhead 1988).

Finally, in red deer, although maternal condition falls with increasing age, hinds over 12 years old allow their calves to suck for longer and their offspring show improved condition and overwinter survival compared to those of middle-aged hinds. As predicted, the costs of breeding to the mother's survival increase with her age (see Figure 10.4). In both the last two cases, the improvement in reproductive performance occurs toward the end of the mother's life span, and it seems unlikely that this is caused by improved reproductive or

Figure 10.4   Changes in
reproductive performance
with maternal age in red deer
(from Clutton-Brock 1984).
(a) Mean duration of suckling
bouts;
(b) calf condition (kidney fat
index) at the onset of winter
(solid line) and overwinter calf
survival (broken line);
(c) difference in age-specific
mortality rates between milk
and yeld hinds.
Suckling durations based on
samples of 5–10 mothers
sampled per age category;
juvenile survival based on
samples of 10–30 calves per
maternal age category; female
survival based on samples of
20–100 females per age
category.

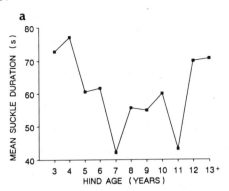

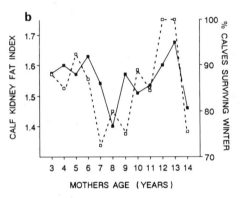

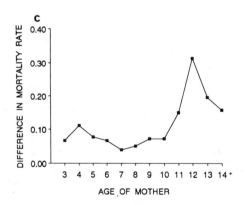

feeding skills. However, here, too, it is important to demonstrate that similar changes occur within individuals, for positive correlations between reproductive success and age could occur because better-quality animals both live longer and breed more successfully (see above).

Comparisons of parental care between successive broods in the same season also suggest that parental expenditure may change in an adaptive fashion (Montgomerie and Weatherhead 1988). In temperate animals that can breed more than once per season, breeding periods are often followed by increased adult mortality during the winter months. As a result, the residual reproductive value of breeding adults declines during the course of breeding seasons, and parents might be expected to invest more heavily in late broods than in early ones (Barash 1975; Pianka and Parker 1975; Curio, Regelmann, and Zimmermann 1984). In lace bugs (*Gargaphia solani*) females attack approaching predators more often, more quickly, and more aggressively while guarding their second clutch than when guarding their first (Tallamy 1982, 1984). Females of some multibrooded bird species show a similar tendency to defend late broods more intensely than early ones (Greig-Smith 1980; Biermann and Robertson 1981, 1983; East 1981; Curio and Regelmann 1987), though in others studies have found no consistent seasonal increase in nest defense (Weatherhead 1979b; Montgomerie and Weatherhead 1988). However, a number of other variables may change during the course of the breeding season, including the availability of resources, the parents' experience of predators, and the age and experience of the predators themselves and these trends could be responsible for changes in parental behavior (see Harvey and Greenwood 1978; Nur 1983; Clutton-Brock 1984; Gochfeld 1984; Knight and Temple 1986a,b).

10.5 Parental Investment by Males and Access to Mating Partners

The fundamental problem faced by attempts to demonstrate age-related changes in parental investment is that reproductive efficiency is likely to change with breeding experience (see Dolhinow, McKenna, and Laws 1979; Knight and Temple 1986a; Montgomerie and Weatherhead 1988). The most convincing evidence that parents adjust their expenditure of time or energy in relation to fitness costs comes from studies that have manipulated the costs of paternal care to the mating rate of males (see Robertson and Biermann 1979; Carlisle 1982; Armstrong and Robertson 1988). For example, in the biparental cichlid, *Herotilapia multispinosa*, males guard single clutches and must desert their previous offspring before they can breed again (Keenleyside 1983). By changing the ratio of breeding females to breeding males in different ponds,

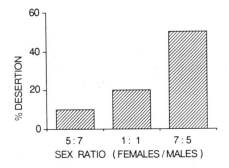

Figure 10.5 Proportion of male *Herotilapia multispinosa* deserting their broods in relation to changes in the adult sex ratio (females:males) (from Keenleyside 1983).

Keenleyside was able to vary the probability that males would mate again and, hence, the costs of remaining to guard their previous brood. As predicted, male desertion increased as the adult sex ratio became progressively biased toward females (see Figure 10.5).

The relative attractiveness of males to females and of females to males may also affect their level of investment (Burley 1988). For example, male and female zebra finches (*Poephilia guttata*) whose attractiveness to the opposite sex has been increased by changing the color of their leg rings (Burley 1981) reduce their expenditure on parental care (Burley 1988). Partners given "unattractively" colored leg rings not only show higher levels of parental expenditure but also have shorter life spans and lower long-term breeding success. Attractive birds may benefit from reducing expenditure because this raises their ability to get extra matings or their long-term reproductive output, while unattractive birds may benefit from increasing their level of expenditure because it permits them to get attractive mates and to retain them (Burley 1988). Burley's results provide a good demonstration of the conflicts of interest that are likely to arise between care-givers.

10.6   Parental Investment and Conflicts of Interest between Care-givers

In animals where both sexes care for the young, the benefits that a parent's offspring gain from parental expenditure are likely to be influenced by the amount of care they are already receiving from the other parent (Chase 1980; Davies 1985; Houston and Davies 1985; Winkler 1987; Lazarus 1989). In many circumstances, the benefits of investing to the parent's fitness will decline in relation to expenditure by the other parent, though there will also be cases where they should increase (Chase 1980; Regelmann and Curio 1986).

How should investment by one parent change in relation to the level of expenditure provided by its mate? Assuming that each parent's survival is reduced in relation to its total expenditure, Houston and Davies (1985) predict that where parents treat all members of the brood in the same way and offspring fitness increases as an asymptotic function of total parental expenditure, once parental expenditure exceeds some threshold level, each parent should respond to increases in care by its partner by *reducing* its own expenditure (see Figure 10.6a,b). Conversely, parents would be expected to respond to reductions in care by their mates by *increasing* their own expenditure. An equilibrium (ESS) will be reached when the male's expenditure E_m is the male's best value given that the female expends E_f, and where E_f is the female's best value if the male expends E_m (a Nash equilibrium). If curves for E_m plotted on E_f and E_f plotted on E_m ("reaction" curves) intersect and each has a slope of less than -1, the intersection point should be an ESS for both partners (see Figure 10.6c). In such cases, parents should respond to reductions in care by their partners by increasing their own effort while not fully compensating for the reduction, so that total expenditure is reduced. Where two partners are making alternate investments in parental care, their expenditures are likely to converge on the ESS through a sequence of smaller and smaller changes (Chase 1980).

If reaction curves do not intersect or the slope of either is greater than -1 at the region of intersection, other outcomes are possible (Chase 1980; Winkler 1987). If the female's reaction curve is completely above that of the male (Figure 10.6d), then the male should desert and the female's best strategy would be to take over care of the young. Similarly, if the male's reaction curve is above that of the female (Figure 10.6e), then the male should be responsible for parental care. Finally, if reaction curves have slopes greater than -1 at the region of intersection (Figure 10.6f), the intersection point is unstable. Under these conditions, if one parent reduces its expenditure, the response of the other parent is to increase its own expenditure by a *larger* amount, more than compensating for the reduction. Reactions then get larger and larger until one parent is responsible for all expenditure (Houston and Davies 1985). Which parent this will be depends on the initial conditions.

In practice, the responses of partners to changes in the level of investment by their mates vary widely—which may help to account for the different effects of removing one partner (see Chapter 8). In some cases, increased assistance by the male has little or no effect on the level of parental care provided by its mate (Slagsvold and Lifjeld 1988; Whittingham 1989). In others, the remaining parent compensates partly or fully for the reduction in input by its mate, leading to negative correlations between the expenditure of partners (S.G. Martin 1974; Weatherhead 1979a; J.M.N. Smith, Yom-Tov, and Moses 1982; Breitwisch 1988; Wanless, Harris, and Morris 1988; L. Wolf Ketterson, and Nolan 1988, 1989). Most experimental demonstrations of this effect involve the removal of one mate and have the disadvantage that the consequences of mate removal may be different from those of changes in investment by a parent that continues to help, especially if the level of expenditure is determined by a "bargaining" process (see Figure 10.6c). One recent study of

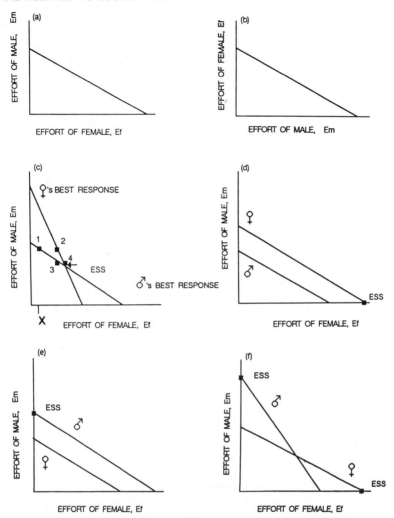

Figure 10.6 Optimal parental expenditure for males and females plotted on the level of expenditure by the other sex (from Houston and Davies 1985). (a) Optimal male effort plotted on female effort; (b) optimal female effort plotted on male effort. (c) refers to the third figure, which shows the first two curves (a, b) superimposed, both with slopes of <−1 so that the point of intersection of the two lines is stable. Under these conditions, if the female expends X, the male's best response is to expend 1. The female then replies with 2, the male with 3, the female with 4, and so on until the efforts reach the ESS. At the stable point, it does not pay either individual to change its effort. (d)–(f) show three other possible outcomes suggested by Chase (1980). In (d), the female curve lies completely above that of the male; here the ESS is for the female to do all the work. In (e), the male curve lies above that of the female; here the ESS is for the male to do all the work. In (f), the curves intersect but the intersection point is unstable. Either "all female work" or "all male work" is an ESS.

starlings (*Sturnus vulgarus*) avoids this problem by varying the rate of feeding by partners by attaching weights to their tails (J. Wright and Cuthill 1989). Weighted partners (either males or females) reduce their feeding rate, and their unweighted mates compensate for this effect by increasing their own rate, though they do not do so fully. Finally, positive correlations between levels of investment have been found in a few cases. For example, great tits increase the intensity of their mobbing response to an artificial predator in relation to the intensity of their mate's behavior (Regelmann and Curio 1986).

As Chase (1980) predicted, similar effects of the level of input by other parents occur among cooperative and communal breeders. For example, in the dunnock, *Prunella modularis*, some females are polyandrously mated to two males, an Alpha male who is responsible for most of the mating and a Beta male who mates less often (Davies 1985; Houston and Davies 1985). Beta males generally expend less effort on the broods than Alphas, and their expenditure on care is related to the frequency with which they copulate with the female (Burke, Davies, et al. 1989). For a given brood size, females feeding chicks on their own feed them more frequently than when they are helped by a single male. The assistance of a second male reduces the work load of the Alpha male but has little effect on that of the female, who generally works harder than either male (Table 10.1).

In some biparental species, one sex may desert before parental care is complete (e.g., Beissinger 1987; Ezaki 1988). This situation is likely to lead to conflicts of interest over who will care for the young (see Section 8.2). The benefits of desertion to the deserter will depend on whether the deserted partner takes over all care of the brood or deserts in its turn (Dawkins 1976; Lazarus 1989). A parent will be particularly likely to desert its mate at times when the potential payoffs of desertion are high, when reductions in care have

Table 10.1

Mean provisioning rates of dunnocks, *Prunella modularis*, when working alone, in pairs, or in trios (from Houston and Davies 1985).

| Brood Size | Single Female | Feeds to Brood per Hour by | | | | |
| | | Pair | | Trio | | |
		Female	Male	Female	Alpha Male	Beta Male
1	5.2	3.9	3.8	3.4	3.1	0.3
2	—	5.6	4.6	4.1	4.8	3.3
3	9.4	6.6	5.9	6.2	4.6	5.6
4	—	6.9	8.2	7.1	6.4	5.1
5	—	8.6	11.9	8.1	8.3	6.1

relatively little effect on the fitness of its brood, or when its partner is least likely to desert in response (Lazarus 1989).

Situations of this kind can lead to complex interactions between the best strategies of the two sexes (Lazarus 1989). For example, parents may sometimes be selected to continue parental care because desertion would cause their mates to desert, reducing the original deserter's fitness below the level it would achieve under biparental care (Dawkins 1976; Lazarus 1989). It is even conceivable that one or even both sexes may be selected to desert "preemptively" in order to avoid being placed in a cruel bind by their mates at a later stage of parental care (Lazarus 1989). In practice, these complexities are likely to be difficult to distinguish respectively from cases where selection favors parental care in both sexes or where one sex is selected to desert.

Empirical studies of desertion in species where both sexes usually care for the young are scarce but go some way to support theoretical predictions. In the great reed warbler (*Acrocephalus arundinaceus*) males defend territories used by one or more females and commonly desert their last partners of the season, entering moult while their mates are still feeding nestlings (Ezaki 1988). In this case, desertion may not be favored in males until continued defense of the breeding territory is unlikely to yield further mating partners. In the monogamous snail kite, *Rostrhamus sociabilis*, either sex may desert before the end of parental care. The frequency of desertion increases when food is plentiful (see Figure 10.7), presumably either because costs to existing broods are low or mating seasons are prolonged (Beissinger and Snyder 1987). Parents that are going to desert also feed the young less frequently from the second week after hatching. Beissinger and Snyder (1987) suggest that this may occur either because individuals adjust their reproductive expenditure in relation to their perceived opportunities to remate, or because partners are testing their mates to check whether they are likely to increase parental expenditure if deserted.

Figure 10.7 Changes in the percentage of snail kite (*Rostrhamus sociabilis*) nests that were deserted by one parent in relation to the farthest distance from the nest traveled by adults in search of snails (an index of food scarcity) (from Beissinger and Snyder 1987).

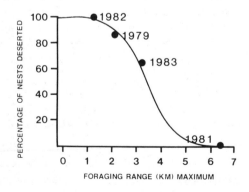

10.7   Parental Decision Rules

As the previous section describes, individuals that have already invested heavily in a brood are often less likely to desert than individuals that have invested less heavily. In his original (1972) paper, Trivers proposed that parents should adjust current expenditure in relation to past investment in order to minimize wastage of reproductive effort—an argument subsequently nicknamed the "Concorde fallacy" following the justifications produced by British politicians for spending further funds on developing supersonic aircraft in order to minimize wastage of previous expenditure (Dawkins 1976).

Trivers's argument is incorrect, for parents should adjust their current expenditure to prospective (net) benefits, not past costs (Dawkins and Carlisle 1976), as economists have recognized for some time (Boucher 1977). However, past investment will commonly reduce the parent's capacity for further expenditure and may consequently provide parents with a reliable index of the prospective benefits of future expenditure (see Maynard Smith 1977; Sargent and Gross 1985; Curio 1987; Coleman and Gross, in press). If so, they may behave *as if* they were committing the Concorde fallacy, although their levels of expenditure are in fact related to the prospective benefits of parental care. Behavior of this kind is sometimes called "Concordian" because of its resemblance to the adjustment of expenditure in relation to past investment.

Several studies have now shown that current expenditure is correlated with past expenditure. For example, in savannah sparrows (*Passerculus sandwichensis*), the intensity of nest defense is more closely predicted by the amount of time for which parents have been defending the nest than by the time remaining in the breeding season to renest if the current nest fails (Weatherhead 1979b, 1982).

In an elegant experiment on bluegill sunfish (*Lepomis machrochirus*), Coleman, Gross, and Sargent (1985) tested four alternative decision rules that males might be using to determine how intensively to defend their broods: (1) invest according to brood size alone; (2) invest according to past expenditure; (3) invest according to both brood size and past expenditure; (4) invest according to neither brood size nor past expenditure. If parents were behaving in a Concordian fashion, investment would be expected to vary with past expenditure (2).

To test which rule the fish used, Coleman et al. measured the response of males to a model of an intruding male presented on the sixth day of the brood cycle, having manipulated broods to produce three experimental samples:

1. Families where brood size was artificially reduced by 50% after the first day of the brood cycle ("Early"). In these, the value of the brood was small but so, too, was past expenditure on fanning the eggs, which varied with brood size.

2. Families where brood size was artificially reduced by 50% on the fifth day of the brood cycle ("Late") with the result that, while the value of the brood was small, past expenditure on fanning was high.

3. Families where broods were not reduced and both brood value and past expenditure were high ("Control").

If the intensity of defense depended on brood size alone, they predicted that values for Controls should be higher than for either Late or Early samples, which should not differ from each other. In contrast, if it depended on past expenditure, values for Control and Late samples should be similar and both should exceed values for Early samples. If both brood size and past expenditure were involved, values for Control samples should be higher than for Late samples, which should be higher than for Early samples. Finally, if neither brood size nor past investment were involved, values for the three groups should be similar. Their results (see Figure 10.8) show that intensity of nest

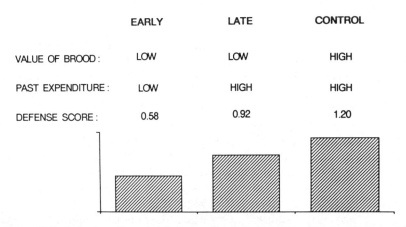

	EARLY	LATE	CONTROL
VALUE OF BROOD :	LOW	LOW	HIGH
PAST EXPENDITURE :	LOW	HIGH	HIGH
DEFENSE SCORE :	0.58	0.92	1.20

Figure 10.8 Intensity of brood defense on the sixth day of the brood cycle by male bluegill sunfish against a model intruder (from Coleman, Gross, and Sargent 1985). Defense scores are shown for these samples. *Early*: where brood size was reduced by 50% early in the brood cycle and both the value of the brood and parental expenditure on fanning (which is correlated with brood size) were low; *Late*: where brood size was reduced by 50% late in the brood cycle with the result that while the value of the brood was low, past expenditure on fanning was high; *Control*: where brood size was not reduced and both the value of the brood and past expenditure were high.

defense was higher for Control than Late samples, which were higher than Early, indicating that parents adjusted their defense in relation to brood size and past expenditure.

While this experiment suggests that past expenditure may be one of the factors affecting current expenditure, it does not indicate that male blue-gills were committing the Concorde fallacy—for their expenditure on fanning may have affected their subsequent capacity for expenditure and thus the relative value of future broods (Coleman, Gross, and Sargent 1985). Correlations between past and present expenditure do not necessarily indicate that any causal link exists, for a variety of other factors are likely to generate positive correlations between past and present expenditure, including variation in environmental factors and individual differences in the fitness costs of parental expenditure.

So far, none of the putative examples of the Concorde fallacy have shown that animals make the mistake identified by Dawkins (Coleman, Gross, and Sargent 1985; Curio 1987; Coleman and Gross, in press). Moreover, evidence that humans do so is founded on the naïve assumption that politicians base their actual (as against their expressed) calculations on value for money rather than on the likely effects of conspicuous wastage on future votes.

10.8 Summary

Parental expenditure should vary with changes in the fitness costs of care to the parent. However, firm evidence that it does so is scarce. Parental expenditure is commonly related to resource availability, but adaptive strategies of parental expenditure are difficult to distinguish from the ecological consequences of reduced food availability. Similarly, lower expenditure by young breeders may occur because young animals are minimizing the risks of breeding as a response to their high reproductive value—or because of their inexperience in finding resources or their lower body condition. Increased parental expenditure at the end of the lifespan is apparently uncommon but has been found in a number of species. However, no study has yet conclusively shown that this is not a consequence of improvements in the parent's ability to find resources, though this appears unlikely in some cases. The firmest evidence that parental expenditure is adjusted to the fitness costs of care to the parent come from studies of male care in fish, which show that increased access to other breeding partners reduces the extent or duration of male care.

Where two parents care for the young, a parent's optimal level of expenditure will also depend on the level of expenditure of its mate. Empirical evidence shows that parents do commonly increase their expenditure in response to reductions in expenditure by their partners and reduce it in relation to increases by their partner. In theory, a stable situation will be one where each partner is expending its optimum level given the level of expenditure by its partner.

Conflicts of interest between partners also occur where one adult deserts the brood before the completion of parental care. Here, too, the benefits of desertion will depend on the likely response of the partner to being deserted as well as on access to alternative mates and are likely to vary during the course of parental care. Some evidence suggests that partners which have invested less in broods are more likely to desert them. This could occur because past investment affects an individual's ability to invest at a later stage or because individuals are testing their partner's response to a reduction in care on their part. However, relationships of this kind could also occur because the fitness costs of parental expenditure to some individuals are consistently higher than to others.

Finally, "decisions" about parental investment are sufficiently complex that parents may sometimes use some simple "rules of thumb" that give a rough but reliable indication of the optimal level of investment. One possible rule may be to invest in relation to past investment. This is unlikely to be (as was originally suggested) so as to reduce the chance that previous investment will be wasted, but because past expenditure commonly predicts the parent's future capacity for expenditure. Several studies have now shown that an individual's past expenditure on parental care predicts its likely expenditure in the future, but there are usually several possible interpretations of these relationships.

11 Parent-Offspring Conflict

11.1 Introduction

Most predictions about optimal levels of parental care are based on the assumption that parents are free to adjust expenditure in their offspring in relation to their own interests. However, this may not always be the case, for the parent's level of expenditure may be constrained by the behavior of its own offspring. Whenever parents are not genetically identical to their offspring, conflicts of interest between parents and offspring are likely to arise that may influence the level of parental investment and depress the mean fitness both of parents and offspring (Trivers 1974; Parker 1985).

The first section of this chapter reviews theoretical predictions about conflicts of interest between parents and their young, while the second examines the empirical evidence that conflicts modify parental expenditure.

11.2 Parent-Offspring Conflict in Theory

Interbrood Conflict under Uniparental Care

In a characteristically original paper, Trivers (1974) was the first to point out that because the degree of relatedness between parents and their offspring is only 0.5 in outbred, diploid animals, parents and their progeny should "disagree" about the level of parental investment. Trivers envisaged a situation where mothers produced a single young each year, where subsequent offspring were full sibs, and where fathers did not invest in their offspring after conception. He measured the benefits of parental expenditure in terms of the fitness of current offspring and the costs in comparable units of the number or fitness of the mother's future offspring.

Trivers's graphical model envisages a situation where the amount of parental expenditure per day is fixed and the benefits of parental care to the current offspring decline as it grows older, while the costs to the parent either remain constant or increase, so that the ratio of benefit to cost (B/C) shows a progressive decline (see Figure 11.1). Basing his argument on Hamilton's (1964) con-

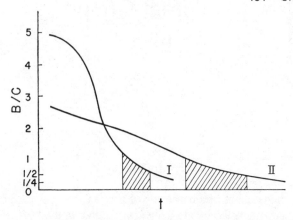

Figure 11.1   The benefit/cost ratio (B/C) of a parental act (such as nursing) toward an offspring as a function of time (from Trivers 1974). Benefit is measured in units of reproductive success of the offspring, and cost in comparable units of reproductive success of the mother's future offspring. Two species are plotted: in species I the benefit/cost ratio decays quickly; in species II, it does so more slowly. Shaded areas indicate times during which parent and offspring are in conflict over whether the parental care should continue. Future sibs are assumed to be full-sibs. If future sibs were half-sibs, the shaded areas would have to be extended until B/C = 1/4.

cept of inclusive fitness, Trivers argued that because the coefficient of relatedness between the offspring and its future sibs will be 0.5 while the offspring's relatedness to itself is 1.0, the offspring should favor the mother's continuing parental care until the costs of care to the mother's fitness exceed twice the benefit to itself. Mothers and offspring should consequently "disagree" over the continuation of parental care from the time when the B/C ratio equals 1.0 to the time when it equals 0.5 (see Figure 11.1). Depending on the rate at which B/C declines with increasing offspring age, this may be either a relatively short period (Curve I) or a relatively long one (Curve II). Where offspring can raid parental resources directly or can deceive parents into exceeding their own optimal level of investment (for example, by begging frequently for food), Trivers predicted that they should do so until the B/C ratio falls to 0.5 if future offspring are full sibs or 0.25 if they are half-sibs. Periods of evolutionary conflict between parents and their offspring might be expected to give rise to overt behavioral conflict, such as the apparent "disagreements" over weaning between female mammals and their young (Trivers 1974).

In a subsequent model, Trivers showed that evolutionary conflicts between parents and offspring are likely to arise over levels of investment at any stage of the period of parental care. Trivers argued that since the cost of parental expenditure to the adult will be double the cost that is suffered by its offspring, the latter will be selected to favor a level of parental expenditure that maximizes the difference between B and C/2, while parents will be selected to favor a level that maximizes the difference between B and C (see Figure 11.2a). Since both the parent and the current offspring will be related to the parent's future progeny by 0.5 (assuming that they have the same father), subsequent treatments of parent-offspring conflict have argued that the *costs* of expenditure to the parent and its current offspring will be identical and that it is the *benefits* to parents and offspring that will differ. Lazarus and Inglis (1986) argue that benefits to the current offspring will be twice as high as to the parent, on the grounds that offspring receiving investment are 100% related to themselves while the benefits to the parent of investing in its progeny must be devalued by their average relatedness to their offspring (Dawkins 1976; Hartung 1977, 1980; Robinson 1980; Lazarus and Inglis 1986; see Figure 11.2b). However, Trivers's fundamental point that offspring can usually be expected to favor a level of parental expenditure that is higher than their parents' optimum remains unchanged.

Implicit in Trivers's theory is the suggestion that where offspring are able to modify their parents' behavior, they may reduce the parent's fitness by shifting the level of investment away from the parent's optimum toward their own. Shortly after Trivers's original (1974) paper appeared, this suggestion was challenged by Alexander (1974), who argued that selection operating on the

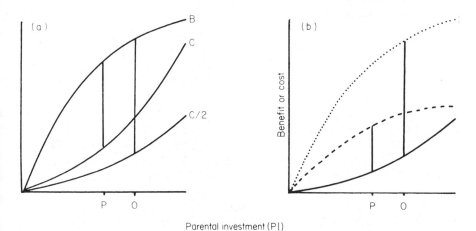

Parental investment (PI)

Figure 11.2 (a) The benefit, cost, and half the cost of a parental act toward an offspring at one moment in time as a function of the amount of fitness the parent invests in the act (PI) (from Trivers 1974). B = benefit to parent and offspring; C = cost to parent; C/2 = cost to offspring. At P the parent's inclusive fitness (B–C) is maximized; at O the offspring's inclusive fitness (B–C/2) is maximized. Parent and offspring disagree over whether P or O should be invested. The offspring's future siblings are assumed to be full-siblings. (b) The reformulation of parent-offspring conflict proposed by Lazarus and Inglis (1986). B = benefit to parent; 2B = benefit to offspring; C = cost to both parent and offspring. Parental benefit is shown as a dashed line and offspring benefit as a dotted line (from Lazarus and Inglis 1986).

parent will always result in a level of expenditure that is adjusted to the parent's optimum. A gene that reduced parental fitness by encouraging its carriers to extract more resources from their parents than the latter were selected to give could not, he argued, spread to fixation, because individuals that benefited as offspring by extorting additional investment would suffer the consequences of the same trait once they became parents.

Alexander's challenge was taken up in a series of papers by Parker and Macnair (Parker and Macnair 1978; Macnair and Parker 1978, 1979; Parker 1985). Parker and Macnair argued that the selection pressures leading to parent-offspring conflict are likely to be frequency-dependent since the best level of investment for an offspring to take depends on the current level being taken by the rest of the offspring in the population. They consequently used Game Theory models rather than simple optimization procedures to investigate the conditions under which a rare "conflictor" gene, causing its carriers to extort more resources from their parents than the latter were selected to give, could spread through populations of "nonconflictors."

Their first model explores the evolution of conflictor genes under the circumstances that Trivers originally envisaged, where a single parent invests in a single offspring that is related to its future sibs by half. Parker and Macnair (1978) show that a rare conflictor gene will spread if $f(m) > 1/2\,(m + 1)$, where $f(m)$ is the fitness gained by a conflictor compared with a nonconflictor offspring $(f(m) > 1)$, and m is the amount of parental investment taken by a conflictor relative to $m = 1$ for a nonconflictor. Where sibships consist of nonconflictors as well as conflictors, the nonconflictor allele will lose to the conflictor allele if m is not too large, and if m is small there is no bar to the fixation of the conflictor gene. As the cost of the conflictor gene to the parent increases, the probability that it will spread declines, but it can still be expected to reduce parental fitness by causing parents to exceed their optimal level of investment (Parker 1985). A variety of similar models have come to similar conclusions (see Blick 1977; Stamps, Metcalf and Krishnan 1978; Metcalf, Stamps and Krishnan 1979; Hartung 1980; Robinson 1980; Stamps 1980; Feldman and Eshel 1982; Bull 1985).

Though Alexander was clearly wrong in arguing that "conflictor" genes which reduced parental fitness could not spread in sexually reproducing diploid organisms (and has since modified his views: Alexander 1979), his argument is correct for asexually reproducing species where the genetic interests of parents and offspring are identical. Moreover, in sexual organisms, the spread of conflictor alleles can be inhibited if their cost to the parent is extremely high, or if offspring are unable to persuade or force their parents to exceed their own optimal level of investment—for example, where the amount of parental expenditure is not related to cues derived from the offspring (Parker and Macnair 1978; see also Harper 1986). It is also important to remember that where there are large differences in reproductive value between offspring and parents, the *realized* benefits of extracting additional parental investment to current offspring may be reduced, minimizing the difference between the level of expenditure favored by the parent and the level favored by its offspring.

PARENT-OFFSPRING CONFLICT IN OTHER BREEDING SYSTEMS

Subsequent models by Macnair and Parker (1978, 1979) and Parker (1985) explore the effects of differences in the number of care-givers and in the relatedness of progeny within and between broods on the evolution of conflict. Though they confirm that, under most circumstances, selection will favor "conflictor" genes, both these aspects of the breeding system have an important influence on the conditions under which "conflictor" genes can be expected to spread.

If, instead of a single parent rearing the young, two parents invest equally in their offspring and are constrained to breeding with a single partner during their lifespan ("true monogamy"), the parents' optimal investment per young will be the same as under uniparental care by a monogamous female, but because two parents contribute to care, twice as many offspring can be reared (Parker 1985). However, where both parents invest equally but pair only for the duration of a single brood so that each partner can "save" expenditure for subsequent breeding attempts with other mates, the responses of both parents to the level of investment provided by their mates have to be considered. Conflict between parents (see Section 10.6) favors a reduction in the level of investment per offspring, so that, paradoxically, investment per progeny should be less when two parents invest than when only one does (Parker 1985). However, whether or not this is translated into a reduction in *expenditure* per offspring will depend on the effects of biparental (versus uniparental) care on fitness costs per unit expenditure by the parent.

Where conflicts occur between parents and offspring over the allocation of a fixed amount of investment to members of the *same brood* (as against between successive offspring or successive broods), conflictor genes can be expected to spread whether they are dominant or recessive and whatever the mating system, provided that there are no direct costs of attempting to extort additional investment or of the presence of other conflictors in the same brood (Macnair and Parker 1979; Metcalf, Stamps, and Krishnan 1979; see also Harper 1986). Where direct costs to progeny are involved, the spread of conflictor genes and the intensity of conflict will depend on the magnitude of these costs; on whether they are visited only on their carriers ("individual" costs) or on all members of the broods they belong to ("shared" costs); and on whether conflict affects the fitness of one or both parents (Macnair and Parker 1979; Parker 1985; Lazarus and Inglis 1986; Harper 1986).

For all mating systems, conflict will be higher where costs of solicitation are low or where they are shared than where they are paid by conflictors alone, since nonconflictors will suffer some of the costs of solicitation in mixed broods (Macnair and Parker 1979). Where only one parent invests, conflict levels should be higher if competition among siblings for investment occurs within rather than between broods. In contrast, if both parents invest, inter-brood competition may be associated with higher levels of conflict. Increased levels of conflict would also be expected where a single parent invests if mating is promiscuous or broods are of mixed paternity. Where both parents invest and are equally susceptible to solicitation, conflict will generally be higher than under uniparental care (Macnair and Parker 1979; Metcalf, Stamps, and Krishnan 1979; Parker 1985).

Like previous models, these arguments suggest that parents respond to solicitation by their young, otherwise conflictor genes would be unable to spread (see above). They consequently raise the question as to why parents should respond at all. Harper (1986) argues that this is because the relative frequency of begging by different chicks within a brood allows parents to allocate investment to chicks that need it most, thereby maximizing the success of the brood. Once parents respond to begging behavior, he suggests, chicks will be selected to increase their rate of begging until the costs of further increases outweigh their benefits. The resulting increase in begging rate is likely to reduce the average fitness of parents and young, but will not necessarily affect the way in which parents allocate investment within broods, since individual variation in nutritional requirements and in begging rate is likely to persist.

EFFECTS OF BROOD SIZE ON PARENT-OFFSPRING CONFLICT

Brood size is also likely to affect the level of conflict between parents and their offspring. Whether investment is nondepreciable or depreciable, the level of conflict should decline with increasing brood size (Lazarus and Inglis 1986). This is because parental gains increase in relation to the size of the brood, justifying increased investment, while, at larger brood sizes, gains to offspring through collateral relatives (to which parents and young are equally related) represent a larger proportion of their inclusive fitness—with the result that differences between optimal levels of investment for parents and offspring decline. A similar decline is expected where increased investment only benefits a single offspring (for example, because a predator seldom takes more than one from a brood) but for different reasons (Lazarus and Inglis 1986). Here, parental optima are independent of brood size, while the chances that each offspring will be taken decline as brood size increases. As a result, the optimum level of investment for offspring declines toward the parental value as brood size increases and the level of conflict is likely to fall.

Differences in the way in which the costs of conflict are distributed among brood members may influence the extent of overt conflict between parents and offspring. For example, where the chicks of altricial birds can increase their allocation of food by begging, begging frequency might be expected to decline with brood size where the chance that an entire brood will be predated rises in proportion to total begging frequency (Harper 1986). Conversely, where the costs of begging are confined to the beggar, individual begging frequency might be expected to level off at some intermediate value of brood size (Figure 11.3). In both cases, the zero frequency of begging in brood sizes of one is

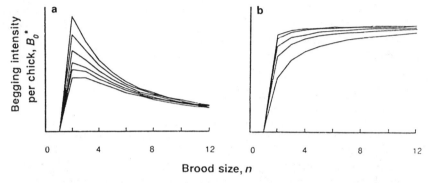

Figure 11.3   The evolutionary stable typical begging intensity (B) for different brood sizes in circumstances where (a) the costs of begging affect all brood members equally ("predation costs only") and (b) where they affect only the individual who begs ("energetic costs only") (from Harper 1986). Different lines represent variation in the effects of begging rate on chick survival; vertical scales are not necessarily comparable between the two graphs.

unrealistic and results from Harper's assumption that begging affects only the distribution of food within the brood and has no influence on the total level of parental investment.

EFFECTS OF PARENT-OFFSPRING CONFLICT ON BROOD SIZE

Parents are also likely to be in conflict with their offspring over the size of broods (Eickwort 1973; O'Connor 1978; Parker and Mock 1987). If resource availability declines during the breeding season, it may be to the parent's advantage to reduce brood size by sacrificing one or more of its offspring if this increases the number of young that are likely to survive or reduces the costs of breeding (see Section 9.6). Adults will favor a reduction in brood size whenever this increases the number of young that they rear, while a potential victim should accept its own demise only where its indirect fitness through sibs, if it dies, exceeds the sum of its own fitness plus its indirect fitness if it survives (O'Connor 1978). Since increasing brood size is likely to be associated with an increase in the relative importance of gains in indirect fitness, parents and potential victims are more likely to be in agreement where brood size is large (see above).

Adults are also likely to disagree over optimal brood size with their older or better-quality offspring, which are most likely to survive if broods are reduced (O'Connor 1978; Parker and Mock 1987). "Survivors" are more likely to favor brood reduction than their parents (see Chapter 9), who may be selected to intervene to prevent siblicide by stronger offspring. In fact, conflicts between parents and offspring over clutch or brood size are likely to be a general phenomenon. The magnitude of the conflict (measured by the difference between the clutch size that is optimal for the parent and the size that is optimal for the offspring) will increase in relation to the degree of multipaternity of the brood and the extent to which additional offspring affect the fitness of their sibs as a result of intrabrood competition (Parker and Mock 1987).

WHO WINS? PREDICTING THE OUTCOME OF PARENT-OFFSPRING CONFLICT

Whenever selection favors conflictor genes that cause offspring to extort more investment than the parent is selected to give, counteracting selection pressures will favor counterstrategies in the parent. These may lead to a variety of parental responses, ranging from capitulation to chick demands (e.g., Zahavi 1977) to punishment of "conflicting" chicks by their parents (Feldman and Eshel 1982).

Subsequent models by Parker and Macnair (1979) and Parker (1985) explore the likely outcome of interactions between two loci: a "conflictor" locus, which determines whether or not offspring solicit their parents for extra investment, and a "suppressor" locus, which determines how parents retaliate to solicitation (see also Feldman and Eshel 1982). Solicitation is assumed to have a variable cost, which may affect either the solicitor alone or all members of its brood. Parker and Macnair consider two forms of parental response: where parents invest in proportion to offspring demands, and where they ignore solicitation altogether and invest a fixed amount.

When parents invest in proportion to offspring demands, pure ESS's are likely to evolve where parents invest at some level intermediate between their own optimum and that of their offspring (a "pro-rata" ESS: see Parker and Macnair 1979). However, an ESS where the offspring wins the conflict (i.e., a level of investment close to the offspring's optimum) can also occur where conflict affects future sibs, while a "parent wins" ESS is possible where there are high costs affecting contemporary sibs (Parker and Macnair 1979; Parker 1985). When parents retaliate by ignoring offspring solicitation, they can win conflicts and maintain levels of investment close to their own optimum where the costs of ignoring solicitation are small. However, where costs are substantial (as well as in some "fixed payment" models), there is no stable outcome.

Parker and Macnair (1979) conclude that the most likely outcome of evolutionary conflicts between parents and offspring are "pro-rata" ESS's, where the level of investment is intermediate between parental optima and offspring optima. In comparable circumstances, it is always more difficult to maintain a "parent wins" or an "offspring wins" solution when there is investment by both parents, or a "parent wins" solution where there are shared costs of solicitation and conflict affects the fitness of both parents (Parker 1985).

One additional way in which parents may retaliate to progeny that attempt to extort additional investment is to punish them by reducing their fitness (Trivers 1974; Alexander 1974; Parker and Macnair 1979). Behavior of this kind is likely to have immediate costs to parental fitness and should only develop where the costs of successful extortion to the parent are high or the costs of punishment are low—for example, where offspring survival is negatively related to brood size. Nevertheless, behavior of this kind does appear to occur (see Section 9.5).

Where parents and their offspring disagree over clutch size and offspring are able to kill each other, offspring will generally "win" conflicts with parents if they can kill sibs that are in excess of their own optimum brood size, and the costs of siblicide are confined to the indirect component of offspring fitness (Parker and Mock 1987). Under these circumstances, the mother's best strategy may be to lay the offsprings' optimal clutch size even though her fitness is correspondingly reduced (Parker and Mock 1987; Godfray 1986).

11.3 Parent-Offspring Conflict in Practice: Predictions

Reliable estimates of parent-offspring conflict should be based on measures of the extent to which optimal levels of investment differ between parents and their offspring (see Stamps, Clark, et al. 1985). Unfortunately, it is seldom possible to identify these optima or to measure the extent to which observed levels of investment diverge from them. We know virtually nothing about the costs to offspring of attempting to extort additional investment or the costs of counterstrategies to their parents (see Stamps, Clark, et al. 1985), and many of the assumptions of current models of parent-offspring conflict are clearly unrealistic. In particular, it seems unlikely that the costs of solicitation to offspring will ever be entirely confined to the solicitor or evenly shared among all brood members (see Section 11.2) and very likely that they will vary systematically with the offspring's birth order, relative size, and age. In few, if any, species do both parents invest equally so that the costs of additional parental expenditure may commonly differ between care-givers (see Section 8.1).

Though measures of overt behavior (such as begging rates and weaning tantrums) are often assumed to be related to the intensity of conflict between parents and offspring, there is no guarantee that this is the case. Changes in begging rate, for example, may occur for a wide variety of reasons unrelated to the degree of conflict (see Henderson 1925; Harper 1986). Measuring changes in the ratio of solicitation to parental expenditure, as Lazarus and Inglis (1986) suggest, fails to solve the problem since these ratios need not reflect changes in conflict. As a result, models of parent-offspring conflict are generally difficult to test. For this reason, it may be most useful to concentrate on the simplest and most robust predictions arising from these models. The next three subsections briefly review evidence supporting three qualitative predictions.

1. *That overt conflicts should be evident between parents and their offspring* (Trivers 1974; Parker and Macnair 1978; Parker 1985). Both in birds and mammals, offspring commonly attempt to persuade their parents to give them additional food (Trivers 1974). In birds, fledglings beg for food (e.g., Burger 1981) and begging can affect both the allocation of resources to particular chicks (Lockie 1955; Ryden and Bengtsson 1980; Stamps, Clark, et al. 1985) and total expenditure on the brood (Hudson 1979; Bengtsson and Ryden 1981 1983). There are also indications that begging by nestlings can attract predators (e.g., Perrins 1965), and Zahavi (1977) has even suggested that the begging of young passerine birds may represent an attempt by offspring to blackmail their parents into feeding them because begging calls raise the chance that the nest will attract the attention of a predator. The young of some rodents actively resist being removed from the breeding burrow by their mother, beating her off with their forepaws and refusing to be picked up (Daly and Daly 1975). Similarly, young chimpanzees whose attempts to suck have been persistently frustrated will throw weaning tantrums in which they kick, scream, and even attempt to attack their mothers (Goodall 1986). Tantrums appear to make mothers nervous and tense and they commonly give in to their offspring's demands.

Parents may respond to solicitation by giving in, by ignoring offspring demands and maintaining a feeding schedule that maximizes their own fitness (e.g., Hudson 1979), or by attempting to deter their offspring from continuing to solicit additional investment. Responses commonly vary between parents. For example, while male budgerigars that assist their mates respond to variation in begging rate and usually feed whichever chicks approach and beg first, females ignore begging frequency by older nestlings and feed smaller offspring preferentially (see Section 9.5). As a result, offspring in nests where males assist females in feeding the young are fed around three times as frequently as in nests where only the female feeds the young. However, even here, the frequency of feeding may not reach the offspring's optimum, for the frequency of begging exceeds feeding frequency, suggesting that parents and offspring reach a "pro-rata" compromise. Attempts by one or both parents to prevent the largest and most vociferous offspring from monopolizing food brought to the nest have been observed in a variety of other species (Ryden and Bengtsson 1980; Bengtsson and Ryden 1981, 1983; see Sections 9.5 and 9.6).

In some cases, parents may actively discourage attempts to solicit additional food. For example, coots will "tousle" chicks that persistently attempt to follow them after they have been fed in order to obtain a disproportionate share of resources. These attacks can be savage and, in some cases, have been seen to cause chick mortality (Horsfall 1984; Leonard, pers. comm.).

Where parental care is terminated by the onset of the parents' next breeding attempt, parents and offspring may disagree over the timing of the next attempt. Early resumption of breeding by the parent, combined with an early termination of investment in the previous offspring, can reduce the survival and eventual breeding success of previous offspring substantially (T. H. Brown 1959, 1964; Hobcraft, McDonald, and Rutstein 1983; Blurton-Jones 1986; Clutton-Brock, Albon, and Guinness 1988). Offspring are consequently likely to favor a later date for the resumption of breeding by their mother than the latter's optimum. In a variety of primates, frequent suckling delays the resumption of ovulatory cycles and prolongs interbirth intervals (vervet monkeys: Lee 1987; baboons and macaques: Nicolson 1982; Pope, Gordon, and Wilson 1986; gorillas: Stewart 1988; humans: Short 1976a,b, 1983; McNeilly 1988). Some evidence suggests that offspring may attempt to manipulate this effect to prolong interbirth intervals (Worlein, Eaton, et al. 1988). For example, juvenile rhesus macaques increase the frequency of their attempts to suck at the onset of the next mating season when their mothers are likely to conceive again (Gomendio 1989). If females do come into estrus, juveniles may also attempt to prevent males from copulating with their mothers (Clutton-Brock and Harvey 1976). The demands of human infants to suck throughout the night may have much the same function (Blurton-Jones and da Costa 1987).

Conflicts between parents and offspring are not confined to the period of parental care. Especially in species that live in stable groups, parents and offspring may disagree over whether or not offspring should remain in the group (Dobson 1982; Cockburn, Scott, and Scotts 1985; Lyon, Goldman, and Hoage 1985; Liberg and von Schanz 1985) and, if they do so, over their access to resources and over whether they should breed themselves or assist their parents (Emlen 1984, pers. comm.; Brown 1987; Smith and Ivins 1987; Cockburn 1988). All of these circumstances are commonly associated with overt, behavioral conflict that varies in intensity with ecological or social circumstances (Hauser and Fairbanks 1988; Worlein, Eaton, et al. 1988).

2. *That solicitation by offspring should cause parents to exceed their optimal level of investment and reduce parental fitness* (Trivers 1974; Parker 1985; Lazarus and Inglis 1986; Stamps, Clark, et al. 1985). Studies of the begging rate of chicks in altricial birds provide some evidence that offspring may persuade parents to increase their level of investment. In budgerigars, for example, it seems likely that the threefold increase in feeding rate extracted from parents in male-assisted broods (see above) would have costs to parental fitness in natural populations (Stamps, Clark, et al. 1985). Female counter-

strategies, too, may have costs: selective feeding of smaller chicks by female budgerigars is associated with a reduction in the rate of food delivery compared to males, who feed chicks on demand (Stamps 1985).

Both in birds and in insects, there is suggestive evidence that competition between offspring may reduce the fitness of adults. In many viviparous species, developing young feed on subsequent eggs or embryos, while in some oviparous species newly hatched larvae regularly eat unhatched eggs or larvae smaller than themselves (Ho and Dawson 1966; Mertz and Robertson 1970; Eickwort 1973). In parasitoid Hymenoptera, females usually either lay a single egg or a considerable number of eggs per host (Le Mesurier 1987). The larvae of species of the first kind often have large mandibles, which they use to kill any other larvae in the host, while in the second kind larvae seldom compete directly (Ives 1987). Godfray (1986, 1987) shows that, where clutch size is small, individual larvae can increase their fitness by killing competing sibs and suggests that selection operating on larvae may constrain clutch size to a single egg. His findings closely resemble the outcome of parent-offspring conflict over brood size predicted by Parker and Mock (1987).

The clearest evidence that offspring can manipulate investment to the parent's detriment comes from studies of sex ratio variation in social Hymenoptera, for here it is possible to identify the optimal patterns of investment for parents versus workers (see Trivers and Hare 1976; Oster, Eshel, and Cohen 1977; Benford 1978; Craig 1980; Noonan 1981; Uyenoyama and Bengtsson 1981; Charnov 1982; Page and Metcalf 1982; Pamilo 1982). In haplo-diploid species, queens are related to their offspring by 0.5, whereas workers are related to their sisters by 0.75 and to their brothers by 0.25 (Hamilton 1964). Since queens are equally closely related to their sons and daughters, they should aim to divide their total investment equally between their reproductive offspring (see Chapter 13). In contrast, workers (who, on average, are related to their reproductive sisters by 0.75 but to brothers by 0.25) would maximize their inclusive fitness by investing more heavily in reproductive sisters (i.e., young queens) than in drones. Since the average fitness of males is negatively related to their relative frequency in the population, the ESS sex ratio for workers is to allocate three times as much investment to young queens as to drones. Trivers and Hare (1976) tested this prediction by comparing investment ratios (the sex ratio of young for different degrees of sexual dimorphism in size) for twenty-one ants. Though there was considerable scatter, their results suggested that most species showed investment ratios of around 3:1 (Figure 11.4) whereas in two species of slave-making ants, where workers are unrelated to the young that they rear, 1:1 investment ratios were found (see also Bourke 1989). Trivers and Hare concluded that workers won the conflict by manipulating the sex ratio of the offspring that they were rearing. A number of other cases where workers appear to win conflicts over the sex ratio have since been discovered (Charnov 1982; Ward 1983).

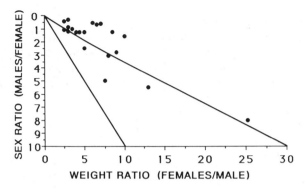

Figure 11.4 The sex ratio (male/female) of reproductives (alates) is plotted as a function of adult dry weight ratio (female/male) for monogynous ant species (from Trivers and Hare 1976). Lines showing 1:1 and 1:3 ratios of investment are drawn for comparison.

In other Hymenoptera, queens appear to win conflicts, achieving a 1:1 investment ratio. One way in which the queen can avoid subversion of her optimum sex ratio by workers is to limit the supply of diploid eggs or to produce only haploid eggs early in the season (Bulmer and Taylor 1981). The temperate social wasp, *Polistes metricus*, appears to show a modified version of the latter strategy. Nests are founded in spring by one or more mated queens who produce males relatively early in the summer, followed by queens at a later stage, whereas workers are produced throughout the season (Metcalf 1980). Metcalf suggests that situations where the queen wins may be commoner in polistine wasps than in ants because queens are involved in caring for eggs and young and are consequently in a better position to control the investment ratio. Alternatively, in tropical species with long breeding seasons and extensive overlap between generations, temporal separation of male and female production may be difficult to achieve and workers may be more likely to win conflicts of interest for this reason (see West-Eberhardt 1969, 1975).

A variety of other factors can also affect the optimal sex ratios for queens and workers. Where queens mate more than once, workers will be related to young queens by less than 0.75; this will favor a more equal investment ratio than 3:1 (Alexander and Sherman 1977). More equal investment ratios will also be favored by workers where they lay a proportion of male eggs themselves and are consequently more closely related to males than 0.25 (e.g., Ward 1983). And in some species, "orphaned" colonies (those where the founding queen has died) continue to rear males produced by workers, while colonies with resident queens produce more females (e.g., Owen, Rodd, and Plowright 1980; Forsyth 1981). Predictions about investment ratios also assume that the costs of rearing sons and daughters are directly proportional to their relative weights and will be affected by any sex difference in survival (see Metcalf 1980) or in the efficiency of growth during the rearing period (see

Section 12.3). The queen's optimum sex ratio will also be influenced by the extent of inbreeding by her sons (local mate competition: Hamilton 1967). Where a queen's sons compete to mate with their sisters, a strongly female-biased sex ratio will be favored, both because the average fitness of sons declines as more are produced and because the production of daughters increases the mating success of their brothers (see Grafen 1984).

Multiple mating, reproduction by workers, and local mate competition are probably present to some degree in many social insects (see Metcalf 1980; Page and Metcalf 1982) and may vary considerably between colonies of the same species (e.g., Owen, Rodd, and Plowright 1980; Ward 1983), so that investment ratios as extreme as 1:1 or 3:1 would not always be expected. This agrees with the increasing evidence that many investment ratios lie between these extremes (see Charnov 1982; Ward 1983).

Conflicts of interest over the sex ratio probably occur in many animals other than haplo-diploid Hymenoptera. In diploid organisms, selection on gametes may oppose parental attempts to manipulate the sex ratio before hatching or birth and may constrain the evolution of adaptive variation in sex ratios (Williams 1979; Reiss 1987a). However, it is rarely possible to identify contrasting optima, and the interpretation of other sex ratio trends presents a number of problems (see Chapter 13).

3. *That levels of solicitation by young should rise to a point at which they have appreciable costs to the young* (Harper 1986; Lazarus and Inglis 1986). So far, there have been virtually no attempts to investigate the costs to offspring of competing directly with siblings or of soliciting additional investment from parents. One indication that solicitation may have appreciable costs is provided by evidence that increased feeding rates resulting from increased begging rates in altricial birds are not associated with any significant increase in growth rate, fledging age, or fledging size (Ryden and Bengtsson 1980; Stamps, Clark, et al. 1985). However, an alternative explanation of these results is that uncontrolled factors obscure effects of increased feeding rate on growth, and it is not yet clear that higher begging rates are necessarily associated with substantial energetic costs to offspring.

11.4 Summary

In all organisms where parents and their offspring are not genetically identical, conflicts of interest are likely to arise between them over the level of parental investment. Offspring will commonly attempt to extract higher levels of investment than it is in the parents' interests to provide, lowering the parents' fitness where they are successful. Parents may respond by setting fixed limits to the level of investment or by raising the costs of attempts to extract additional investment.

The intensity of conflict between parents and offspring is likely to depend on the breeding system and may be affected by the size of the brood, the relatedness of sibs, the number of caregivers, and the costs of conflict to the offspring. Evolutionary conflicts may lead to "victories" for parents or their offspring or to "pro-rata" ESS's where parents invest at some level intermediate between their own optimum and that of their offspring. Models suggest that compromises of the latter kind may be the commonest outcome of conflict.

Empirical evidence confirms that behavioral conflict between parents and offspring is not uncommon and several lines of evidence suggest that the behavior of offspring can cause parents to increase their level of investment. However, it is seldom possible to test more specific predictions because optima for parents and offspring can rarely be identified. In the future, it may be worth concentrating on the most robust predictions of theoretical models, including predictions that conflict should be most intense where

- the costs to the offspring of attempts to extort additional investment to the offspring are low;
- the costs fall on all members of the brood rather than on the extortioner alone;
- conflict occurs within broods and only one parent invests;
- the coefficient of relatedness between siblings is relatively low and fathers do not invest;
- brood size is small and the benefits of care are nondepreciable.

12

Parental Investment
in Sons
and Daughters

12.1   Introduction

Most of the previous chapters have ignored the fact that, in sexually reproducing animals, offspring can be divided into two fundamental categories: males
and females. Willson and Pianka (1963) first suggested that parents might be
expected to invest more heavily in offspring of one sex, and a number of
subsequent papers have qualified or added to their predictions (Trivers and
Willard 1973; Reiter, Stinson, and Le Boeuf 1978; Dittus 1979; Maynard
Smith 1980; Clutton-Brock, Albon, and Guinness 1981; Stamps, in press). The
basis for expecting sex biases in parental expenditure is straightforward: where
resources allocated to offspring of one sex provide a greater return in terms of
parental fitness, parents might be expected to invest more heavily in that sex.

This chapter reviews the evidence for differential investment in sons and
daughters. Section 12.2 examines the evidence that parental expenditure affects the fitness of sons and daughters to different extents, while Sections
12.3–12.5 review four kinds of evidence that parental investment differs between sons and daughters. Section 12.6 considers an alternative explanation of
differences in the costs of raising sons and daughters—that they are a consequence of sex differences in juvenile behavior arising from sexual selection.
Section 12.7 examines the consequences of differential costs for the population sex ratio, and Section 12.8 summarizes the main conclusions of the
chapter.

12.2   Does Parental Expenditure Have a Greater Effect
on Fitness in Males or Females?

The relative effects of parental expenditure on the fitness of males and females
may often differ between ectotherms and endotherms. In oviparous ectotherms, where the reproductive success of females increases with their egg
production, it is commonly suggested that adult size and early growth exert a
stronger influence on the breeding success of females than on males (Gross
and Sargent 1985). For example, the close association between female body
size and fecundity in many fish and amphibia (see Figure 2.2) may explain

why females are commonly the larger sex. However, there is little firm evidence that body size affects the fitness of females more than males (see Gross and Sargent 1985; B. H. King 1987, 1988). In some invertebrates at least, juvenile growth and adult size have a stronger effect on the reproductive success of males than females (e.g., Crespi 1988). It would not be surprising if the direction and magnitude of these differences prove to vary with the mating system, and further studies are badly needed.

In endotherms where male success depends on fighting ability and body size, the breeding success of sons may often be more strongly influenced by early growth and parental expenditure than that of daughters for at least three reasons. First, variance in male breeding success is likely to be relatively greater than variance in female breeding success (Clutton-Brock 1988). Second, adult size and, hence, early growth may have a stronger influence on breeding success in males than in females (Trivers and Willard 1973). And third, parental expenditure and resource availability commonly have a stronger influence on growth and survival in males than in females (see Chapter 14). However, there is little direct evidence that the effects of parental expenditure differ between the sexes. One of the few investigations of the effects of parental phenotype on the relative fitness of sons and daughters is the long-term study of red deer on the Isle of Rhum (Clutton-Brock, Guinness, and Albon 1982). Here, dominant mothers (those above median rank in the female population) show superior body condition, breed more successfully than subordinates, live longer and are probably able to expend more resources on each reproductive attempt. Both within and across cohorts of mothers, the breeding success of sons increases more sharply with their mother's rank than that of daughters (see Figure 12.1).

As among ectotherms, the effects of parental investment on the fitness of sons probably vary widely. In some mammals, adult size is not closely related to weight at weaning (e.g., McCann, Fedak, and Harwood 1989) or mating success depends on factors other than body size (Fedigan 1983; Schwagmeyer 1988, 1989). The effects of parental investment on the fitness of daughters probably differ, too. In some species, food shortage before birth or during lactation can permanently impair the growth and breeding success of daughters while, in others, females that initially show low growth rates can catch up at a later stage (Huck, Labov, and Lisk 1986, 1987; Albon, Clutton-Brock, and Guinness 1987).

In at least some endotherms, parental investment and parental quality may have a greater effect on the fitness of daughters than on that of sons. In some social primates where females normally remain in their natal group throughout their lives, a female's rank and breeding success depend on the support of her mother and other matrilineal relatives (Fedigan 1983; Harcourt 1989a,b). Competition between female kin groups can be intense. In some species of baboons and macaques, unrelated females selectively attack one anothers' fe-

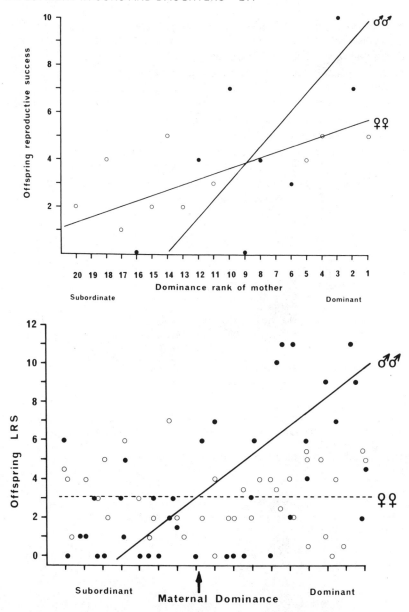

Figure 12.1   Lifetime reproductive success (LRS) of sons and daughters of red deer hinds in relation to their mother's social rank (from Clutton-Brock, Albon, and Guinness 1984). (a) Data for progeny born to the 1972 cohort of hinds; (b) combined data from six different cohorts. LRS estimates for males based on measures of harem size, weighted by the probability of conception and checked by DNA fingerprinting.

male offspring, possibly because they represent potential competitors (toque macaques: Dittus 1977, 1979; Barbary macaques: Paul and Thommen 1984; rhesus macaques: Gomendio 1989; baboons: Pereira, in press). As a result, the daughters of subordinate mothers show higher mortality than other juveniles (Figure 12.2; see also Dittus 1979; Silk, Clark-Wheatley, et al. 1981; van Schaick and van Noordwijk 1983; Altmann Hausfater, and Altmann 1988). There is even a suggestion that females carrying female fetuses are more frequently attacked by conspecifics than those pregnant with sons and may be more likely to abort as a result. In captive groups of pigtail macaques (*M. nemestrina*), mothers pregnant with female fetuses were almost three times as likely to require medical treatment for bite wounds than mothers pregnant with male fetuses (Sackett, Holm, et al. 1975; see Figure 12.3).

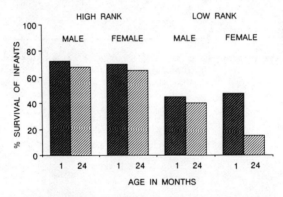

Figure 12.2 The proportion of male and female bonnet macaque (*Macaca radiata*) infants born to high- and low-ranking females that survived to one month and 24 months (from Silk 1983). Daughters of low-ranking females are significantly less likely to survive than their sons or than the daughters of high-ranking females.

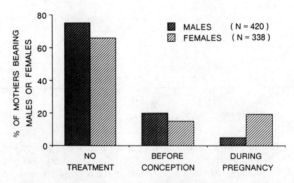

Figure 12.3 Percentage of male- and female-bearing pigtail macaques (*Macaca nemestrina*) in a sample of 758 mothers that received no medical treatment; were treated before conception; or were treated during pregnancy (from Sackett, Holm, et al. 1975). 80% of medical treatments were for bite wounds.

In some of these species, a mother's rank may have a relatively weak effect on the rank and reproductive success of her sons, who disperse to breed in other groups (Altmann 1980; Silk 1983). In particular, among baboons and macaques, a male's rank and reproductive success depends to a large extent on alliances with other males who will support him in competition for mates (Packer 1977). As a result, it may be that, in contrast to red deer, maternal rank has a stronger effect on the fitness of daughters than on that of sons. However, direct evidence that this is the case is lacking, and the only study that has tried to estimate the comparative effects of maternal rank on the relative fitness of sons and daughters in free-ranging primates concluded that it affected the fitness of sons more than that of daughters (Meikle, Tilford, and Vessey 1984).

These results underline the importance of investigating the comparative effects of parental investment on the lifetime fitness of males and females. Without empirical studies of these relationships, explanations of parental behavior are bound to be insecure.

12.3   Parental Expenditure and Sex Differences in Early Growth

Claims that parents invest more heavily in offspring of one sex depend on three kinds of evidence: sex differences in juvenile growth and body size; differences in the energetic costs of rearing males and females; and differences in the reproductive costs of producing or rearing sons and daughters. The next three sections review each of these in turn.

Among invertebrates, as well as in birds and mammals, growth rates are commonly faster or the period of development is longer in juveniles of the larger sex (Charnov 1982). In birds these differences can appear as early as two days after hatching, and by fledging weight dimorphism is usually nearly as great as in adulthood (Howe 1976, 1979; Willson 1966; Patterson, Erckmann, and Orians 1980; Fiala 1981b; Richter 1983; Røskaft and Slagsvold 1985). In many dimorphic mammals, sex differences in weight are already evident by birth (Figure 12.4), though in some cases they do not persist through lactation (McCann, Fedak, and Harwood 1989). Sexual dimorphism subsequently develops more slowly than in birds, seldom peaking before both sexes reach breeding age (e.g., Sackett, Holm, et al. 1975; Laws, Parker, and Johnstone 1975; B. Mitchell, Staines, and Welch 1977; Gosling, Baker, and Wright 1984; Kovacs and Lavigne 1986; Trillmich 1986; S. S. Anderson and Fedak 1987; Stamps 1990).

Unfortunately, sex differences in size and growth provide unreliable estimates of differences in parental expenditure, since growth priorities and body composition often differ between the sexes (Stamps, in press). In some cases, sex differences in juvenile weight overestimate differences in expenditure. For example, in social Hymenoptera, males and reproductive females can differ in

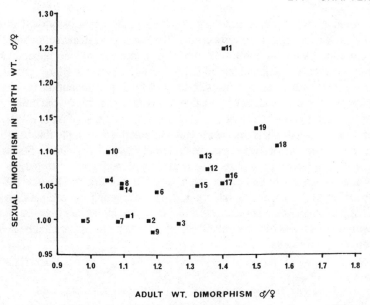

Figure 12.4   Sexual dimorphism in birth (or pouch) weight (male/female weight) plotted against sexual dimorphism in adult weight for different mammals (from Clutton-Brock, unpubl.). (1) and (2) *Onchomys*; (3) *Proechimys*; (4) coypu, *Myocastor coypus*; (5) ringtail possum, *Pseudocheirus peregrinus*; (6) pig; (7) horse; (8) roe deer, *Capreolus capreolus*; (9) moose, *Alces alces*; (10) Chinese water deer, *Hydropotes inermis*; (11) mouflon, *Ovis ammon*; (12) Soay sheep; (13) pigtail macaque, *Macaca nemestrina*; (14) Rhesus macaque, *Macaca mulatta*; (15) fallow deer, *Dama dama*; (16) red deer, *Cervus elaphus*; (17) wapiti, *Cervus canadensis*; (18) reindeer/caribou, *Rangifer* sp.; (19) *Odocoileus*.

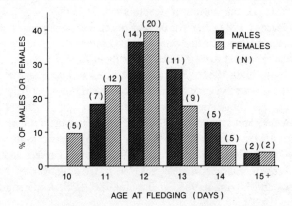

Figure 12.5   Frequency distribution of fledging ages for male and female yellow-headed blackbirds (from Richter 1983).

water and caloric content, and female-to-male weight ratios may often over-estimate the difference in energy costs of the two sexes (Boomsma and Isaaks 1985; Boomsma 1989). In birds, the smaller sex may channel a higher proportion of its resources into feather growth. For example, in dimorphic icterids, male juveniles increase in body size and weight more quickly than females, but the latter show accelerated feather development and fledge at a younger age (Holcomb and Twiest 1970; Richter 1983; Stamps, in press: see Figure 12.5). An opposite tendency is found in the European sparrow-hawk, where females are nearly twice as heavy as males by fledging, and males are fatter, more active, and fledge earlier. The rate of food intake by male and female nestlings is apparently identical (Newton 1978, pers. comm.; see also Bortolotti 1986).

In other cases, sex differences in juvenile body size may underestimate differences in parental expenditure. In some mammals, juvenile males tend to be more active than females, have higher rates of heat loss, and convert crude protein into growth less efficiently and may consequently have greater nutritional requirements per unit body weight (see Slee 1970, 1972; Trivers 1972; Glucksman 1974; Pratt and Anderson 1979; Clutton-Brock, Guinness, and Albon 1982). For example, in horses, which show little size dimorphism, male foals are more active than females and spend more time sucking (Duncan, Harvey, and Wells 1984).

12.4   Energy Expenditure on Sons and Daughters

In several dimorphic vertebrates where males are larger than females, the energy intake of juvenile males has been shown to be higher than that of females. In the great-tailed grackle, *Quiscalus mexicanus*, measurements of oxygen and food consumption under controlled conditions show that males, which are around 50% heavier than females by the twelfth day after hatching, require around 20% more food (Teather and Weatherhead 1988). In red-winged blackbirds, where males are around 30% heavier at fledging than females, they require approximately 27% more energy (Fiala 1981b; Fiala and Congdon 1983: see Figure 12.6), and parents increase feeding rates to broods with a high proportion of males (Yasukawa, McClure, et al., in press). Greater food intake by male nestlings has also been found in rooks (Slagsvold, Røskaft, and Engen 1986) and giant petrels (C. M. Brooke, pers. comm.). However, as yet, there is no firm evidence of greater nutritional requirements among female nestlings in species where females are the larger sex (see Newton 1979; Collopy 1986).

In several dimorphic mammals, too, the energetic costs of producing or rearing males exceed those of rearing females. In fur seals, *Arctocephalus*, measurement of milk intake shows that males (which are substantially larger than females) take more than 30% more milk than females, though they ingest

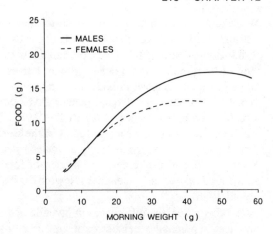

Figure 12.6 Daily food consumption (grams wet weight) as a function of body weight in hand-raised nestling red-winged black-birds, *Agelaius phoeniceus* (from Fiala, 1981b).

less relative to their body mass (Trillmich 1986; Costa and Gentry 1986; Costa, pers. comm.). In the less dimorphic gray seal, *Halichoerus grypus*, the energy costs of rearing sons are around 10% higher than those of daughters (S. S. Anderson and Fedak 1987). However, in the highly dimorphic Southern elephant seal, *Mirounga leonina*, there appears to be no difference in maternal expenditure on sons and daughters during lactation, although males are born heavier than females (McCann, Fedak, and Harwood 1989).

Less reliable evidence of the increased energetic costs of rearing sons is available from a variety of other mammals. In coypu (*Myocastor coypus*) male pups spend more time sucking from the highest-yielding teats (Gosling, Baker, and Wright 1984). Sex differences in suckling frequency and/or in the duration of suckling bouts have also been found in red deer, goats, American bison, and African elephants (Clutton-Brock, Albon, and Guinness 1981; Pickering 1983; Lee and Moss 1986; Wolf 1988; see Figure 12.7), though they are apparently absent in a number of ungulates that show unusually fast juvenile growth rates, including pronghorn, *Antilocapra americana* (Byers and Moodie 1990). Female juveniles are also weaned earlier than males in elephant seals and zebu cattle (Reiter, Stinson, and Le Boeuf 1978; Reinhardt and Reinhardt 1981).

However, measures of sucking duration or frequency are not necessarily closely related to milk delivery since weaker or less well-developed young may take longer to drain the mother's milk supply or may be prepared to suck longer for lower levels of delivery (Mendl and Paul 1989). The offspring of malnourished rats, for example, nurse more than those of well-fed mothers (Hall, Leakey, and Robertson 1979), while differences in time spent nursing by juvenile mice and domestic cats are negatively related to growth rates (Mendl and Paul 1989).

So far, parental expenditure on sons and daughters has not yet been compared in sexually monomorphic mammals or in species where females are larger than males (Ralls 1976).

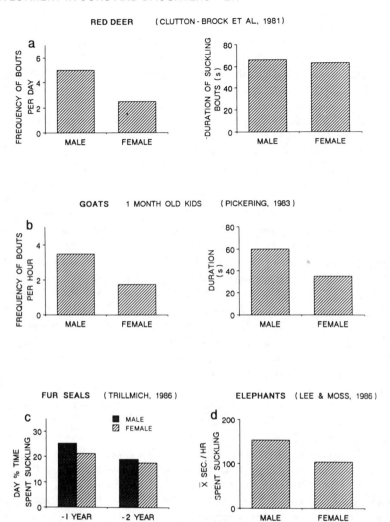

Figure 12.7   Sex differences in suckling behavior in (a) 1–2-month-old red-deer calves; (b) 1-month-old goat kids; (c) juvenile and yearling fur seals; and (d) elephant calves of up to 24 months.

12.5 Reproductive Costs of Sons and Daughters

Few studies have compared the reproductive costs of rearing sons and daughters. In red deer, mothers that have reared a male calf are more likely to die the following winter and, if they survive, are almost twice as likely to fail to produce a calf compared to subordinate mothers that have reared female calves (see Figure 12.8). The increased costs of sons do not affect all females equally and fall primarily on subordinate mothers (see Section 13.1).

A similar reduction in subsequent reproductive performance occurs in female bison that have reared sons versus daughters (Wolf 1988), and consistent tendencies for mothers that have produced male calves to show longer intervals before the next conception have been found in African elephants and Indian cattle (Singh, Singh, and Srivastava 1965; Dhillon, Acharya, et al. 1970; Lee and Moss 1986). In reindeer, mothers that rear males lose more weight during lactation (Kojola and Eloranta 1989). And in Mongolian gerbils (*Meriones unguiculatus*) females that have raised all-male litters show longer periods of vaginal closure (preventing copulation) after birth and subsequently produce smaller litters than females that have raised all-female litters (Clark, Bone, and Galef 1990). In contrast, the longer lactation periods of male pups in northern elephant seals (see above) are not associated with increased costs to the mother's survival or subsequent breeding success (Le Boeuf, Condit, and Reiter 1989), possibly because mothers feed their pups from stored reserves, which they can replenish as soon as lactation is complete.

While differences in the energetic costs of rearing sons and daughters may often be associated with differences in the fitness costs of rearing the two sexes, differences in fitness costs are not necessarily associated with increased energy expenditure on juveniles of one sex. In a number of multiparous species, hormones emanating from males during gestation may depress the growth of their litter mates so that the sex of litter mates affects an individual's birth weight: in sheep, for example, lambs born co-twin with females have higher birth weights, grow faster, and are more likely to survive than those born co-twin with males (Burfening 1972; Clutton-Brock, unpublished data). Other mechanisms may be responsible for differences in the fitness costs of raising sons and daughters in cercopithecine primates. In some populations of macaques, subordinate mothers that have reared female infants do not conceive again as quickly as those that have reared males and are more likely to lose their next offspring (Simpson, Simpson, et al. 1981; Simpson and Simpson 1985; Gomendio 1989; but see also, Crockett and Rudran 1984; Small and Smith 1984; Berman 1988; Silk 1988). Differences in suckling patterns are apparently slight (Gomendio 1988, 1989) and probably favor males. Where they occur, the higher costs of females appear to be associated with high rates of aggression received by the daughters of subordinates from other

Figure 12.8 (a) Proportion of female red deer that had previously raised male and female calves that became pregnant the same year, based on a sample of 447 calves reared successfully; (b) conception dates of red deer hinds (estimated by back-dating from known birth dates) following years in which they had failed to rear a calf, reared a female calf, or reared a male calf (from Clutton-Brock, Albon, and Guinness 1981); (c) mortality of mothers that had reared sons and daughters in the previous summer. Expected values calculated on the assumption that mothers rearing sons and daughters were equally likely to die.

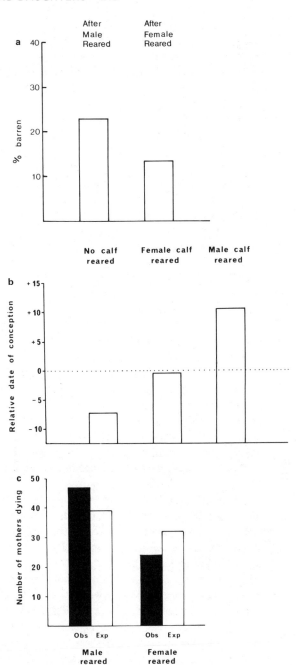

females belonging to the same group (see above). Mothers comfort offspring that have been attacked by other group members by allowing them access to the nipple, with the consequence that daughters of subordinate mothers have more frequent access to their mother's nipples at older ages than other juveniles. The effects of these differences in nipple stimulation on the mother's gonadotrophin secretion may be the cause of the lower fecundity of mothers that have raised daughters (McNeilly 1987).

Where female offspring are allowed to share their mother's home range throughout the rest of their lives, sex differences in parental care after offspring have reached nutritional independence may also have substantial costs. Not only may continuing social support for daughters have some cost to the mother, but the daughter's presence may reduce her mother's subsequent breeding success by increasing competition for resources (Clark 1978; Caley, Boutin, and Moses 1988). In some species, offspring of one sex associate more closely with their mothers as juveniles (e.g., Kojola 1984), and this, too, can have measurable costs to parental fitness. For example, in red-necked wallabies (*Macropus rufogriseus*), sons spend more time close to their mothers before dispersing than daughters, and mothers with subadult sons are less likely to be successful at their next reproductive attempt than those with subadult daughters (C. N. Johnson 1985, 1987). Conversely, the presence of mature offspring can *enhance* their parents' subsequent breeding success. For example, both in red-cockaded woodpeckers, *Picoides borealis* (Gowaty and Lennartz 1985), and in African wild dogs, *Lycaon pictus* (Malcolm and Marten 1982), sons commonly remain and help to provision the mother's subsequent offspring. Results from other cooperative breeders suggest that the presence of helpers may have an important effect on the reproductive success of the breeding female (J. L. Brown 1987). Here, sons can be viewed as "repaying" part of the costs of their production to the parent, and the net costs of rearing daughters may exceed those of rearing sons (Emlen, Emlen, and Levin 1986).

Where offspring of one sex associate with their parents after nutritional independence, it is usually difficult to combine estimates of their costs and benefits to their parents at different stages of their life histories to produce a single measure of the relative costs of sons and daughters. For example, the increased investment in male red-deer calves before weaning may be balanced by postweaning investment in daughters (see Figure 12.9, curve b); but differences in postweaning costs between sons and daughters could also be smaller (curve c) or larger (curve a) than differences in investment before weaning. In effect, this means that while it is possible to compare differential investment in sons and daughters at particular stages, reliable estimates of the total costs of sons and daughters are only possible where neither male nor female offspring are philopatric.

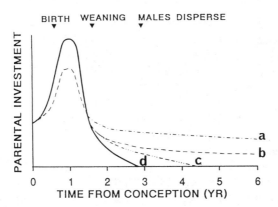

Figure 12.9 Hypothetical plot of changes in the distribution of parental investment by red-deer hinds in their male and female offspring (from Clutton Brock, Albon, and Guinness 1982). The plot shows levels of parental investment in sons (solid line) and daughters (interrupted lines) at different stages of the reproductive cycle. Curves a–c illustrate different costs of raising daughters beyond weaning. If postweaning costs of daughters are high (a), the total costs of raising daughters may exceed those of sons; if postweaning costs are small (d or c), sons may be the more expensive sex to rear.

12.6   Sex Allocation or Differential Extraction?

Most studies that have demonstrated differences between the costs of rearing sons and daughters have interpreted these as adaptations resulting from selection on the parent to make the best use of reproductive resources. However, an alternative possibility is that sex differences in investment are a consequence of differences in the behavior of male and female juveniles. If so, they may have arisen as a result of selection pressures operating on juveniles: increased expenditure on sons may commonly be a consequence of strong selection favoring male offspring that maximize their early growth rates and hence their body size as adults. These pressures are likely to be stronger than those on parents to withold resources with the result that the amount of resources extracted by sons may be shifted beyond the parental optimum (see Chapter 11).

What evidence is there that differences in the costs of rearing sons and daughters are a result of parental discrimination? In many parasitic Hymenoptera, it is clear that parents vary the resources available to sons and daughters by adjusting the sex ratio of progeny in relation to the size of the host (see Chapter 13). However, in birds and mammals, sex differences in parental expenditure are commonly associated with sex differences in juvenile behavior, and there is little evidence that parents treat their sons and daughters differently. For example, in African elephants, the increased sucking times of males (see Figure 12.7) are caused by higher frequencies of suckling attempts by male calves (Lee and Moss 1986). Mothers are equally tolerant of sucking by male and female calves for the first two years of calf life, subsequently terminating a higher proportion of sucking attempts by males. In red deer, too, young males attempt to suck more frequently than females, and the proportion of sucking attempts by males that are rejected by the mother is higher than for females (Clutton-Brock, Albon, and Guinness, unpublished). In rhesus macaques, the increased costs of daughters to subordinate females apparently arise from increased aggression directed at juvenile females and their mothers by other group members (see above).

Recent studies of birds suggest parents may sometimes vary their expenditure in relation to brood sex ratios (Stamps, Clark, et al. 1985; Stamps 1987; Stamps, Kus, and Arrowood 1987; Stamps, in press). In captive budgerigars, Stamps found that although neither fathers nor mothers discriminated between sons and daughters within broods, fathers (and, to a lesser extent, mothers) fed female-biased broods more frequently than male-biased broods before fledging and also started feeding them at an earlier stage (Figure 12.10a). These differences were not associated with higher begging rates by chicks in female-

Figure 12.10   Parental expenditure on male- and female-biased broods of budgerigars (from Stamps, Kus, and Arrowood 1987).
(a) Average paternal feeding rate (regurgitations/nestling/hr) during the later stages of chick development versus the percentage of females in the brood;
(b) average age at fledging of brood members versus the percentage of females in the brood.

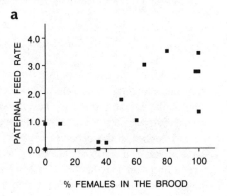

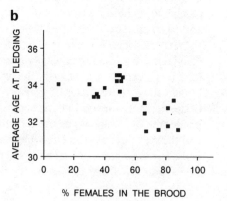

biased families but with an increased response of fathers to begging. As a result of this bias, female-biased broods obtained nearly three times more regurgitations in the final prefledging period than male-biased broods, and fledged earlier (Figure 12.10b). Early-fledging females showed higher reproductive success at their first breeding season than late-fledging ones, while similar correlations were not apparent for males. Stamps et al. suggest that parents may feed female-biased broods more frequently than male-biased ones because parental expenditure affects the reproductive success of females (which are primarily responsible for the acquisition and defense of nesting sites) more than that of males. Parents may not differentiate between offspring within broods because this would delay food delivery. Similar results are reported for bluebirds (Gowaty, unpublished, quoted by Stamps, in press), but a study of rooks found no correlation between the sex ratio of broods and feeding frequency (Røskaft and Slagsvold 1985). In mammals, there are some indications that parents, as well as other conspecifics, sometimes treat male and female juveniles differently (e.g., Dittus 1979; Moore and Morelli 1979; Mori 1979b; Moore 1982; Gomendio 1989), but there is no clear evidence that they discriminate against the more expensive sex when resources are short (see Section 13.3).

12.7   Sex Differences in Parental Investment and the Population Sex Ratio

Differences in rearing costs between male and female offspring might be expected to affect the average sex ratio produced by all parents in the population as well as the sex ratios produced by individual parents. As Fisher argued (1930), parents should, on average, divide their total reproductive effort equally between the sexes so that extra costs of rearing one sex should lead to the sex ratio at the end of the period of parental investment being biased against the more expensive sex. The rationale underlying Fisher's theory is intuitively obvious: if total reproductive effort is not equally divided between the sexes in a randomly mating population, fitness returns per unit investment will be greater for the cheaper sex. This will lead to selection favoring either a reduction in the number of the more expensive sex or a reduction in per capita expenditure on that sex.

A substantial number of specific exceptions to Fisher's theory have been identified, most arising from circumstances where the fitness returns per unit investment differ between sons and daughters (see Charnov 1982; Frank 1987). These include cases where siblings of one sex compete more intensely for mates or resources (Hamilton 1967; Clark 1978; Bulmer and Taylor 1980; Taylor 1981; Charnov 1982; Frank 1987; Bulmer 1986; Caley, Boutin, and Moses 1988); where offspring of one sex assist their parents' breeding attempts or increase the fitness of their brothers and sisters (Malcolm and Marten 1982; Gowaty and Lennartz 1985; Avilés 1986; Emlen, Emlen, and Levin 1986; Lessells and Avery 1987; Frank and Crespi 1989); where selection favors increased per capita investment in one sex and it is either impossible or too costly to vary the sex ratio during the period of parental investment (Maynard Smith 1980); and where sex is environmentally determined and one sex develops under more favorable conditions than the other (Bull 1983; Frank and Swingland, in press). In addition, a number of circumstances may favor temporary biases in the division of total parental effort between the sexes, including fluctuations in the adult sex ratio (Burley 1982, 1986; Werren and Charnov 1978), changes in the intensity of competition or cooperation between offspring of one sex (see Werren and Taylor 1984), and variation in the costs or benefits of producing either sex (Charnov 1982).

A general model by Frank (1987) provides a framework that incorporates most of these exceptions. Frank shows how the division of parental effort between the sexes (population allocation) may be affected by the length of investment periods, the number of offspring produced per period, the shape of resource distribution curves, and the form of relationships between investment in male and female offspring. The last two effects are particularly likely to be important in organisms where fecundity is low, and, under some conditions, the way in which parental effort should be divided among males and females cannot yet be predicted. For example, Frank emphasizes that where parents produce sequential broods and there is a negative correlation between parental expenditure in successive periods of investment (a relatively common situation in long-lived organisms with low fecundity, like red deer), there is no adequate quantitative theory yet available for predicting how parental effort should be divided between the sexes.

The clearest tests of Fisher's prediction of equal allocation are provided by studies of sex ratio variation in social or eusocial Hymenoptera, where the costs of raising daughters exceed those of rearing sons (Trivers and Hare 1976; Charnov 1982; Nonacs 1986; van der Have, Boomsma, and Menken 1988; Frank and Crespi 1989). As Section 11.3 describes, the higher costs of raising females are commonly compensated by a proportional reduction in the number of females produced—with the added complexity of conflicts between workers and queens.

Though the problems of predicting how parental effort should be divided rob us of firm theoretical predictions for vertebrates showing low fecundity and negative correlations between expenditure in successive periods (Frank 1987), several lines of evidence suggest that differences in the costs of raising males and females are sometimes compensated by variation in the number of the two sexes that are reared. For example, parents might compensate for higher mortality in males during the period of parental care by producing a male-biased sex ratio at birth or hatching, which should subsequently change to a female bias by the end of the period of care (Fisher 1930; Leigh 1970). As

Fisher himself pointed out, the human sex ratio follows this pattern, while data for ungulates suggests that the highest birth sex ratios occur in sexually dimorphic species where males show higher preweaning mortality than females (Clutton-Brock and Albon, in prep.). A little evidence suggests that the situation may be reversed in mammals where females are larger than males. For example, in spotted hyenas, fetal sex ratios for ten litters gave a sex ratio of 47% male while the juvenile sex ratio for fifteen litters was 55% male (van Jaarsveld, Skinner, and Lindeque 1988). Though the difference is not significant, cases where litter sex ratios are higher than fetal sex ratios are rare in mammals where males are the larger sex (see Section 13.2).

In many dimorphic vertebrates, the situation is complicated by the higher energy costs of raising one sex—usually males (see Section 12.4). In some species where males cost more to rear, weaning or fledging ratios show a significant female bias (Røskaft and Slagsvold 1985; Skogland 1986; Teather and Weatherhead 1989), but in others weaning or fledging ratios are equal or even male-biased (Clutton-Brock, Guinness, and Albon 1982; Trillmich 1986; Kojola and Eloranta 1989). Moreover, an alternative interpretation of female-biased fledging or weaning ratios in species where males cost more is that they are a consequence of a birth or hatching sex ratio that is constrained close to parity combined with lower viability among males (see Chapter 13).

Biased sex ratios at hatching or birth have also been found in a number of vertebrates where sex is chromosomally determined, and these biases have sometimes been interpreted in adaptive terms. For example, in three different species of lemmings, females produce nearly three times as many females as males (Kalela and Oksala 1966; Bengtsson 1977; Fredga, Gropp, et al. 1977). In all three lemming species, unusual sex determining mechanisms are involved. In wood lemmings (*Lemmus schisticolor*) the bias is caused by the presence of two types of X chromosomes, X^o and X^*, which have different effects; X^oY is male, but X^*Y is female. X^*Y females are fertile but produce only X^* ova, owing to nondisjunction of fetal oocytes, and Bengtsson (1977) has shown that the equilibrium sex ratio could be as low as 25% male. However, this does not explain why selection has not produced autosomal modifiers that would cause the X^*Y females to be males and thus increase the fitness of their carriers in populations with low sex ratios (Stenseth 1978; Williams 1979; Bull and Bulmer 1981). Many lemming populations are cyclical, and one possible adaptive explanation is that, when population density is low and groups are widely separated, inbreeding and local mate competition may favor a female-biased sex ratio (Stenseth 1978).

In some species where relatives of one sex compete more intensely for resources, birth sex ratios appear to be biased toward the sex that is less likely to compete. This idea was first suggested by Clark (1978) to account for a strong male bias in sex ratios in bushbabies, *Galago crassicaudatus*. More extensive data provide little evidence that the birth sex ratio is strongly biased in this species (Masters, Centner, and Caithness 1982; see also Hoogland 1981), but male-biased birth sex ratios have now been found in several mammals where females are philopatric (Caley, Boutin, and Moses 1988; Stuart-Dick and Higginbottom, in press). Across a small sample of primate species, birth sex ratios tended to be most strongly male-biased where related females are most likely to compete for resources (C. N. Johnson 1988). Similar effects have been reported within species. For example, among populations of the semelparous marsupial mouse, *Antechinus*, the proportion of males among pouch young increases in relation to the proportion of females in the population that breed more than once (Cockburn, Scott, and Dickman 1985). In these animals, daughters are philopatric and are likely to compete for resources with their female relatives. Cockburn (ibid.) suggests that the decline in the proportion of females produced in populations where mothers commonly breed more than once represents an adaptation to the increased postweaning costs of raising daughters.

Additional evidence that vertebrate sex ratios may be adjusted to compensate for differences in the costs of raising sons and daughters comes from studies of species where offspring of one sex remain for part or all of their lives in their natal group and assist their parents' subsequent breeding attempts, thus "repaying" the costs of their own production (Emlen, Emlen, and Levin 1986; Lessells and Avery 1987). Sex ratios in these species are commonly biased toward the cooperative sex. For example, in red-cockaded woodpeckers, where sons commonly remain in their natal group and assist in their parents' subsequent breeding attempts, nestling sex ratios show a significant male bias (59%) (Gowaty and Lennartz 1985). A consistent male bias has also been found in African wild dogs, where sons again usually remain in their natal groups and assist their mother in rearing subsequent litters while daughters disperse (Malcolm and Marten 1982). Similarly, in the social spider, *Anelosimus eximius*, where immature females cooperate with their mother to build webs and capture prey, juvenile sex ratios show a strong female bias (Avilés 1986; Elgar and Godfray 1987).

12.8 Summary

Parents might usually be expected to invest more heavily in offspring of one sex where additional investment has greater fitness returns, calculated in terms of grandoffspring, than investment in offspring of the other sex. In ectotherms where female fecundity increases in relation to body size, selection may commonly favor increased investment in daughters. In contrast, in some birds and mammals where males compete for harems of females, body size and early growth affect the fitness of males more than females, and mothers would be expected to invest more heavily in sons. However, the direction of these biases probably varies among species. For example, in some polygynous primates, social support after weaning may be more likely to influence the rank and fitness of a female's daughters (who usually remain in their natal group throughout their lives) than that of their sons (who disperse to other groups at adolescence). In addition, the relative effects of parental investment on sons and daughters may vary at different stages of development.

In a variety of social and eusocial insects, there is evidence of increased investment in female offspring, while in some polygynous birds and mammals where males are the larger sex, sons grow faster than daughters, require more food, and depress their parents subsequent breeding success or survival to a greater extent.

In most vertebrates where sons cost more to rear than daughters, these differences appear to be a consequence of sex differences in the behavior of offspring rather than of differences in parental behavior. This could suggest that they are a by-product of sexual selection favoring sex differences in juvenile growth rates rather than a consequence of evolved parental strategies, though it is possible that parents have been selected to permit one sex to extract more resources. Ideally, the way to distinguish between explanations would be to compare observed levels of investment with optima for parents versus their offspring, but, in practice, this is rarely possible (see Chapter 11). Further studies of the way in which parents treat their sons and daughters in species where rearing costs vary with offspring sex are badly needed. Like studies of differential investment in large and small offspring (see Section 9.5), these need to control for differences in offspring behavior that may influence the parents' access to their sons and daughters.

Where offspring of one sex cost more to rear, parents might, on average, be expected to divide their total effort equally between the sexes by producing an excess of the cheaper sex, though the division of parental effort may be affected by the distribution of resources available for reproduction and by the effects of investment on the fitness of sons and daughters. In social Hymenoptera, sex ratios are commonly adjusted to differences in the net costs of raising males and females. There is some evidence of similar trends in vertebrates: in mammals, birth sex ratios appear to be most heavily male-biased in species where males are most likely to die before the end of parental investment; where females are more likely to compete for resources than males; and where males assist their parents' subsequent breeding attempts.

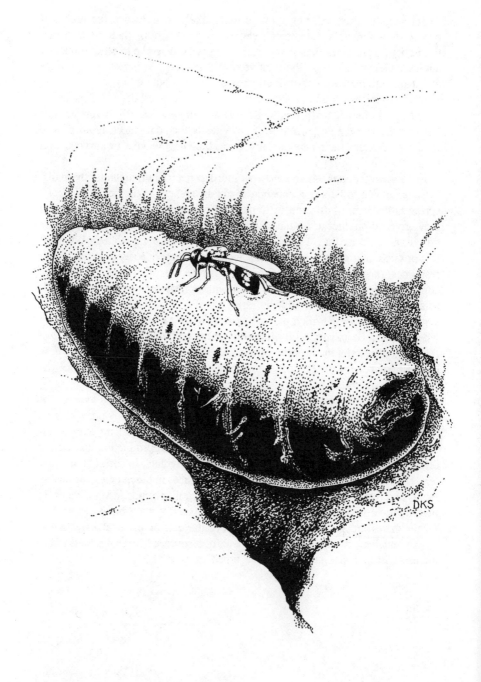

13
Sex Ratios and Differential Juvenile Mortality

13.1   Introduction

Where offspring of one sex are more costly to rear, parents might be expected to vary the sex ratio of their offspring in relation to their ability to expend resources, prematurely terminating investment in offspring that have little chance of surviving to breed successfully (Trivers 1972; Trivers and Willard 1973). For example, in polygynous birds and mammals where breeding success in males is more strongly influenced by body size than in females, parents might rear sons when resources were plentiful and daughters when they were scarce. Similarly, the sex ratio might vary with parental quality, superior parents producing the more expensive sex and inferior ones the cheaper sex.

Sex ratio variation in response to the parent's ability to expend resources might be achieved by several different mechanisms. Parents might manipulate the sex ratio at conception if the relative profitability of sons and daughters can be assessed in advance. Alternatively, they might initially produce an equal sex ratio and modify it at some subsequent stage of development by prematurely terminating investment in offspring of one sex. For example, Trivers (1972) suggested that parental discrimination of this kind may account for the common tendency for juvenile mortality in mammals to be male-biased when resource availability is low. Though Myers (1978) challenged this suggestion, arguing that it is better to produce young with a low chance of breeding successfully than none at all, her argument assumes that reproduction has no effects on the parent's residual reproductive value (Okansen 1981), thus removing the potential benefits of infanticide. Where reproduction has fitness costs, it is possible for infanticide to increase parental fitness (see Chapter 9).

Trivers and Willard's theory of sex ratio variation needs to be qualified in several ways. First, the way in which resources available for reproduction are distributed among parents may affect the optimal sex ratio for different individuals. The importance of this effect may vary widely between breeding systems and is likely to be most pronounced in species with low fecundity (Frank 1987).

Second, several mechanisms other than those that Trivers and Willard envisaged can lead to variation in fitness returns on investment in male and female progeny. In particular,

(1) Where one sex is more expensive to rear, the costs of raising the more expensive sex may vary between parents, and these differences may favor sex ratio variation. For example, in red deer, the increased costs of raising sons are confined to subordinate mothers (see Figure 13.1), while in rhesus macaques the increased costs of raising daughters are confined to subordinate mothers (Gomendio, Clutton-Brock, et al. 1990). In the absence of any relationship between maternal rank and the relative fitness of sons and daughters, these effects could favor the production of females by subordinate red-deer hinds and of males by subordinate macaques.

(2) In multiparous species, there may be trade-offs between the sex and number of offspring (Williams 1979; Burley 1982; McGlinley 1984; Gosling 1986). For example, mothers in very poor condition might produce a single offspring of the cheaper sex, those in slightly better condition a singleton of the more expensive sex, those in good condition twins of the cheaper sex, and those in very good condition twins of the more expensive sex (see Figure 13.2).

(3) Sex ratio trends need not be confined to cases where offspring of one sex are more costly to rear. For example, where the relative attractiveness (and hence the reproductive value) of sons varies between parents, females mated to attractive males might bias the sex ratio of their offspring toward males, while females mated to unattractive partners might tend to produce daughters (Burley 1977, 1981, 1986).

(4) A number of other factors may affect the fitness returns of producing sons and daughters to different mothers. These include the chance of successful dispersal by the two sexes (McCullough 1979; Armitage 1987); variation in the extent of competition or cooperation between siblings for mates or resources (Hamilton 1967; Clark 1978; Bulmer and Taylor 1980; Taylor 1981; Charnov 1982; Werren and Taylor 1984; Elgar and Godfray 1987; C. N. Johnson 1988) or between parents and their offspring (Malcolm and Marten 1982; Emlen, Emlen, and Levin 1986; Cockburn, Scott, and Dickman 1985; Lessells and Avery 1987) and local fluctuations in the adult sex ratio (Werren and Charnov 1978; Charnov 1982).

Finally, conflicts between parents and their offspring may occur over the sex ratio (see Chapter 11). In particular, where selection favors the premature termination of investment in one sex, conflicts of interest are likely to arise, and offspring would be expected to take countermeasures. In some cases, parents may win these conflicts, but in others offspring may force parents to diverge from their optimal sex ratio. In biparental species where rearing offspring of one sex affects the fitness of one parent more than that of the other or where they subsequently assist the breeding attempts of parents of one sex, optimal sex ratio strategies may also be affected by conflicts of interest between caregivers (see Section 10.6).

The rest of this chapter examines the empirical evidence that parents vary the sex ratio of their offspring in relation to the availability of resources and

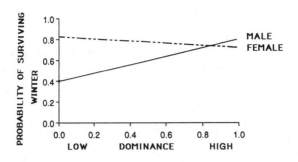

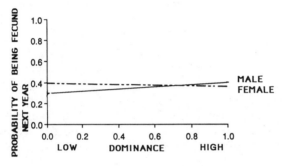

Figure 13.1   Costs of raising sons and daughters to the subsequent survival of red-deer hinds (from Gomendio, Clutton-Brock, et al. 1990). Curves show the probability of mothers that have reared male or female offspring surviving the winter and bearing a calf the following year, plotted against their dominance rank. Dominant mothers are equally likely to survive and to breed again after rearing sons or daughters, while subordinates are less likely to survive or to breed again after rearing sons.

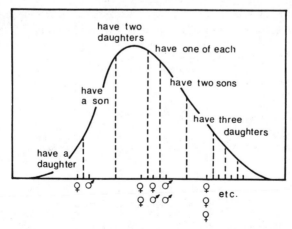

Figure 13.2   Litter sex ratio optimization in a species that usually has two but sometimes one or three young per clutch (from Williams 1979).

their capacity to invest in their offspring. Section 13.2 describes evidence of relationships between resource availability and parental quality and variation in the sex ratio at birth or hatching. This topic has been recently reviewed in invertebrates (Charnov 1982), reptiles (Bull 1983), birds (Clutton-Brock 1986), and mammals (Clutton-Brock and Iason 1986); more extensive evidence of sex ratio variation can be found in these references. Section 13.3 reviews the distribution of sex differences in juvenile mortality and examines whether or not sex ratio trends are likely to be a consequence of parental manipulation.

13.2   Sex Ratio Variation at Hatching or Birth

INVERTEBRATES

In some invertebrates, the sex of developing progeny is determined directly by resource availability or local population density. For example, in some nematodes, females are more than twice the length of males, and sex is environmentally determined (Charnov 1982). The sex ratio (% males) rises with the level of infestation of the host, declining with the amount of food available to the growing larvae (Figure 13.3; see also Christie 1929; A. A. Johnson 1955; Petersen 1972; Charnov 1982).

In other cases, parents vary the sex ratio of their offspring in relation to the immediate availability of resources. In some parasitoid wasps, females lay female (fertilized) eggs in large hosts and male (unfertilized) eggs in small ones (Charnov 1982; B. H. King 1987, 1988). In many of these species, sex ratio control is highly sophisticated. For example, female *Lariophagus*, small parasitic wasps that attack the larvae of granary weevils, not only adjust the sex ratio of the eggs laid in relation to host size but also vary the magnitude of the bias in relation to the distribution of host sizes (see Figure 13.4; Charnov,

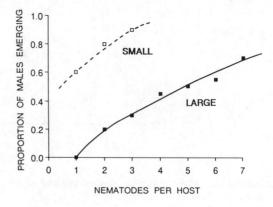

Figure 13.3   Sex ratio in the nematode *Romanomeris* as a function of its infection level, in large and small hosts. (Petersen, Chapman, and Woodward 1968; redrawn from Charnov 1982.)

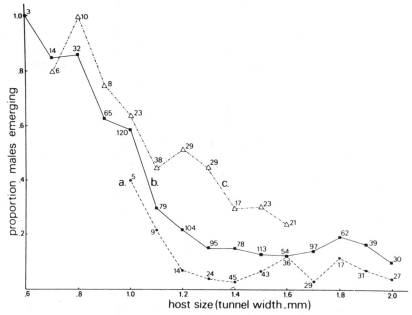

Figure 13.4   Sex ratio shift in *Lariophagus* (from Charnov 1982). The proportion of males emerging is plotted against the size of hosts (weevil larvae) that were offered in sequence. Curve *b* resulted when female wasps were presented sequentially with twenty hosts of a single size. Curves *a* and *c* resulted when wasps were presented in alternating sequence with two host sizes differing by 0.4 mm. Curve *c* is where the host of interest was the smaller of the two. For example, when hosts of 1.4 mm were offered alone, they gave a sex ratio of 15% male (curve *b*). When they were offered alternatively with 1.8 mm hosts, they gave a sex ratio of 30% male (curve *c*) while, when offered with 1.0 mm hosts, this dropped to 2% male (curve *a*).

Los-den Hartogh, et al. 1981). Similarly, female fig wasps adjust the sex ratio of their offspring both in relation to the intensity of local mate competition and the degree of inbreeding (Herre 1985).

In some thrips (Thysanoptera) females either produce all-male broods by viviparity or all-female broods by oviparity (Crespi 1988, and in press). Males apparently gain more from large size than females, and viviparity and the production of males are associated with circumstances that allow offspring to grow into large adults. In spring, viviparous and oviparous females do not differ in size, but females tend to breed viviparously where food supplies are abundant. In summer, viviparous females breed earlier than oviparous ones and their offspring develop into larger adults (Crespi 1988, 1989).

FISH, REPTILES, AND AMPHIBIA

There have been relatively few studies of individual variation in the hatching sex ratio in fish (Charnov 1982), though environmental sex determination occurs in some species, and sex change is common (Bull 1983). In contrast, sex determination (and hence the sex ratio at hatching) is affected by incubation temperature in a considerable number of oviparous reptiles, including lizards, turtles, and crocodiles (Bull 1980, 1983). Biases are frequently extreme and occur over quite small temperature ranges: for example, in some turtles, only males are obtained if eggs are incubated at 23–28°C, both sexes at 28–30°C, and only females at temperatures above 30°C (Bull and Vogt 1979; Bull, Vogt, and McCoy 1982). The direction of effects varies between species: in lizards and alligators, warm temperatures produce males and cool ones females, while in most turtles this pattern is reversed (Bull 1980, 1983). In a few species, extreme temperatures produce females and those in the center of the range produce males. As yet, the functional significance of these differences and the extent to which different temperatures favor males and females is still unclear (Bull 1983).

BIRDS

There is relatively little evidence of systematic variation in hatching sex ratios in birds (Clutton-Brock 1986). Within species, the sex ratio does not appear to vary systematically with egg weight (Fiala 1981b; Harmsen and Cooke 1983; Bancroft 1984; Weatherhead, in press), and most large data sets show that hatching ratios are very close to parity (Clutton-Brock 1986).

However, a number of cases have been reported where the hatching sex ratio varied significantly in relation to factors that might be expected to affect the availability of resources and the growth and survival of young. For example, in a sample of yellow-headed blackbird (*Xanthocephalus xanthocephalus*) nests, the sex ratio of nestlings was significantly male biased (64%) in primary broods where males usually assist females in feeding young, but did not differ from parity in second and third broods where females usually rear young unassisted (Patterson and Emlen 1980). In some birds, sex ratios vary consistently with egg order within clutches (Fiala 1981a; Ryder 1983; Weatherhead 1983). For example, in bald eagles, *Haliaeetus leucocephalus*, first-hatched eggs are usually female (63%), and male-female hatching sequences within two-egg broods are rare (Bortolotti 1986). Males may grow faster in the first week of life and, by placing males second, parents may minimize prefledging mortality (but see Section 13.3).

MAMMALS

In mammals, there is more extensive evidence that birth sex ratios vary in relation to resource availability (Clutton-Brock and Iason 1986). Trends include associations between the sex ratio at or shortly after birth and birth date

(Coulson and Hickling 1961; S. S. Anderson and Fedak 1987); litter size (McShea and Madison 1986); sex of previous offspring (Stuart-Dick and Jarman, unpublished); maternal age or parity (Verme 1969, 1983; Dapson, Ramsey, Smith, and Urbston 1979; Huck, Pratt, et al. 1988; Stuart-Dick and Jarman, unpublished); specific dietary components (Bird and Contreras 1986); overall food quality or availability (Rivers and Crawford 1974; Labov, Huck, et al. 1986; Wright, Crawford, and Anderson 1988; Stuart-Dick and Higginbottom, in press); environmental quality (Verme 1969, 1983; C. N. Johnson and Jarman 1983; Pederson and Harper 1984); maternal condition (Rutberg 1986; Skogland 1986; Kojola and Eloranta 1989); maternal dominance rank (Altmann 1980; Simpson and Simpson 1982; Clutton-Brock, Albon and Guinness 1986; Symington 1987; Pratt, Huck, and Lisk, in press); litter size (Packer and Pusey 1987); and timing of insemination relative to ovulation (Guerrero 1970, 1974; Verme and Ozoga 1981).

A number of experimental studies have now induced changes in the sex ratio. In some species, food deprivation or other forms of environmental stress generate significant reductions in the birth sex ratio associated with reduced litter sizes, while animals fed on ad lib diets produce larger litters and approximately equal numbers of sons and daughters. For example, laboratory mice maintained on low-fat diets produce significantly female-biased sex ratios (24%), while those maintained on control diets produce larger litters with approximately equal sex ratios (Table 13.1; Lane and Hyde 1973; Rivers and Crawford 1974; Moriya and Hiroshige 1978; see also Meikle and Drickamer 1986; Wright, Crawford, and Anderson 1988; Pratt, Huck, and Lisk, in press).

Table 13.1

The effect of diet on the number of male and female offspring of albino mice (from Rivers and Crawford 1974).

	Maternal Diet		
	Low Fat	Control	P
No. litters born	26	28	
No. litters stillborn	6	1	*
No. litters totally or partially cannibalized	12	2	**
No. litters examined for sex ratio	8	15	
No. pups per litter	5.13	7.94	*
No. male pups per litter	1.25	4.00	***
No. female pups per litter	3.88	3.94	N.S.
Sex ratio (% males)	24.40	50.40	*

NOTES: The low-fat diet contained 6 g kg^{-1} lipid, the control diet 66 g kg^{-1}. *.05; **.01; ***.001.

Figure 13.5 Effects of manipulation of food availability during early development on birth sex ratios in golden hamsters, *Mesocricetus auratus* (from Huck, Labov, and Lisk 1986, 1987).
(a) Sex ratios from birth to 28 days of litters produced in adulthood by females raised on ad lib diets (•) or food-restricted during their first 25 days of life (▲), their second 25 days of life (o), or their first 50 days of life (Δ) and subsequently maintained on ad lib diets; (b) sex ratios from birth to 25 days of litters produced in adulthood by daughters of females who were themselves raised on ad lib diets (o) versus daughters of females food-restricted during the first 50 days of life.

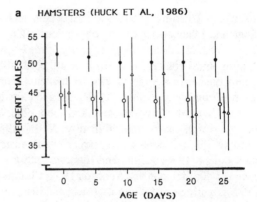

a   HAMSTERS (HUCK ET AL, 1986)

• DAMS NOT FOOD RESTRICTED AS JUVENILES.

o DAMS FOOD RESTRICTED AS JUVENILES DAYS 26-50

▲ DAMS FOOD RESTRICTED DAYS 1-25.

Δ DAMS FOOD RESTRICTED AS JUVENILES DAYS 1-50.

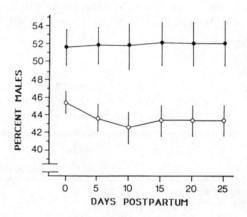

b   HAMSTERS (HUCK ET AL, 1987)

• SUBJECTS, DAMS & GRANDDAMS FED AD LIB.

o SUBJECTS & DAMS FED AD LIB;
  GRANDDAMS FOOD RESTRICTED DAYS 1-50.

Some of the most remarkable work on mammalian sex ratios has been carried out on golden hamsters, *Mesocricetus auratus*. As in several other rodents, food restriction during pregnancy leads to a significant reduction in litter size and a decline in the sex ratio at birth, from 49.6% to 40.7% (Labov, Huck, et al. 1986). In addition, restricting food given to virgin females during their first fifty days of life causes them to produce smaller litters and female-biased sex ratios during adult life, even when they are subsequently replaced on ad lib diets (Huck, Labov, and Lisk 1986; see Figure 13.5a). The effects of food restriction during early development even span generations: the daughters of food-restricted females, themselves reared on ad lib diets, produce smaller litters and relatively fewer sons than the daughters of control females that were not food restricted (Huck, Labov, and Lisk 1987; see Figure 13.5b).

In other cases, improved conditions are associated with male-biased sex ratios. For example, in wild populations of common opossums (*Didelphis*

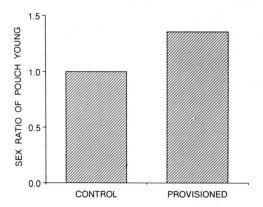

Figure 13.6 Pouch sex ratios in provisioned and control samples of common opossums (from Austad and Sunquist 1986).

marsupialis), experimental feeding of young females led to an increase in the weight and sex ratio of young in the pouch (see Figure 13.6). Recent work shows that both responses can occur in the same population. For example, while severely food-deprived domestic mice produce female-biased litters and females fed ad lib produce an approximately equal sex ratio, competitively successful females under moderate levels of food availability produce significantly male-biased ratios (Wright, Crawford, and Anderson 1988).

Though some sex ratio trends have been found in several species, few are consistent across species (Clutton-Brock and Iason 1986). On these grounds, it is sometimes suggested that all examples of systematic variation in birth sex ratios represent cases where the null hypothesis has been wrongfully rejected.

However, if birth sex ratios are manipulated adaptively, differences in the direction of trends should be expected, since the effects of parental expenditure on sons and daughters evidently vary (see Section 12.2).

Relationships between maternal rank and the sex ratio in red deer and rhesus macaques provide a good example. In these two species, both the relative fitness and the relative costs of raising sons and daughters vary with maternal rank, but they may do so in opposite directions (see Section 12.2). Subordinate female red deer are likely to maximize their fitness by producing a female-biased sex ratio because their daughters will have higher fitness than their sons, and the costs of rearing sons exceed those of rearing daughters. In contrast, subordinate macaques may benefit from producing male-biased sex ratios because their sons are likely to have higher fitness than their daughters and the costs of raising daughters, are higher than those of raising sons (see Section 12.5). As might be expected, the birth sex ratio (% males) increases with maternal rank in red deer (see Figure 13.7) and declines with maternal rank in at least some populations of macaques and baboons (Figures 13.8 and 13.9: Altmann 1980; Simpson and Simpson 1982; Silk 1983, 1986; Altmann, Hausfater, and Altmann 1988; but see also Meikle, Tilford, and Vesey 1984; Berman and Rawlins 1985; Small and Smith 1985; Rawlins and Kessler 1986; Small and Hrdy 1986; Berman 1988).

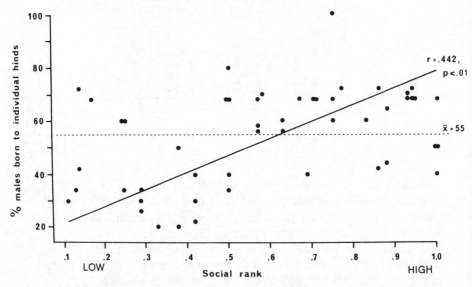

Figure 13.7   Birth sex ratios produced by individual red deer hinds differing in social rank over their lifespans (from Clutton-Brock, Albon, and Guinness 1986). Measures of maternal rank were based on the ratio of animals that the subject threatened or displaced to animals that threatened or displaced it, weighted by the identity of the animals displaced. Values of this ratio range from 0.1 (low ranking) to 1.0 (high ranking).

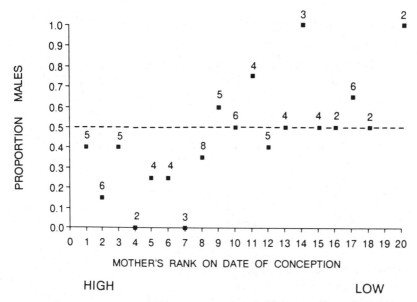

Figure 13.8   Birth sex ratios in relation to maternal rank in wild yellow baboons in the Amboseli National Park, Kenya (from Altmann 1980; Altmann, Hausfater, and Altmann 1988). Numbers of offspring produced per female are shown above each square. Mothers arranged in linear rank order from 1 (highest ranking) to 20 (lowest ranking).

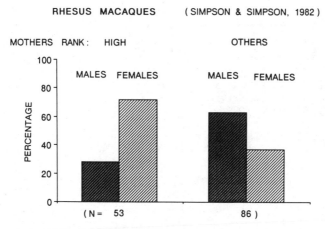

Figure 13.9   Percentage of male and female offspring produced by high-ranking and "other" females in a colony of rhesus macaques at Madingley, Cambridge (Simpson and Simpson 1982). Similar trends were found in two temporal subdivisions of the data set.

MECHANISMS OF SEX RATIO VARIATION BEFORE HATCHING
OR BIRTH

While haplo-diploid animals can manipulate sex ratios by choosing to fertilize or not to fertilize their eggs, we still know little about the causes of sex ratio variation in species where sex is chromosomally determined. In mammals, a wide variety of different mechanisms could generate variation in birth sex ratios, including differential production of X- and Y-bearing sperms, differential motility or mortality of sperm affecting the rate of fertilization, and differential implantation or survival of male and female zygotes (see Parkes 1926; Beatty 1971; James 1987; Muehleis and Long 1976; Bengtsson 1977; Roberts 1978; Clutton-Brock and Iason 1986; Gosling, 1986).

However, only for the last of these mechanisms is firm evidence available. In a wide variety of species, deprivation or other forms of environmental stress lead to female-biased birth sex ratios associated with reductions in litter size, while unstressed animals produce approximately equal numbers of males and females or a slight male bias (see above). The most likely explanation of these trends is that environmental stress causes higher mortality among male fetuses. In a variety of species, there is direct evidence that prenatal mortality is commonly male-biased. The sex ratio of aborted fetuses is usually male-biased in macaques (Sackett, Holm, et al. 1975; Digiacomo and Shaughnessy 1979) and man (see Figure 13.10; Parkes 1926; McMillen 1979), and studies of several mammals show that fetal sex ratios decline with increasing fetal age (Figure 13.11). It is interesting to note that the latter trend is found both in cattle and in elk (where males are larger than females), and in the baleen whales (where females are larger than males in some species) (Ralls 1976; Trivers 1985).

Though male fetuses are generally less likely to survive gestation than females, this is not always the case. In large samples of coypu, Gosling (1986)

Figure 13.10 Mean sex ratio of fetal deaths by month of gestation calculated from vital statistics data for the United States from 1922 to 1936 (solid squares) and 1950 to 1972 (open squares) (from McMillen 1979).

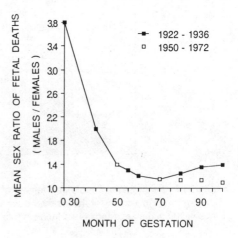

Figure 13.11 Sex ratios
among fetuses of different
sizes or ages:
(a) domestic pig: Parkes
1925;
(b) elk: Greer and Howe
1964;
(c) blue, fin, and sei whales:
Trivers 1985; Trivers and
Seger, unpubl.

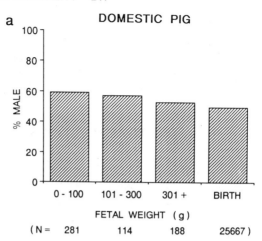

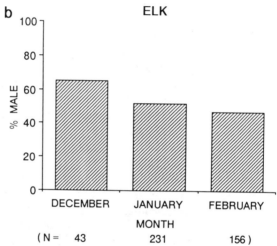

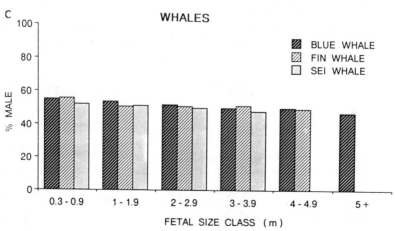

found that young females in superior body condition selectively abort small, female-biased litters. Gosling suggests that they do so because this allows them to reconceive a larger litter and that they refrain from aborting small male-biased litters because, in this case, mothers can express their reproductive potential by producing a litter of large, successful sons (but Section 9.7).

SEX RATIO VARIATION: PARENTAL STRATEGY OR BY-PRODUCT OF SEXUAL SELECTION?

Variation in the sex ratio within populations is commonly interpreted in two different ways. First, changes in the sex ratio may occur because parents actively manipulate the sex ratio, either at conception or at some subsequent stage of offspring development, so as to maximize their fitness. Alternatively, physiological or behavioral differences between male and female juveniles may render one sex (usually males) more susceptible to food shortage or other forms of environmental harshness, leading to consistent variation in sex ratios. These two explanations are often difficult to distinguish, for many of their predictions are similar. For example, both arguments predict that inferior parents should produce a lower proportion of the more expensive sex and that sex ratio biases should increase as food availability declines.

In haplo-diploid invertebrates, the precision and magnitude of variation in hatching sex ratios clearly indicate that they are a consequence of adaptive parental strategies (Charnov 1982; Crespi, in press). However, in most other animals, the situation is less clear. For example, the simplest explanation of cases where mammals living under stressful conditions produce small litters with low sex ratios is that sexual selection for rapid growth has lowered the viability of males under adverse conditions. In contrast, where improved conditions are associated with sex ratios at birth or hatching that are significantly higher than parity while controls produce sex ratios significantly lower than parity, it would seem more likely that adaptive manipulation is involved (Clutton-Brock and Iason 1986). At the moment, we know too little about the pattern of sex ratio variation to draw firm conclusions, but the available evidence suggests that more than one mechanism is probably involved.

13.3   Sex Differences in Juvenile Mortality

PATTERNS OF DIFFERENTIAL MORTALITY

The second way in which parents might vary the sex ratio of their offspring is by manipulating the survival of male and female offspring after birth. In energetic terms, this is a relatively wasteful method, but this may be unimportant if the costs of egg or neonate production are low and the parent's reproductive output is limited by the number of young they can raise.

There is evidence of higher mortality among juvenile males in a substantial number of birds and mammals (Table 13.2), most of them species in which males are larger than females. For example, in one study of common grackles, an approximately equal sex ratio at hatching fell to a ratio of 38% males by fledging (Howe 1977). Evidence of increased mortality among male nestlings or of a decline in the sex ratio after hatching has also been found in capercaillie (Wegge 1980; Moss and Oswald 1985), black grouse (Linden 1981; Angelstam 1984), rooks (Røskaft and Slagsvold 1985), and red-winged blackbirds (Cronmiller and Thomson 1980, 1981), though not in a number of other species where the sexes are of similar size (Kessel 1957; Pinkowski 1977; Harmsen and Cooke 1983, Gowaty and Lennartz 1985). As a result, when the average change in sex ratios between hatching and fledging is plotted against adult sexual dimorphism for the small number of species for which data is available, there is a significant tendency for the extent to which the sex ratio declines to be greatest in dimorphic species (Clutton-Brock, Albon, and Guinness 1985; see Figure 13.12).

In mammals where males are larger than females, neonatal mortality is again slightly higher among males (Parkes 1926: see Table 13.2) but the most pronounced sex differences in survival commonly occur after weaning (Ralls, Brownell, and Ballou 1980). Many of the most pronounced differences have been found in ungulates, including roe deer (Andersen 1953; Borg 1971), black-tailed and mule deer (Taber and Dasmann 1954; Mohler, Wampole, and Fichter 1951), wapiti (Flook 1970), reindeer (Bergerud 1971; Leader-Williams 1988; Skogland 1985), musk ox (Tener 1954), and Soay sheep (Grubb 1974), and, as in birds, the degree of differential mortality appears to increase with the degree of sexual dimorphism in body size (Clutton-Brock, Albon, and Guinness 1985). However, the extent and, in some cases, the direction of differential mortality varies among populations. For example, studies of *Odocoileus* suggest that juvenile mortality can be either female-biased, equal in both sexes, or male-biased (Mohler, Wampole, and Fichter 1951; Taber and Dasmann 1954; Robinette, Gashwiler, et al. 1957; Dapsen, Ramsey, and Smith 1979; Woolf and Harder 1979).

In contrast to ungulates, juvenile mortality is typically female-biased in some populations of baboons and macaques (see above). Here the bias appears to have quite different causes and is associated with increased rates of aggression directed at juvenile females by other group members (Section 12.2). However, in some primates, higher mortality in males has been found at some stages of juvenile development. In a declining population of baboons, Altmann found a tendency for male juveniles to be more likely to die than females (see Table 13.2), while in a colony of Barbary macaques, male juveniles were more likely to die during their first year of life (Paul and Thommen 1984). In addition, food shortage or other forms of environmental stress appear to affect the growth of males more than that of females, as they do in many other mammals (see next section).

Table 13.2
Sex differences in hatching and juvenile mortality in some birds and mammals.

BIRDS

Domestic pigeons: Cole and Kirkpatrick 1915.

Hatching Live		% Dead in Shell		% Mortality to Day 28	
Males	Females	Males	Females	Males	Females
866	810	4.8	4.4	39.6	37.9

Domestic chickens: McArthur and Baillie 1932.

Hatching Live		% Mortality to Day 120	
Males	Females	Males	Females
527	503	28.1	24.9

Pheasants: Haig-Thomas and Huxley 1927.

Hatching Live		% Mortality within 6 Months	
Males	Females	Males	Females
514	252	70.4	50.8

MAMMALS

Ungulates

Red deer: Clutton-Brock et al. 1982, unpublished ($n = 227$).

% Born		% Mortality, 0–3 Months		% Mortality, 3–12 Months		% Mortality, 12–24 Months	
Males	Females	Males	Females	Males	Females	Males	Females
57.0	43.0	9.80	8.90	39.0	30.5	28.9	19.4

Domestic sheep: from Burfening 1972.

Born		% Mortality, 0–60 Days		% Mortality, 60–100 Days	
Males	Females	Males	Females	Males	Females
1819	1730	18.6	18.6	4.4	3.8

Feral goats: Pickering 1983.

	% Mortality, 0–1 Year		% Mortality, 1–2 Years	
	Males	Females	Males	Females
	25	14	19	7
$n =$	48	36	36	30

Thoroughbred horses: from Platt 1978.

Born		% Stillborn		% Mortality before 70 Days	
Males	Females	Males	Females	Males	Females
1,400	1,445	2.1	1.9	3.6	2.1

Table 13.2 (*continued*)

Primates

Yellow baboons: Altmann 1980.

	% Mortality 0–1 Year	
	Males	Females
	43.4	28.5
n =	23	21

Toque macaques: Dittus 1977, 1979.

	% Mortality					
	0–1 Year		1–2 Years		2–5 Years	
	Males	Females	Males	Females	Males	Females
	39.5	52.6	13.1	33.3	14.1	17.4

N: based on a population of 446 animals.

Barbary macaques: Paul and Thommen 1984.

	% Mortality 0–12 Months	
	Males	Females
	10.4	6.3
n =	144	160

Pigtailed macaques: from Sackett et al. 1974.

Born		% Stillborn/ Aborted		% Perinatal Mortality	
Males	Females	Males	Females	Males	Females
420	380	14.0	13.0	18.0	5.0

Rhesus macaques: Small and Smith 1986.

Born		% Mortality before 30 Days		% Mortality after 30 Days – 1 Year	
Males	Females	Males	Females	Males	Females
398	399	6.3	3.8	7.2	8.1

Man, England and Wales, 1913: from Parkes 1926.

Percentage Mortality							
0–3 Months		3–6 Months		6–9 Months		9–12 Months	
Males	Females	Males	Females	Males	Females	Males	Females
6.8	5.2	2.2	1.8	1.7	1.4	1.4	1.3

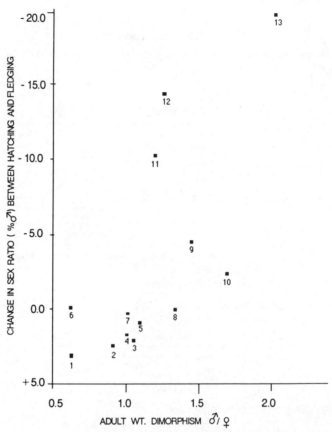

Figure 13.12   Changes in the sex ratio (% males) between hatching and fledging in bird species showing different degrees of adult weight dimorphism (male weight/female weight) (from Clutton-Brock, Albon, and Guinness 1985). (1) peregrine falcon; (2) American kestrel; (3) eastern bluebird; (4) starling; (5) snow goose; (6) European sparrow-hawk; (7) red-cockaded woodpecker; (8) blue grouse; (9) red-winged blackbird; (10) yellow-headed blackbird; (11) rook; (12) Capercaillie.

In man, sex differences in juvenile mortality are also not uncommon but may have different causes again (Chen, Huq, and D'Souza 1981). Neonatal mortality is usually male-biased, but this is reversed between the ages of one and five in some developing countries with high infant mortality (Cowgill and Hutchinson 1963a,b). There is some evidence that differential juvenile mortality at this age may be caused by less careful nurture of daughters or even by differential infanticide (Cowgill and Hutchinson 1963a,b; Voland 1984).

Finally, in altricial birds where females are the larger sex, it is commonly assumed that males may have difficulty in obtaining food and are likely to show higher mortality than females. However, there is little firm evidence that this is the case. Detailed studies of European sparrow-hawks (*Accipiter nisus*) and blue-footed boobies (*Sula nebouxii*), where females are substantially heavier than males, have found no sex differences in nestling mortality (Newton and Marquiss 1979; Newton 1979; Drummond, Osorno, et al., in press; though see also Bortolotti 1986; Edwards and Collopy 1983). In blue-footed boobies, the assumption that males are subordinate to females because they are smaller is evidently wrong: when males hatch first, they are able to maintain dominance over their sisters even after the latter exceed them in size. Even when females hatch first, males are often able to grow and survive despite a large disparity in size (Drummond, Osorno, et al., in press). These results raise the interesting possibility that pronounced sex differences in juvenile mortality may be confined to species where sexual size dimorphism arises through sexual selection.

DIFFERENTIAL JUVENILE MORTALITY AND FOOD SHORTAGE

Sex differences in juvenile mortality among dimorphic birds appear to increase when food is short. In one study of common grackles (*Quiscalus quiscula*), sex differences in mortality were greatest among chicks hatching from the last egg (Howe 1977, 1979). In red-winged blackbirds, experimental addition of two chicks per nest led to an increase in starvation mortality from less than 1% to over 20% of nestlings and a fledgling sex ratio of 37% male compared to 52% in natural nests (Cronmiller and Thomson 1980, 1981). In rooks, too, experimental manipulation of brood size caused an increase in the mortality of sons compared to daughters (Røskaft and Slagsvold 1985).

Food shortage or competition often also has stronger effects on the growth of males where they are the larger sex (Slagsvold 1982; Bortolotti 1986). In yellow-headed blackbirds, variance in the weight of male nestlings is greater than in females, male nestlings are lighter relative to females in secondary clutches (where the father does not assist in rearing the young) than in primary ones, and male juveniles lose weight more rapidly when food is short (Willson 1966; Patterson, Erckmann, and Orians, 1980). In common grackles, the growth of male nestlings is more strongly affected by brood size and hatching synchrony than female growth, and egg weight is a better predictor of the weight of females than males (Howe 1976, 1979). And in rooks, experimental enlargement of brood size combined with sex ratio manipulation led to increased variability in the weight of male chicks compared to females (Røskaft and Slagsvold 1985).

In mammals, too, sex differences in juvenile mortality are most pronounced when food is scarce—which may help to explain why mortality differentials often vary within species. For example, in the red deer population of Rhum, no sex differences in juvenile mortality were found when the population was expanding, but as it approached carrying capacity, higher mortality of males

Table 13.3

Percentage mortality of male and female red-deer calves in relation to the dominance rank and group size of their mothers.

	% Mortality in First Year of Life			
	Females		Males	
Mother's rank				
Dominant (above median rank)	36	(125)	38	(184)
Subordinate (below median rank)	30	(109)	51	(102)
Mother's group size				
Small (below median size for population)	24	(87)	16	(117)
Large (above median size)	22	(95)	36	(118)

NOTE: Sample size is in parentheses.

occurred among calves and yearlings (Clutton-Brock, Guinness, and Albon 1982). These differences were more pronounced among the offspring of subordinate mothers and juveniles born into matrilines of above average size (Table 13.3). Several studies of catastrophic mortality in ungulates also show that sex differences in mortality are most pronounced during periods of acute food shortage. For example, when a population of six thousand reindeer on St. Matthew Island crashed to forty-two in the course of a single winter, males of all ages were almost totally eliminated (Klein 1968). Similar sex differences in juvenile mortality have been found in die-offs of white-tailed deer (Woolf and Harder 1979), wapiti (Flook 1970), mule deer (Taber and Dasmann 1954; Robinette, Gashwiler, et al. 1957), wildebeest (Child 1972), and Soay sheep (Grubb 1974).

Experimental manipulation of food availability has been shown to affect juvenile males more than females in several mammals. When lactating female wood rats (*Neotoma floridana*) were maintained on 70–90% of the maintenance food requirements of nonreproductive females of equivalent size, the sex ratio of their pups declined from around 50% males at birth to 29% by 20 days (McClure 1981: see Figure 13.13). This trend was remarkably consistent: in all but one of the litters, every male pup died before there were any deaths among females.

Food shortage is also commonly associated with a reduction in male growth relative to female growth. For example, though juvenile male wood rats typically grow faster than females, female juveniles in McClure's food-restricted litters put on more weight than males (see Figure 13.14). In coypu, the faster growth rates of males are negatively influenced by increasing litter size, while the slower growth rates of females are unaffected (Gosling, Baker, and Wright 1984). Among captive reindeer maintained on a high level of nutrition, male calves gain weight faster than females, while in wild populations living on an inferior diet, there is little difference in growth between males and females

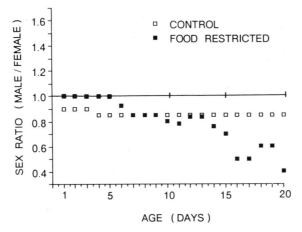

Figure 13.13 Changing ratio of males to females in 29 control (open squares) and 32 food-restricted (solid squares) litters of wood rats, *Neotoma floridana* (from McClure 1981). Food-restricted mothers were given 70–90% of the maintenance requirement of a nonreproductive female of equivalent body mass.

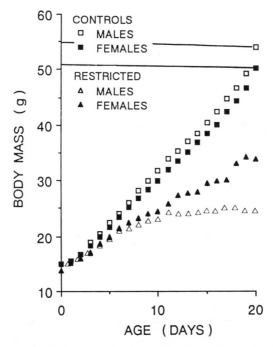

Figure 13.14 Average body mass of male and female nestlings in food-restricted and control litters of wood rats, *Neotoma floridana* (McClure 1981).

Figure 13.15 Differential growth in male and female reindeer:
(a) under captive conditions on a high plane of nutrition;
(b) in natural populations where food was scarce (McEwan 1968).

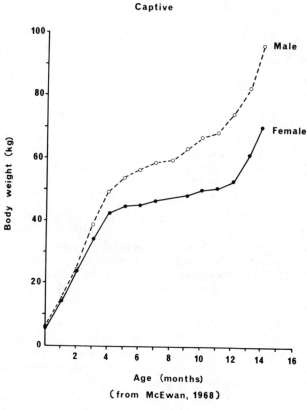

Captive

Body weight (kg)

Male

Female

Age (months)

(from McEwan, 1968)

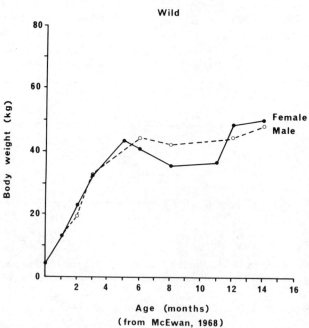

Wild

Body weight (kg)

Female
Male

Age (months)

(from McEwan, 1968)

during the first months of life (see Figure 13.15). Similarly, while infant male pigtailed macaques reared in breeding colonies are typically slightly heavier than females of the same age, male infants rejected by their mothers and raised on an artificial diet grow more slowly than females (Sackett, Holm, et al. 1975). Among human orphans who suffered from acute undernourishment during the later stages of World War II, boys were considerably more affected than girls (Widdowson 1976b).

Do Parents Selectively Deprive One Sex of Resources?

Like variation in sex ratios at birth, sex differences in juvenile mortality may either occur because parents manipulate the sex ratio of their progeny, or they may be caused by differences in the susceptibility of male and female juveniles to adverse environments arising from sexual selection operating on one sex (Clutton-Brock and Iason 1986). Several sex differences may contribute to the reduced viability of males but, in species where males are substantially larger than females, the faster growth rates, lower fat reserves, and greater nutritional requirements of males probably play an important role (Clutton-Brock, Albon, and Guinness 1985).

Two predictions offer some hope of discriminating between these two explanations of differential mortality. First, if sex differences in survival occur because parents selectively eliminate offspring of one sex whose chances of surviving to breed are low, they should do so early in the period of parental investment to minimize wastage of their resources. This prediction is rather less than secure since there may be delays in the effects of cutbacks in investment on mortality of offspring. However, the available evidence indicates that sex differences in mortality are often more pronounced late in the period of parental investment or even after investment is completed. For example, in red deer, sex differences in neonatal mortality are small, and increased mortality among males is most pronounced well after weaning, toward the end of the calf's first or second winter (see Table 13.2). Much of the evidence of differential mortality in other species of birds and mammals shows that it is greatest after parental care is terminated (Clutton-Brock 1986; Clutton-Brock and Iason 1986).

Second, if parents manipulate the sex ratio of their progeny after hatching or birth, we might expect to find some evidence of active discrimination against offspring of one sex. It is clear that sex differences in juvenile mortality can occur in the *absence* of parental discrimination. The results of feeding experiments where male and female juveniles are raised on different planes of nutrition in the absence of their parents show that the sexes often differ in their susceptibility to food shortage. For example, male piglets raised on an inadequate diet in the absence of their parents were four times more likely to die than females (Widdowson 1976b; McCance and Widdowson 1962). The experimental exposure of young pheasants, turkeys, ruffed grouse, and mallards (all dimorphic species) to extreme temperatures showed that males consistently died before females, whereas little or no sex difference in mortality was found in monomorphic species, including bob-white quail and Hungarian partridges (Latham 1947). Higher mortality in juvenile males also occurs in several nidifugous birds that do not feed their young (Wegge 1980; Angelstam, in prep.). Firm evidence that parents actively manipulate sex ratios after hatching or birth must consequently demonstrate that parents treat sons and daughters differently and that these differences are responsible for sex differences in survival. While a number of studies of vertebrates have claimed to demonstrate parental manipulation of the sex ratio after birth or hatching, none have yet been able to meet these criteria. For example, although McClure's (1981) experiments with wood rats are widely cited as evidence of active manipulation of brood sex ratios by the mother, McClure and her coworkers collected no data that allowed them to determine the cause of the lower survival of males (McClure, pers. comm.), and this bias could equally well have arisen because male pups were more susceptible to starvation than females. Similarly, while Howe (1977, 1979) and Burley (1986) interpret differential mortality in chicks (see above) as a consequence of active parental manipulation, in neither case is evidence of parental discrimination yet available.

The absence of evidence of parental discrimination does not necessarily indicate that selection operating on parents has not prevented the evolution of mechanisms to compensate for the increased susceptibility of one sex. However, it is clearly unsafe to assume that sex-biased mortality before or after birth is a consequence of adaptive parental behavior. Unfortunately, the usual practice of describing parents as manipulating the sex ratio easily gives the impression that changes occur as a result of active parental intervention. In the future, studies would do well to avoid any form of words that carries this implication—unless there is clear evidence that changes in the sex ratio are a consequence of parental discrimination between sons and daughters.

13.4 Summary

The extent to which rearing sons and daughters benefits the parent's fitness is likely to vary with environmental conditions as well as with the quality of the parent. Parents might consequently be expected to vary the sex ratio of their progeny by manipulating the sex ratio either at conception or at some later stage of offspring development.

Variation in the sex ratio at hatching or birth is not uncommon among invertebrates. Especially among haplo-diploid species, where unfertilized eggs produce males, there is extensive evidence of adaptive sex ratio control. Variation in hatching sex ratios is also common in some groups of reptiles, where sex determination is temperature-dependent, but the adaptive significance of these trends is not yet clear. Sex ratio variation is apparently commoner in mammals than in birds, but trends are inconsistent and often vary both within and among species.

Except in some invertebrates and reptiles, the causes of underlying variation in the sex ratio at hatching or birth are usually unknown. In mammals, several lines of evidence suggest that harsh environmental conditions can lead to higher mortality of male fetuses and to female-biased sex ratios at birth. Mechanisms of this kind are most likely to account for cases where resource deprivation is associated both with sex ratios below parity and with reduced litter sizes.

Sex differences in juvenile mortality have been documented in a wide variety of birds and mammals. In most species, juvenile males are more likely to die than juvenile females. Like variation in sex ratios at hatching or birth, sex differences in mortality during development have often been interpreted as a consequence of adaptive parental manipulation of the sex ratio: where sons cost more to rear than daughters, parents might prematurely terminate investment if they cannot afford the expenditure necessary to rear them successfully. As predicted, juvenile mortality is often higher among offspring of the more expensive sex. Sex differences increase as food availability declines and are most pronounced in polygynous species showing marked sexual dimorphism in body size. However, except in humans, there is no clear evidence of parental discrimination against brood or litter members of one sex. In addition, sex differences in juvenile mortality commonly extend beyond the period of parental investment and occur among juveniles raised away from their parents. These results suggest that higher mortality rates among male juveniles may commonly be caused by their faster growth rates and higher energetic requirements, which parents may not be able to fulfill.

14 Conclusions

14.1   Introduction

In this chapter, I return to the five questions raised in the Introduction and summarize the principal findings of the relevant chapters. Finally, I assess the prospects for future research, suggesting some lines of work that need to be developed.

14.2   What Are the Benefits and Costs of Parental Care?

The principal benefits of most forms of parental care to the care-giver lie in its effects on the survival, growth, and eventual breeding success of its progeny. There is abundant evidence that these effects can be substantial and that differences in reproductive performance are often associated with variation in early development. Field studies that have been able to follow individuals throughout their lives (Clutton-Brock 1988) show that the effects of variation in parental investment are rarely confined to a single stage of the offspring's life history or a single component of its fitness and can frequently be identified throughout its entire breeding career. Indeed, some studies show that the benefits of parental investment can be transferred across generations, influencing the phenotype and fitness of grandchildren through their effects on the mothers phenotype—a finding that needs to be considered in any attempt to measure the heritability of fitness components in natural populations. Since most studies of the benefits of parental investment are confined to particular stages of the lives of offspring, estimates of the effects of parental care that are based on single components of fitness will often underestimate benefits. Conversely, variation in parental quality may generate (or reinforce) positive correlations between parental expenditure and offspring fitness, leading to overestimates of the importance of parental expenditure.

Though the benefits of parental care are evidently large, we currently know little of the form of relationships between parental expenditure and offspring fitness. It seems likely that these will commonly be nonlinear and that they may sometimes incorporate one or more threshold effects. In addition, we know little about the extent of variability in the effects of parental expenditure in relation to offspring sex, age, or quality or to environmental factors such as temperature, population density, or food availability.

The benefits of parental care are opposed by substantial costs of expenditure to the parent's fitness. Here, too, expenditure can affect many different components of a female's subsequent fitness in addition to her immediate survival—including her subsequent fecundity, the survival and breeding success of subsequent offspring, and, in some cases, the fitness of older sibs. The costs of parental care evidently vary widely, in relation to the quality of the environment and the quality of the parent. For example, comparisons of the costs of raising sons and daughters show that the fitness costs of increased expenditure on one sex may be restricted to certain categories of parents (Figure 13.1).

One of the most obvious conclusions to be drawn is that there are many pitfalls in attempting to measure the costs of reproduction to parents. In particular, individual variation in phenotypic or genetic quality may obscure the effects of parental expenditure on subsequent fitness. Selection can lead to changes in the average phenotypic quality of breeding adults, which can generate correlations between age or breeding experience and reproductive performance. Experimental manipulation of parental expenditure is needed, but both laboratory and field experiments have their own problems: it is difficult to avoid affecting other aspects of the environment that are likely to affect the fitness of parents or offspring or to include measures of all the likely consequences of parental expenditure. In addition, past experience of attempts to measure reproductive costs emphasizes the importance of understanding the physiological mechanisms involved in order avoid misinterpreting correlations between reproductive expenditure and subsequent fitness.

All theoretical models of parental investment necessarily incorporate assumptions about the form of cost and benefit functions. Because of our inability to measure these functions adequately, few predictions concerning the form or extent of parental investment in practice have a secure basis. For example, we do not currently know whether parents should generally invest more heavily in offspring of superior quality because they are more likely to survive or in inferior ones because parental expenditure will have a greater influence on their fitness (see Chapter 9). Almost certainly, parental decisions should vary with environmental factors of which we know virtually nothing. These problems should cause us to be skeptical of predictions based only on theoretical principles, however sensible they appear. Where data fail to conform to predictions, we should be wary of rejecting the empirical results rather than the theory. More importantly, where results confirm predictions, it is important to consider how many other explanations would predict similar trends.

14.3   Why Does the Extent of Parental Care Vary
So Widely Among Species?

Though the absence or extent of parental care in particular cases will continue
to puzzle us, the general answer to this question is clear. Parental care is most
highly developed in circumstances where eggs or young face adverse environments, high rates of predation, parasitism, or intense competition with conspecifics. Relatively large egg size, viviparity or other forms of bearing, guarding of eggs, and feeding of young all appear to be responses to environmental
problems of this kind (Chapters 4 and 5). Environmental conditions also affect
the relative duration of developmental rates, though the size of parents, eggs,
and neonates may constrain the rate of development (Chapter 6).

Although this answer is clear in outline, the factors governing the evolution
of parental care are complex. For example, a considerable number of factors,
including the seasonality of environments and the length of breeding seasons,
the predictability of resources and the degree of relatedness among competitors, are all likely to affect the evolution of parental care and can interact in a
complex fashion.

All other things being equal, organisms would be expected to minimize the
amount of time they spend in the most dangerous stages of development. Here,
predictions have been complicated by unrealistic assumptions (see Chapter 4).
In particular, it has been assumed that embryos which spend a relatively long
time in the egg progress rapidly through the next growth stage, thus enhancing
the advantages of relatively large eggs in species where juveniles show low
survival. However, in many animal groups, large egg size and long developmental periods before hatching are associated with relatively slow rates of
subsequent development.

Viviparity and other forms of bearing among ectotherms are most commonly found in species where newly hatched juveniles would otherwise face
hostile environments. In mammals, the evolution of viviparity was probably
preceded by the evolution of endothermy and lactation. Parental provisioning
may have allowed reduction in egg size, resulting in the development of egg
retention and eventually in viviparity. The absence of viviparity in long-lived
birds with small clutch sizes, and especially in terrestrial species, remains a
mystery.

Parental expenditure during offspring development and the duration of incubation, gestation, and lactation vary widely among species. So far, research has
concentrated primarily on establishing how measures of parental expenditure
and offspring development vary with body size. However, though correlations
with adult size are usually close, there is considerable variation around allometric relationships, and deviations for subsequent stages of development are

often positively correlated with each other (Harvey and Read 1988; Harvey, Read and Promislow 1989; Harvey, Promislow, and Read 1989). Recent comparative studies show that differences in the relative rates of development are often related to environmental variables, including resource availability and predation rates. In mammals, for example, the relative rates of development are highest in species with the highest age-specific mortality rates (Harvey, Promislow, and Read 1989).

Although it is commonly assumed that allometries of development rate are the result of inevitable physiological or developmental constraints, the evidence that this is the case is often weak, and the precise reasons for the form of many allometric relationships are unknown (Clutton-Brock and Harvey 1979; Harvey and Read 1988). At least some allometries may arise from the independent adaptation of body size and life histories to variation in ecology. This poses comparative studies of life histories with a fundamental problem. Should they examine the ecological correlates of the absolute duration of particular life-history stages, accepting the risk that part of the variation is a consequence of differences in body size? Or should they investigate relationships between ecology and the relative lengths of developmental stages when size effects have been removed, accepting the probability that part of the variation attributed to size is a consequence of the adaptation of life histories to differences in environment? At the moment, no general answer is available.

14.4 Why Do Only Females Care for Eggs and Young in Some Species, Only Males in Others, and Both Parents in a Few?

Many of the most fascinating and far-reaching questions about parental care concern the extent to which males and females are involved in parental care. Any satisfactory explanation of interspecific differences in parental care must initially account for the tendency for uniparental male care to be common in fish and to be well represented in amphibia but to be relatively rare in insects, reptiles, and birds. As discussed in Chapter 7, these differences probably arise partly from variation in the relative costs and benefits of care to males and females associated with their different modes of temperature regulation and partly from differences in the incidence of internal fertilization. In those fish where only males care for eggs or young, fertilization is usually external, males typically defend territories or nests where one or more females lay their eggs, and females may prefer mating with males who already have eggs in their nests or territories. Especially where females continue to lay in the territories of guarding males, the costs to males of caring for eggs are likely to be low. In contrast, the costs of parental care to females are likely to be relatively high, especially where they constrain their movements or access to resources or affect their growth and fecundity (see Gross and Sargent 1985). Similar sex differences in the costs of parental care probably explain the relatively high frequency of uniparental male care in frogs and salamanders (see Chapter 7).

Insects and reptiles differ from fish and amphibia in that fertilization is internal, eggs are often hidden, and mating areas and oviposition sites are often spatially separated. As a result, males are less commonly able to defend oviposition sites used by several females in succession, and paternal care is more likely to interfere with a male's mating success. Moreover, internal fertilization probably provides an easy route for the evolution of female care. In birds and mammals, where some form of care is obligatory, the relative costs of care to females may be reduced by the restrictions on female fecundity imposed by large egg size and feeding young, while the costs of care to male mating success are likely to be high.

A satisfactory explanation of the distribution of parental care also needs to account for the substantial number of exceptions to these trends, including the evolution of uniparental female care in externally fertilizing fish and amphibia, and of uniparental male care in internally fertilizing species, as well as in some birds. Female care in externally fertilizing fish is often associated either with circumstances where females are unable to breed repeatedly and the costs of care to females are consequently reduced, or with the defense by males of oviposition sites used simultaneously by several females, where females may need to guard their eggs against predation by conspecifics. In a minority of cases, uniparental female care has probably evolved from biparental care. Here, the benefits of mate desertion may be higher for males, and uniparental female care may be more likely to develop than care by males alone.

Uniparental male care is unusual in internal fertilizers, both among insects, fish, and higher vertebrates. In fish, uniparental male care apparently occurs in conjunction with internal fertilization in species where the evolution of male care has been followed by the evolution of internal fertilization. Under these circumstances, there is little reason to suppose that selection would necessarily favor maternal care.

In birds, males are the only or predominant care-giver in a small number of monogamous or polyandrous shorebirds, as well as in a number of species where males incubate large clutches of eggs to which several females have contributed. The evolution of these breeding systems is not well understood. It has been suggested that uniparental male care in monogamous species occurs where unusually high costs of egg production or high rates of brood loss favor the emancipation of females from parental care but there is little evidence to support this explanation. No single explanation of avian polyandry is satisfactory, and its evolution continues to puzzle behavioral ecologists. However, a number of characteristics of the smaller shorebirds may facilitate the evolution of polyandry, including the involvement of males in parental care in the likely antecedent state (monogamy with biparental care), the relatively small clutch size, and the fact that females are generally larger than males and dominant to them.

The adaptive significance of breeding systems where several females lay in a nest that is subsequently incubated by a single male is also mysterious. Though breeding systems of this kind are common in fish and occur among insects and amphibia, among birds they are common only in the ratites. In fish, females may prefer to add their eggs to existing clutches because this reduces predation risks. In the larger ratites, it may occur becauses clutch size is large while the rate of egg production is low, so that the time necessary for a single female to assemble a large clutch would be dangerously long. Once several females lay in a single nest, a full clutch may represent a larger fraction of the male's lifetime reproductive output than a female's. Care may also be favored in males in order to limit competition among females that might otherwise endanger the male's investment.

Finally, we need to account for the evolution and distribution of biparental care. This occurs in a minority of invertebrates, fish, amphibia and mammals but in a majority of birds. Biparental care is probably uncommon in ectotherms since one parent may usually be almost as effective as two at guarding eggs. Where it does occur, it is usually associated with parental protection of juveniles as well as eggs. There are even some species where eggs are guarded by a single parent but both parents cooperate to guard fry.

Both in fish and in birds, the removal of one parent in normally biparental species is commonly (but not universally) associated with a reduction in juvenile growth or survival. In birds, biparental care may be the norm because young are dependent on parental feeding, growth seasons are relatively short, young are usually accessible to predators, and food is widely and unpredictably distributed. Compared to most other animals, birds are long-distance foragers exploiting unusually scattered resources. Birds in which only females are involved in care mostly either have precocious young that require little or no feeding or exploit food supplies that are highly nutritious or relatively easy to collect. A high proportion of these species are polygynous, though it is not clear whether the emancipation of males from parental care preceded polygynous breeding or vice versa.

In mammals, biparental care is common only among the carnivores and, to a lesser extent, the primates. It is usually associated with monogamy. Where males breed monogamously, the evolution of some degree of male care is unsurprising, and the central problem is to account for the evolution of monogamy. Recent reviews suggest that this is likely to evolve where males can substantially enhance the reproductive rate of their mates by assisting in caring for the offspring (Dunbar 1988; Clutton-Brock 1989a). This represents a likely explanation of the evolution of monogamy in canids where litter sizes are unusually large. However, as yet there is little evidence that male assistance has an important influence on female fecundity or juvenile survival in other species. An alternative possibility is that monogamy represents a form of mate guarding in species where females are long-lived and widely dispersed—as is apparently the case in a number of reef fish.

14.5   To What Extent is Parental Care Adjusted to Variation in Its Benefits to Offspring and Its Costs to Parents?

The fundamental difficulty faced by attempts to predict how parents should vary expenditure on their offspring is that we know so little about the benefits or costs of parental care or about how these vary in relation to differences in environment or in the quality of parents. As a result, many current arguments may be naïve. Nevertheless, patterns of parental care are in agreement with a number of functional predictions. In many species, females discriminate accurately between their own offspring and other juveniles, while expenditure by males varies in relation to their probability of paternity. Parents take greater risks for larger broods and, in at least some cases, for offspring that are nearing independence. Parental expenditure varies in relation to the resources available to the parents as well as with the parents' opportunities for additional breeding attempts.

In contrast, there is little evidence that parents discriminate actively against qualitatively inferior offspring, though they may do so indirectly by failing to interfere in competition among juveniles. Where active parental discrimination has been observed, parents generally favor weaker or younger offspring, presumably in an attempt to maximize the number of young reared. Similarly, there is little firm evidence that parental investment increases with parental age in species where life expectancy declines toward the end of an adult's lifespan.

Conflicts of interest may often prevent parents from achieving their optimal levels of expenditure. First, parents and their offspring will commonly disagree over the level of investment, and in some cases offspring are probably able to extract higher levels of investment than the parent is selected to provide. The level of conflict between parents and their offspring is likely to vary with the nature of the breeding system, the number of care-givers, the size of the brood, and the relatedness of members of the same and different broods. Unfortunately, it is seldom possible to measure the intensity of conflict or to estimate the extent to which offspring force parents to diverge from their optima. One of the few exceptions concerns conflicts of interest over the sex ratio of progeny between parents and workers in eusocial Hymenoptera. Here, optimal levels of investment in males and females are likely to differ between parents (who are related to both by 0.5) and workers (who are related to their sisters by 0.75 and their brothers by 0.25), and it is possible to compare predicted levels of investment in males and females with those observed in natural populations. However, even here, difficulties arise from simplifying assumptions that are not fulfilled, and the effects of conflict cannot be measured precisely.

Second, in species where more than one adult cares for the young, conflicts of interest will arise between care-givers. In general, each parent should seek to maximize its fitness return from each breeding attempt while minimizing its expenditure. For example, in a variety of birds, parents reduce the rate at which they feed their young if their partner's feeding rate is higher than average, and in some species males commonly desert before the end of parental investment. As a result, their behavior may force their mates to increase their investment beyond the level that would otherwise have maximized their fitness.

14.6 How Do Parents Divide Their Investment between Sons and Daughters?

In many polygamous animals, parental expenditure may have a greater influence on the fitness of offspring of one sex than on those of the other. In insects where females are larger than males and a female's fitness is closely related to her fecundity, parents may obtain greater fitness returns per unit expenditure on daughters than per unit expenditure on sons, though, as yet, there is little empirical evidence that this is the case. Conversely, in mammals where a male's mating success is closely related to his size, mothers may obtain greater returns per unit expenditure on sons. In situations of this kind, parents might be expected to invest more heavily in offspring of one sex and there is now considerable evidence that they do so. For example, in some parasitoid wasps, mothers manipulate the sex ratio so as to lay female eggs in large hosts and male eggs in small ones. Conversely, in several polygynous mammals where males are larger than females, male offspring suck more frequently and depress the mother's subsequent reproductive success more than daughters.

In some cases, it is clear that the mother controls the allocation of resources to her sons and daughters—for example, in parasitoid wasps, mothers clearly control the allocation of fertilized eggs (females) to large hosts and unfertilized eggs (males) to small ones. However, in others, offspring rather than their mothers may be responsible for sex differences in investment. In ungulates, for example, mothers do not appear actively to favor males. Instead, the higher suckling rates of males occur because males commonly attempt to suck more frequently than females. In such cases, it is difficult to determine whether sex differences in investment are an adaptive parental strategy or have arisen as a consequence of sexual selection on juveniles. In theory, the way to distinguish between these alternatives would be to compare the observed levels of investment in males and females with optima for parents versus offspring, but, in practice, this is seldom possible.

A second way in which parents might vary their investment in their offspring is by manipulating the sex ratio in relation to their capacity for parental expenditure. Although it is clear that adaptive manipulation of the sex ratio at hatching is widespread among invertebrates, there is little evidence of similar adaptations in birds. In mammals, the situation is complex. In a number of species, sex ratios vary with environmental or parental quality, but few trends are consistent either within or among species. Several different mechanisms are probably involved, including variation in the sex ratio at or soon after conception and differential mortality of male and female embryos during gestation. Though some sex ratio trends are probably adaptive, others may occur because sexual selection favoring rapid growth in males has increased their susceptibility to food shortage or other forms of environmental stress.

Parents may also manipulate the sex ratio after birth by prematurely terminating investment in offspring that they cannot afford to rear. While sex differences in juvenile mortality are common in birds and mammals, there is little evidence that they are a consequence of adaptive parental behavior, for they occur among juveniles raised in the absence of their parents and are usually most pronounced toward or after the end of parental investment. As yet there is no firm evidence that parental discrimination between sons and daughters is responsible for sex differences in juvenile mortality in any species apart from man. In several sexually dimorphic species, juvenile males require more food for growth and survival than females and these differences may be responsible for higher mortality in males when food availability is low.

14.7   Prospects

Perhaps the most obvious conclusion to be drawn from this review concerns the need for improved measures of the costs and benefits of parental care and of the extent to which these vary with environment and parental quality. Without better measures of costs and benefits, theories will continue to be forced to rely on dubious assumptions. At the moment, research on parental investment is impeded by an adherence to models based on first principles. Though logically sound, many of these may bear little relationship to reality.

A second need is for a redistribution of theoretical investigations. In a number of areas, models have rapidly progressed beyond the capacity of empirical studies to check their assumptions or test their predictions. In contrast, other areas have been almost totally neglected. Topics where theoretical investigations are now needed include the evolution of egg size in species where there are positive correlations between the length of subsequent development periods (see Chapter 4); the evolution of uniparental female care in fish and uniparental male care in birds; the reasons why several females may prefer to lay in the same nest; and the evolution of monogamy in species where neither sex cares for eggs or young (Chapters 7 and 8).

Finally, a theme that recurs throughout the book is the need to discriminate clearly between adaptive parental strategies and the by-products of selection for other characteristics. For example, is partial mortality of nestling birds a consequence of adaptive brood reduction by the parent—or of constraints on feeding rate imposed by ecological variation? Are the increased costs of raising males in sexually dimorphic mammals a consequence of adaptive differences in the treatment of sons and daughters by the parent—or of differences in the behavior of male and female juveniles arising from sexual selection for increased growth rates in males? And are higher rates of mortality in male juveniles a consequence of parents prematurely terminating investment in offspring that are unlikely to survive to breed successfully—or of the greater susceptibility of young males to adverse conditions, arising from their faster growth rates and increased energy requirements?

The identification of evolutionary adaptations—Williams's "special and onerous concept"—is often difficult. As Williams makes clear (1966b), it is necessary but not sufficient that traits should increase the fitness of their carriers. Adaptations should also show "design" features whose form and distribution reveal that they have evolved because they increase particular components of fitness. For example, the close relationship between infant killing by males and the probability that they are the father of the young (see Figure 9.2) clearly indicates that infanticide is an adaptive strategy rather than a form of pathological behavior, as is sometimes suggested.

Much of the current evidence of adaptive parental strategies falls short of Williams's criteria because it relies on demographic data (i.e., variation in birth and death rates) alone. Demographic trends can usually arise for a variety of different reasons, and an understanding of the physiological or behavioral mechanisms underlying them is often necessary to interpret them correctly. For example, the lack of evidence of parental discrimination against juveniles of one sex suggests to me that sex differences in juvenile mortality in mammals are not a consequence of parental attempts to manipulate the sex ratio, as Trivers (1972) originally suggested (see Section 13.3). Firm evidence that reductions in food availability caused parents to reject feeding attempts by offspring of one sex until they died from starvation would cause me to revise my opinion.

Three developments would help to advance our understanding of the evolution of parental care. First, we need to restrict the use of active verbs—invest, allocate, manipulate—to cases where there is some evidence of active discrimination by the parent. For example, it is misleading to refer to differential juvenile mortality as sex ratio manipulation unless there is evidence that the parent is responsible for the difference in survival. Similarly, we should not refer to partial mortality of broods as brood reduction unless there is evidence that parents are directly responsible for the death of some of their chicks. To do so begs the question that we are asking.

Second, we should avoid testing theoretical predictions with demographic data alone and should place considerable emphasis on the importance of understanding the causes of variation in the growth and survival of juveniles.

Finally, we should not allow the dazzling perfection of some parental adaptations to lead us to assume that all aspects of parental behavior are equally finely adapted, or to blunt our interest in other factors that affect the development and survival of offspring.

References

Alatalo, R. V., A. Carlson, A. Lundberg, and S. Ulfstrand. 1981. The conflict between male polygamy and female monogamy: The case of the pied flycatcher, *Ficedula hypoleuca. Am. Nat.* 117: 738–753.

Alatalo, R. V., A. Lundberg, and K. Stahlbrandt. 1982. Why do pied flycatcher females mate with already mated males? *Anim. Behav.* 30: 585–593.

Albon, S. D., and T. H. Clutton-Brock. 1988. Climate and the population dynamics of red deer in Scotland. In M. B. Usher and D.B.A. Thompson, eds., *Ecological Changes in the Uplands*, pp. 93–107. Blackwell Scientific Publications, Oxford.

Albon, S. D., T. H. Clutton-Brock, and F. E. Guinness. 1987. Early development and population dynamics in red deer. II. Density independent effects and cohort variation. *J. Anim. Ecol.* 56: 69–82.

Albon, S. D., F. E. Guinness, and T. H. Clutton-Brock. 1983. The influence of climatic variation on the birth weights of red deer calves. *J. Zool.* 200: 295–297.

Albon, S. D., B. Mitchell, and B. W. Staines. 1983. Fertility and body weight in female red deer: A density dependent relationship. *J. Anim. Ecol.* 52: 969–980.

Alexander, G. 1960. Maternal behavior in the merino ewe. *Anim. Prod.* 3: 105–114.

Alexander, R. D. 1974. The evolution of social behavior. *Ann. Rev. Ecol. Syst.* 5: 325–383.

Alexander, R. D. 1979. *Darwinism and Human Affairs*. Pitman, London.

Alexander, R. D., and G. Borgia. 1979. On the origin and basis of the male-female phenomenon. In M. S. Blum and N. A. Blum, eds., *Sexual Selection and Reproductive Competition in Insects*, pp. 417–440. Academic Press, New York.

Alexander, R. D., and P. W. Sherman. 1977. Local mate competition and parental investment in social insects. *Science* 196: 494–500.

Allan, J. D. 1984. Life history variation in a freshwater copepod: Evidence from population crosses. *Evolution* 38: 280–291.

Allden, W. G. 1970. The effects of nutritional deprivation on the subsequent productivity of sheep and cattle. *Nutrition Abstracts and Reviews* 40: 1167–1185.

Altmann, J. 1980. *Baboon Mothers and Infants*. Harvard University Press, Cambridge, Mass.

Altmann, J. 1983. Costs of reproduction in baboons. In W. P. Aspey and S. I. Lustick, eds., *Behavioral Energetics: The Cost of Survival of Vertebrates*, pp. 67–88. Ohio State University Press, Columbus.

Altmann, J., S. A. Altmann, and G. Hausfater. 1978. Primate infant's effects on mother's future reproduction. *Science* 201: 1028–1030.

Altmann, J., G. Hausfater, and S. A. Altmann. 1988. Determinants of reproductive success in savannah baboons. In T. H. Clutton-Brock, ed., *Reproductive Success*, pp. 403–418. University of Chicago Press, Chicago.

Altmann, S. A., S. S. Wagner, and S. Lennington. 1977. Two models for the evolution of polygyny. *Behav. Ecol. Sociobiol.* 2: 397–410.

Amoroso, E. C. 1960. Viviparity in fishes. *Symp. Zool. Soc. Lond.* 1: 153–183.

Amoroso, E. C. 1968. The evolution of viviparity. *Proc. Roy. Soc. Med.* 61: 1188–1200.

Anderson, D. J. 1989a. Differential responses of boobies and other seabirds in the Galapagos during the 1986/1987 El Niño/Southern Oscillation event. *Marine Ecology Progress Series* 52: 209–216.

Anderson, D. J. 1989b. The role of hatching asynchrony in siblicidal brood reduction of two booby species. *Behav. Ecol. Sociobiol.* 5: 363–368.

Anderson, D. J., N. C. Stoyan, and R. E. Ricklefs. 1987. Why are there no viviparous birds? A comment. *Am. Nat.* 130: 941–947.

Anderson, J. 1953. Analysis of roe deer population (*Capreolus capreolus* L.) based on extermination of the total stock. *Dan. Rev. Game Biol.* 2: 127–155.

Anderson, S. S., and M. Fedak. 1987. Grey seal *Halichoerus grypus* energetics: Females invest more in male offspring. *J. Zool.* 211: 667–679.

Andersson, M. 1984. Brood parasitism within species. In C. J. Barnard, ed., *Producers and Scroungers: Strategies of Exploitation and Parasitism*, pp. 195–228. Croom Helm, London.

Andersson, M., C. Wiklund, and H. Rundgren. 1980. Parental defence of offspring: A model and an example. *Anim. Behav.* 28: 536–542.

Andersson, S., and C. G. Wiklund. 1987. Sex role partitioning in the rough-legged buzzard *Buteo lagopus*. *Ibis* 129: 103–107.

Angelstam, P. 1984. Sexual and seasonal differences in mortality of the black grouse *Tetrao tetrix* in boreal Sweden. *Ornis. Scand.* 5: 123–124.

Ankney, C. D. 1980. Egg weight, survival and growth of lesser snow goose goslings. *J. Wildl. Mgmt.* 44: 174–182.

Ankney, C. D. 1982. Sex ratio varies with egg sequence in lesser snow geese. *Auk* 99: 662–666.

Ar, A., and Y. Yom-Tov. 1978. The evolution of parental care in birds. *Evolution* 32: 655–669.

Armitage, K. B. 1987. Do female yellow-bellied marmots adjust the sex ratio of their offspring? *Am. Nat.* 129: 501–519.

Armstrong, T., and R. J. Robertson. 1988. Parental investment based on clutch value: Nest desertion in response to partial clutch size in dabbling ducks. *Anim. Behav.* 36: 941–943.

Ashkenazie, S., and U. N. Safriel. 1979. Time energy budget of the semi-palmated sandpiper *Calidris pusilla* at Barrow, Alaska. *Ecology* 70: 783–799.

Askenmo, C. 1977. Effects of addition and removal of nestlings on nestling weight, nestling survival and female weight loss in the pied flycatcher *Ficedula hypoleuca*. *Ornis. Scand* 8: 1–8.

Askenmo, C. 1979. Reproductive effort and the return rate of male pied flycatchers. *Am. Nat.* 114: 748–753.

Austad, S. N., and M. E. Sunquist. 1986. Sex ratio manipulation in the common opossum. *Nature* 324: 58–60.

Austad, S. N., and R. Thornhill. 1986. Female reproductive variation in a nuptial-feeding spider *Pisaura mirabilis*. *Bull. Br. Arachnol. Soc.* 7: 48–52.

Avilés, L. 1986. Sex ratio bias and possible group selection in the social spider *Anelosimus eximius*. *Am. Nat.* 128: 1–12.

Baccetti, B., ed. 1970. *Comparative Spermatology*. Academic Press, New York.

Bagenal, T. B. 1969. The relationship between food supply and fecundity in brown trout *Salmo trutta* L. *J. Fish. Biol.* 1: 167–182.

Bagenal, T. B. 1971. The interrelation of the size of fish eggs, the date of spawning and the production cycle. *J. Fish. Biol.* 3: 207–219.

Bakker, K. 1959. Feeding period, growth and pupation in larvae of *Drosophila melanogaster. Ent. Exp. & Appl.* 2: 171–186.

Bales, K. B. 1980. Cumulative scaling of paternalistic behavior in primates. *Am. Nat.* 116: 454–461.

Balfour, E., and J. C. Cadbury. 1979. Polygyny, spacing and the sex ratio among hen harriers *Circus cyaneus* in Orkney, Scotland. *Ornis. Scand.* 30: 6–12.

Balon, E. K. 1975. Reproductive guilds of fishes: A proposal and definition. *J. Fish. Res. Bd. Can.* 32: 821–864.

Balon, E. K. 1978. Reproductive guilds and the ultimate structure of fish taxocenes: Amended contribution to the discussion presented at a mini-symposium. *Env. Biol. Fish.* 3: 149–152.

Balon, E. K. 1984. Patterns in the evolution of reproductive style in fishes. In G. W. Potts and R. J. Wootton, eds., *Fish Reproduction: Strategies and Tactics*, pp. 35–53. Academic Press, London.

Bancroft, G. T. 1984. Patterns of variation in size of boat-tailed grackle *Quiscalus major* eggs. *Ibis* 125: 496–509.

Bancroft, G. T. 1985. The influence of total nest failures and partial losses on the evolution of asynchronous hatching. *Am. Nat.* 126: 495–504.

Barash, D. P. 1975. Evolutionary aspects of parental behavior: Distraction behavior of the Alpine accentor. *Wilson Bull.* 87: 367–373.

Barlow, G. W. 1972. A paternal role for bulls of the Galapagos Islands sea lion. *Evolution* 26: 307–308.

Barlow, G. W. 1974. Contrasts in social behavior between Central American cichlid fishes and coral-reef surgeon fishes. *Am. Zool.* 14: 9–34.

Barlow, G. W. 1981. Patterns of parental investment, dispersal and size among coral-reef fishes. *Env. Biol. Fish.* 1: 65–85.

Barlow, G. W. 1984. Patterns of monogamy among teleost fishes. *Arch. Fischwiss.* 35: 75–123.

Bart, J., and A. Tornes. 1989. Importance of monogamous male birds in determining reproductive success. *Behav. Ecol. Sociobiol.* 24: 109–116.

Bateman, A. J. 1948. Intrasexual selection in *Drosophila. Heredity* 2: 349–368.

Batra, L. R. 1963. Ecology of ambrosia fungi and their dissemination by beetles. *Transactions of the Kansas Academy of Science* 66: 213–236.

Bauwens, D., and C. Thoen. 1981. Escape tactics and vulnerability to predation associated with reproduction in the lizard *Lacerta vivipara. J. Anim. Ecol.* 50: 733–743.

Baylis, J. R. 1974. The behaviour and ecology of *Herotilapia multispinosa* (Teleostei: Cichlidae). *Z. Tierpsychol.* 35: 114–146.

Baylis, J. R. 1978. Paternal behaviour in fishes: A question of investment, timing or rate? *Nature* 276: 738.

Baylis, J. R. 1981. The evolution of parental care in fishes, with reference to Darwin's rule of male sexual selection. *Env. Biol. Fish.* 6: 223–251.

Beacham, T. D., and C. B. Murray. 1985. Effect of female size, egg size, and water temperature on developmental biology of Chum salmon (*Oncorhynchus keta*) from the Nitirat River, British Columbia. *Can. J. Fish. Aquat. Sci.* 42: 1755–1765.

Beams, H. W., and R. K. Meyer. 1931. The formation of pigeon "milk". *Physiol. Zool.* 4: 486–500.

Beatty, R. A. 1971. Phenotype of spermatozoa in relation to their content. In C. A. Kiddy and H. A. Haafs, eds., *Sex Ratio at Birth—Prospects for Control*, pp. 10–18. American Society of Animal Science, Philadelphia.

Beebe, W. 1925. The variegated tinamou, *Crypturellus variegatus variegatus*. *Zoologica* 6: 195–227.

Beecher, M. D., I. M. Beecher, and S. Lumpkin. 1981. Parent-offspring recognition in bank swallows (*Riparia riparia*). 1. Natural history. *Anim. Behav.* 29: 86–94.

Beehler, B. 1983. Frugivory and polygamy in birds of paradise. *Auk* 100: 1–12.

Beer, C. G. 1970. Individual recognition of voice in the social behavior of birds. In J. Rosenblatt, C. Beer, and R. A. Hinde, eds., *Advances in the Study of Behavior*, vol. 3, pp. 27–74. Academic Press, New York.

Begon, M., and G. A. Parker. 1986. Should egg size and clutch size decrease with age? *Oikos* 47: 293–302.

Beissinger, S. R. 1987. Mate desertion and reproductive effort in the snail kite. *Anim. Behav.* 35: 1504–1519.

Beissinger, S. R., and N.F.R. Snyder. 1987. Mate desertion in the snail kite. *Anim. Behav.* 35: 477–487.

Bekoff, M., T. J. Daniels, and J. L. Gittleman. 1984. Life history patterns and the comparative social ecology of carnivores. *Ann. Rev. Ecol. Syst.* 15: 191–232.

Bekoff, M., J. Diamond, and J. B. Milton. 1981. Life history patterns and sociality in canids: Body size, reproduction and behavior. *Oecologia* 50: 386–390.

Belding, D. L. 1934a. The spawning habits of the Atlantic salmon. *Trans. Amer. Fish. Soc.* 64: 211–216.

Belding, D. L. 1934b. The cause of high mortality in the Atlantic salmon after spawning. *Trans. Amer. Fish. Soc.* 64: 219–222.

Bell, G. 1980. The costs of reproduction and their consequences. *Am. Nat.* 116: 45–76.

Bell, G. 1984a. Measuring the cost of reproduction. I. The correlation structure of the life table of a plankton rotifer. *Evolution* 38: 300–313.

Bell, G. 1984b. Measuring the cost of reproduction. II. The correlation structure of the life tables of five freshwater invertebrates. *Evolution* 38: 314–326.

Bell, G. 1986. Reply to Reznick et al. 1986. *Evolution* 40: 1344–1346.

Bell, G., and V. Koufopanou. 1986. The cost of reproduction. In R. Dawkins, ed., *Oxford Surveys of Evolutionary Biology*, pp. 83–131. Oxford University Press, Oxford.

Bellairs, A. D'A. 1970. *The Life of Reptiles*, vol. 2. Universe Books, New York.

Benford, F. A. 1978. Fisher's theory of the sex ratio applied to the social Hymenoptera. *J. theoret. Biol.* 72: 701–727.

Bengtsson, B. D. 1977. Evolution of sex ratio in the wood lemming. In F. B. Christiansen and T. M. Fenchel, eds., *Measuring Selection in Natural Populations*, pp. 333–343. Springer-Verlag, Berlin.

Bengtsson, H., and O. Ryden. 1981. Development of parent-young interaction in asynchronously hatched broods of altricial birds. *Z. Tierpsychol.* 56: 255–272.

Bengtsson, H., and O. Ryden. 1983. Parental feeding rate in relation to begging behavior in asynchronously hatched broods of the great tit (*Parus major*). *Behav. Ecol. Sociobiol.* 12: 243–251.

Bennett, P. M. 1986. Comparative studies of morphology, life history and ecology among birds. Ph.D diss., University of Sussex, England.

Bennett, P. M., and P. H. Harvey. 1985a. Brain size, development and metabolism in birds and mammals. *J. Zool.* 207: 491–509.

Bennett, P. M., and P. H. Harvey. 1985b. Relative brain size and ecology in birds. *J. Zool.* 207: 151–169.

Bennett, P. M., and P. H. Harvey. 1987. Active and resting metabolism in birds: Allometry, phylogeny and ecology. *J. Zool.* 213: 327–363.

Ben Shaul, D. M. 1962. The composition of the milk of wild animals. *Int. Zoo Yb.* 4: 333–342.

Berger, J. 1979. Weaning conflict in desert and mountain bighorn sheep (*Ovis canadensis*): An ecological interpretation. *Z. Tierpsychol.* 50: 188–200.

Bergerud, A. T. 1971. The population dynamics of Newfoundland caribou. *Wildl. Monogr.* 25: 1–55.

Berglund, A., and G. Rosenqvist. 1986. Reproductive costs in the prawn *Palaemon adspersus*: Effects on growth and predator vulnerability. *Oikos* 46: 349–354.

Berglund, A., G. Rosenqvist, and I. Svensson. 1986. Reversed sex roles and parental energy investment in zygotes of two pipefish (Syngnathidae) species. *Mar. Ecol. Progress* 29: 209–215.

Berglund, A., G. Rosenqvist, and I. Svensson. 1989. Reproductive success of females limited by males in two pipefish species. *Am. Nat.* 133: 506–516.

Berman, C. M. 1980. Early agonistic experience and rank acquisition among free-ranging rhesus monkeys. *Int. J. Primatol.* 1: 153–178.

Berman, C. M. 1988. Maternal condition and offspring sex ratio in a group of free-ranging rhesus monkeys: An eleven-year study. *Am. Nat.* 131: 307–328.

Berman, C. M., and R. Rawlins. 1985. Maternal dominance, sex ratio and fecundity in one social group on Cayo Santiago. *Am. J. Primatol.* 8: 332–333.

Bertram, B.C.R. 1978. Breeding system and strategies of ostriches. *Proc. 17th International Ornithological Congress*, West Berlin.

Bertram, B.C.R. 1979. Ostriches recognize their own eggs and discard others. *Nature* 279: 233–234.

Best, L. B. 1977. Field sparrow nesting biology. *Auk* 94: 188–200.

Biermann, G. C., and R. J. Robertson. 1981. An increase in parental investment during the breeding season. *Anim. Behav.* 29: 487–489.

Biermann, G. C., and S. G. Sealy. 1982. Parental feeding of nestling yellow warblers in relation to brood size and prey availability. *Auk* 99: 332–341.

Biermann, G. C., and R. J. Robertson. 1983. Residual reproductive value and parental investment. *Anim. Behav.* 31: 311–312.

Bird, E., and R. J. Contreras. 1986. Maternal dietary sodium chloride levels affect the sex ratio in rat litters. *Physiology and Behaviour* 36: 307–310.

Bjorklund, M., and B. Westman. 1986. Adaptive advantages of monogamy in the great tit (*Parus major*): An experimental test of the polygyny threshold model. *Anim. Behav.* 34: 1436–1440.

Black, R. 1971. Hatching success in the three-spined stickleback (*Gasterosteus aculeatus*) in relation to changes in behaviour during the parental phase. *Anim. Behav.* 19: 532–541.

Blackburn, D. G. 1982. Evolutionary origins of viviparity in the Reptilia. I. Sauria. *Amph.-Rept.* 3: 185–205.

Blackburn, D. G., and H. E. Evans. 1986. Why are there no viviparous birds? *Am. Nat.* 128: 165–190.

Blancher, P. J., and R. J. Robertson. 1982. Kingbird aggression: Does it deter predation? *Anim. Behav.* 30: 929–930.

Blaxter, J.H.S. 1969. Development: Eggs and larvae. In W. S. Hoar and D. J. Randall, eds., *Fish Physiology*, vol. 3, pp. 177–252. Academic Press, New York.

Blaxter, K. L. 1964. Protein metabolism and requirements in pregnancy and lactation. In H. N. Munro and J. B. Allison, eds., *Mammalian Protein Metabolism*, pp. 172–223. Academic Press, London.

Blick, J. 1977. Selection for traits which lower individual reproduction. *J. theoret. Biol.* 67: 597–601.

Blueweiss, L., H. Fox, V. Kudzma, D. Nakashima, R. Peters, and S. Sams. 1978. Relationships between body size and some life history parameters. *Oecologia* 37: 257–272.

Blumer, L. S. 1979. Male parental care in the bony fishes. *Q. Rev. Biol.* 54: 149–161.

Blumer, L. S. 1985. The significance of biparental care in the brown bullhead, *Ictalurus nebulosus*. *Environ. Biol. Fish.* 12: 231–236.

Blumer, L. S. 1986. Parental care sex differences in the brown bullhead *Ictalurus nebulosus* (Pisces, Ictaluridae). *Behav. Ecol. Sociobiol.* 19: 97–104.

Blurton-Jones, N. G. 1986. Bushman birth spacing: A test for optimal interbirth intervals. *Ethol. Sociobiol.* 7: 91–105.

Blurton-Jones, N. G., and E. da Costa. 1987. A suggested value of toddler night waking: Delaying the birth of the next sibling. *Ethol. Sociobiol.* 8: 135–142.

Boersma, P. D. 1982. Why some birds take so long to hatch. *Am. Nat.* 120: 733–750.

Boersma, P. D., N. T. Wheelwright, M. K. Nerini, and E. S. Wheelwright. 1982. The breeding biology of the fork-tailed storm petrel (*Oceanodroma furcata*). *Auk* 97: 268–282.

Boggs, C. L., and L. E. Gilbert. 1979. Male contribution to egg production in butterflies: Evidence for transfer of nutrients at mating. *Science* 206: 83–84.

Bonner, W. N. 1984. Lactation strategies in pinnepeds: Problems for a marine mammal. *Symp. Zool. Soc. Lond.* 51: 253–272.

Boomsma, J. J. 1989. Sex-investment ratios in ants: Has female bias been systematically overestimated? *Am. Nat.* 133: 517–532.

Boomsma, J. J., and J. A. Isaaks. 1985. Energy investment and respiration in queens and males of *Lasius niger* (Hymenoptera: Formicidae). *Behav. Ecol. Sociobiol.* 18: 19–27.

Borg, K. 1971. On mortality and reproduction of roe deer in Sweden during the period 1948–1969. *Viltrevy* 1971: 121–146.

Borgia, G. 1979. Sexual selection and the evolution of mating systems. In M. S. Blum and N. A. Blum, eds., *Sexual Selection and Reproductive Competition in Insects*. Academic Press, New York.

Bortolotti, G. R. 1986. Influence of sibling competition on nestling sex ratios of sexually dimorphic birds. *Am. Nat.* 127: 495–507.

Boucher, D. H. 1977. On wasting parental investment. *Am. Nat.* 111: 786–788.

Boulenger, G. A. 1912. Observations sur l'accouplement et la ponte de l'Alyte accoucheur, *Alytes obstetricans. Bull. Class. Sci. Acad. Belgique* 1912: 570–579.

Bourke, A.F.G. 1989. Comparative analysis of sex-investment ratios in slave-making ants. *Evolution* 43: 913–918.

Bowen, B. J., C. G. Codd, and D. T. Gwynne. 1984. The katydid spermatophore (Orthoptera: Tettigoniidae): Male nutrient investment and its fate in the mated female. *Aust. J. Zool.* 32: 23–31.

Bowen, W. D., O. T. Oftedal, and D. J. Boness. 1985. Birth to weaning in 4 days: Remarkable growth in the hooded seal. *Can. J. Zool.* 63: 2841–2846.

Boyce, M. S. 1984. Restitution of r- and K-selection as a model of density-dependent natural selection. *Ann. Rev. Ecol.* 15: 427–448.

Boyce, M. S., ed. 1988. *Evolution of Life-Histories: Pattern and Theory from Mammals.* Yale University Press, New Haven, Conn.

Boyd, I. L., and T. S. McCann. 1989. Pre-natal investment in reproduction by female Antarctic fur seals. *Behav. Ecol. Sociobiol.* 24: 377–385.

Bradbury, J. W., and S. L. Vehrencamp. 1977. Social organization and foraging in emballonurid bats. IV. Parental investment patterns. *Behav. Ecol. Sociobiol.* 2: 19–29.

Brattstrom, B. H. 1974. The evolution of reptilian social behavior. *Amer. Zool.* 14: 35–49.

Breder, C. M., and D. E. Rosen. 1966. *Modes of Reproduction in Fishes.* Natural History Press, Garden City, N.Y.

Breitwisch, R. 1988. Sex differences in defence of eggs and nestlings by northern mocking birds, *Mimus polyglottos. Anim. Behav.* 36: 62–72.

Breitwisch, R., P. G. Merritt, and G. H. Whitesides. 1986. Parental investment by the northern mocking bird: Male and female roles in feeding nestlings. *Auk* 103: 152–159.

Brockelman, W. Y. 1975. Competition, the fitness of offspring and optimal clutch size. *Am. Nat.* 109: 677–699.

Brodie, E. D. III. 1989. Behavioral modification as a means of reducing the cost of reproduction. *Am. Nat.* 134: 225–238.

Brody, M. S., and L. R. Lawlor. 1984. Adaptive variation in offspring size in the terrestrial isopod *Armadillium vulgare. Oecologia* 61: 55–59.

Brody, S. 1945. *Bioenergetics and Growth.* Reinhold van Nostrand, New York.

Bro Larsen, E. 1952. On subsocial beetles from the saltmarsh, their care of progeny and adaptation to salt and tide. *Trans. Eleventh Int. Congress of Entomology*, pp. 502–506.

Bronson, F. H. 1979. The reproductive ecology of the house mouse. *Q. Rev. Biol.* 54: 265–299.

Bronson, F. H., and F. A. Marsteller. 1985. Effect of short-term food deprivation on reproduction in female mice. *Biol. Reprod.* 33: 660–667.

Brooks, R. J. 1984. Causes and consequences of infanticide in populations of rodents.

In G. Hausfater and S. B. Hrdy, eds., *Infanticide: Comparative and Evolutionary Perspectives*, pp. 331–348. Aldine, Chicago.

Broom, R. 1895. Note on the period of gestation in echidna. *Proc. Linn. Soc. New South Wales* 10: 576–577.

Brown, J. L. 1987. *Helping and Communal Breeding in Birds*. Princeton University Press, Princeton, N.J.

Brown, T. H. 1959. Parasitism in the ewe and the lamb. *J. Brit. Grass. Soc.* 14: 216–220.

Brown, T. H. 1964. The early weaning of lambs. *J. Agric. Soc.* 14: 191–204.

Browne, R. A. 1982. The costs of reproduction in brine shrimp. *Ecology* 63: 43–47.

Bruning, D. F. 1974a. Social structure and reproductive behavior of the greater rhea. *Living Bird* 13: 251–294.

Bruning, D. F. 1974b. Social structure and reproductive behavior of the Argentine grey rhea, *Rhea americanus albescens*. Ph.D. diss., University of Colorado, Boulder.

Brunson, R. R. 1949. The life history and ecology of two North American gastrotrichs. *Trans. Amer. Microscop. Soc.* 68: 1–20.

Bryant, D. M. 1978. Establishment of weight hierarchies in the broods of house martins *Delichon urbica*. *Ibis* 120: 16–26.

Bryant, D. M. 1979. Reproductive costs in the house martin. *J. Anim. Ecol.* 48: 655–675.

Bryant, D. M. 1983a. Short-term variability in energy turnover by breeding house martins *Delichon urbica*: A study using doubly labelled water. *J. Anim. Ecol.* 52: 525–543.

Bryant, D. M. 1983b. Time and energy limits to brood size in house martins. *J. Anim. Ecol.* 52: 905–925.

Bryant, D. M., and P. Tatner. 1990. Hatching asynchrony, sibling competition and siblicide in nestling birds: Studies of swiftlets and bee-eaters. *Anim. Behav.* 39: 657–671.

Bull, J. J. 1980. Sex determination in reptiles. *Q. Rev. Biol.* 55: 3–21.

Bull, J. J. 1983. *Evolution of Sex Determining Mechanisms*. Benjamin Cummings, Menlo Park, Calif..

Bull, J. J. 1985. Models of parent-offspring conflict: Effect of environmental variance. *Heredity* 55: 1–8.

Bull, J. J., and M. G. Bulmer. 1981. The evolution of XY females in mammals. *Heredity* 47: 347–365.

Bull, J. J., and R. C. Vogt. 1979. Temperature dependent sex determination in turtles. *Science* 206: 1186–1188.

Bull, J. J., R. C. Vogt, and C. J. McCoy. 1982. Sex determining temperatures in emydid turtles: A geographic comparison. *Evolution* 36: 326–332.

Bulmer, M. G. 1986. Sex ratios in geographically structured populations. *Trends Ecol. Evol.* 1: 35–38.

Bulmer, M. G., and P. D. Taylor. 1980. Dispersal and the sex ratio. *Nature* 284: 448–449.

Bulmer, M. G., and P. D. Taylor. 1981. Worker-queen conflict and sex ratio theory in social Hymenoptera. *Heredity* 47: 197–207.

Bunnell, F. L. 1987. Reproductive tactics of Cervidae and their relationships to habitat.

In C. W. Wemmer, ed., *Biology and Management of the Cervidae*, pp. 145–167. Smithsonian Institution Press, Washington, D.C.

Burchard, J. 1965. Family structure in the dwarf cichlid *Apistogramma trifasciatum* Eigenmann and Kennedy. *Z. Tierpsychol.* 22: 150–162.

Burfening, P. J. 1972. Prenatal and postnatal competition among twin lambs. *Anim. Prod.* 15: 61–66.

Burger, J. 1981. On becoming independent in herring gulls: Parent-young conflict. *Am. Nat.* 117: 444–456.

Burke, T., N. B. Davies, M. W. Bruford, and B. J. Hatchwell. 1989. Parental care and mating behaviour of polyandrous dunnocks *Prunella modularis* related to paternity by DNA finger-printing. *Nature* 338: 249–251.

Burley, N. 1977. Parental investment, mate choice and mate quality. *Proc. Nat. Acad. Sci.* 74: 3476–3479.

Burley, N. 1981. Sex ratio manipulation and selection for attractiveness. *Science* 211: 721–722.

Burley, N. 1982. Facultative sex ratio manipulation. *Am. Nat.* 120: 81–107.

Burley, N. 1986. Sex ratio manipulation in color-banded populations of zebra finches. *Evolution* 40: 1191–1206.

Burley, N. 1988. The differential-allocation hypothesis: An experimental test. *Am. Nat.* 132: 611–628.

Buskirk, R. E., C. Frohlich, and K. G. Ross. 1984. The natural selection of sexual cannibalism. *Am. Nat.* 123: 612–625.

Busse, C. D. 1985. Paternity recognition in multi-male primate groups. *Amer. Zool.* 25: 873–881.

Busse, C., and W. Hamilton III. 1981. Infant carrying by male chacma baboons. *Science* 212: 1281–1283.

Busse, K. 1970. Care of the young by male *Rhinoderma darwini*. *Copeia* 1970: 395–402.

Buxton, P. A. 1955. *The Natural History of Tsetse flies*. Lewis & Co., London.

Byers, J. A., and J. B. Moodie. 1990. Sex specific maternal investment in pronghorn and the question of a limit on differential provisioning in ungulates. *Behav. Ecol. Sociobiol.* 26: 157–164.

Calder, W. A. III. 1979. The kiwi egg and egg design: Evolution as a package deal. *Bioscience* 29: 461–467.

Calder, W. A. III. 1984. *Size, Function and Life History*. Harvard University Press, Cambridge, Mass.

Calder, W. A., and B. Rowe. 1977. Body mass changes and energetics of the kiwi's egg cycle. *Notornis* 24: 129–135.

Caley, M.H.J., S. Boutin, and R. A. Moses. 1988. Male-biased reproduction and sex ratio adjustment in musk rats. *Oecologia* 74: 501–506.

Calow, P. 1973. The relationship between fecundity, phenology and longevity: A systems approach. *Am. Nat.* 107: 559–574.

Calow, P. 1977. Ecology, evolution and energetics: A study in metabolic adaptation. *Adv. Ecol. Res.* 10: 1–61.

Calow, P. 1979. The cost of reproduction—a physiological approach. *Biol. Rev.* 54: 23–40.

Calow, P., and A. S. Woolhead. 1977. The relationship between ration, reproductive effort and age-specific mortality in the evolution of life history strategies—some observations on freshwater triclads. *J. Anim. Ecol.* 46: 765–782.

Capinera, J. L. 1979. Qualitative variation in plants and insects: Effect of propagule size on ecological plasticity. *Am. Nat.* 114: 350–361.

Capinera, J. L., and P. Barbosa. 1977. Influence of natural diets and larval density on gypsy moth *Lymantria dispar* (Lepidoptera: Orgyiidae) egg mass characteristics. *Can. J. Entomol.* 109: 1313–1318.

Carey, C., H. Rahn, and P. Parisi. 1980. Calories, lipid, water and yolk in avian eggs. *Condor* 82: 335–343.

Carlisle, T. R. 1982. Brood success in variable environments: Implications for parental care allocation. *Anim. Behav.* 30: 824–836.

Carlisle, T. R. 1985. Parental response to brood size in a cichlid fish. *Anim. Behav.* 33: 234–238.

Carlson, A. 1989. Courtship feeding and clutch size in red-backed shrikes (*Lanio collourio*). *Am. Nat.* 133: 454–457.

Carson, H. L., and A. R. Templeton. 1984. Genetic revolutions in relation to speciation phenomena: The founding of new populations. *Ann. Rev. Ecol.* 15: 97–131.

Case, T. J. 1978. On the evolution and adaptive significance of postnatal growth rates in the terrestrial vertebrates. *Q. Rev. Biol.* 55: 243–282.

Chadwick, A. 1977. Comparison of milk-like secretions found in non-mammals. *Symp. Zool. Soc. Lond.* 41: 341–358.

Chapais, B. 1983. Dominance, relatedness and the structure of female relationships in rhesus monkeys. In R. A. Hinde, ed., *Primate Social Relationships: An Integrated Approach*, pp. 208–217. Blackwell Scientific Publications, Oxford.

Charlesworth, B. 1980. *Evolution in Age-structured Populations*. Cambridge University Press, Cambridge.

Charlesworth, B., and J. A. Leon. 1976. The relation of reproductive effort to age. *Am. Nat.* 110: 449–459.

Charnov, E. L. 1982. *The Theory of Sex Allocation*. Princeton University Press, Princeton, N.J.

Charnov, E. L., and J. R. Krebs. 1974. On clutch size and fitness. *Ibis* 116: 217–219.

Charnov, E. L., R. L. Los-den Hartogh, W. T. Jones, and J. van den Assem. 1981. Sex ratio evolution in a variable environment. *Nature* 289: 27–33.

Chase, I. D. 1980. Cooperative and non-cooperative behavior in animals. *Am. Nat.* 115: 827–857.

Chen, L. C., E. Huq, and S. D'Souza. 1981. Sex bias in family allocation of food and health care in rural Bangladesh. *Population and Development Review* 7: 55–70.

Cheney, D. L., and R. M. Seyfarth. 1986. The recognition of social alliances by vervet monkeys. *Anim. Behav.* 34: 1722–1731.

Child, G. 1972. Observations on a wildebeest die-off in Botswana. *Arnoldia* 31: 1–13.

Christiansen, F. B., and T. M. Fenchel. 1979. Evolution of marine invertebrate reproductive patterns. *Theor. Popul. Biol.* 16: 267–282.

Christie, J. R. 1929. Some observations of sex in the Mermithidae. *J. Exp. Zool.* 53: 59–76.

Clark, A. B. 1978. Sex ratio and local resource competition in a prosimian primate. *Science* 201: 163–165.

Clark, A. B., and D. S. Wilson. 1981. Avian breeding adaptations: Hatching asyn-chrony, brood reduction and nest failure. *Q. Rev. Biol.* 56: 253–277.

Clark, M. M., S. Bone, and B. G. Galef, Jr. 1990. Evidence of sex-biased postnatal maternal investment by Mongolian gerbils. *Anim. Behav.* 39: 735–744.

Clutton-Brock, T. H. 1984. Reproductive effort and terminal investment in iteroparous animals. *Am. Nat.* 123: 212–229.

Clutton-Brock, T. H. 1985. Birth sex ratios and the reproductive success of sons and daughters. In P. J. Greenwood, P. H. Harvey, and M. Slatkin, eds., *Evolution: Essays in Honour of John Maynard Smith*, pp. 221–236. Cambridge University Press, Cambridge.

Clutton-Brock, T. H. 1986. Sex ratio variation in birds. *Ibis* 128: 317–329.

Clutton-Brock, T. H. 1988. Reproductive success. In T. H. Clutton-Brock, ed., *Reproductive Success*, pp. 472–486. University of Chicago Press, Chicago.

Clutton-Brock, T. H. 1989a. Mammalian mating systems. *Proc. Roy. Soc. B.* 235: 339–372.

Clutton-Brock, T. H. 1989b. Female transfer, male tenure and inbreeding avoidance in social mammals. *Nature* 337: 70–72.

Clutton-Brock, T. H., and S. D. Albon. 1982. Parental investment in male and female offspring in mammals. In Kings College Sociobiology Group, eds., *Current Problems in Sociobiology*, pp. 223–247. Cambridge University Press, Cambridge.

Clutton-Brock, T. H., and P. H. Harvey. 1976. Evolutionary rules and primate societies. In P.P.G. Bateson and R. A. Hinde, eds., *Growing Points in Ethology*, pp. 195–237. Cambridge University Press, Cambridge.

Clutton-Brock, T. H., and P. H. Harvey. 1984. Comparative approaches to investigating adaptation. In J. R. Krebs and N. B. Davies, eds. *Behavioural Ecology: An Evolutionary Approach*, pp. 7–29. Blackwell Scientific Publications, Oxford.

Clutton-Brock, T. H., and P. H. Harvey. 1979. Comparison and adaptation. *Proc. R. Soc. Lond. B* 205: 547–565.

Clutton-Brock, T. H., and G. R. Iason. 1986. Sex ratio variation in mammals. *Q. Rev. Biol.* 61 : 339–374.

Clutton-Brock, T. H., S. D. Albon, and F. E. Guinness. 1981. Parental investment in male and female offspring in polygynous mammals. *Nature* 289: 487–489.

Clutton-Brock, T. H., S. D. Albon, and F. E. Guinness. 1982. Competition between female relatives in a matrilocal mammal. *Nature* 300: 178–180.

Clutton-Brock, T. H., S. D. Albon, and F. E. Guinness. 1984. Maternal dominance, breeding success, and birth sex ratios in red deer. *Nature* 308: 358–360.

Clutton-Brock, T. H., S. D. Albon, and F. E. Guinness. 1985. Parental investment and sex differences in juvenile mortality in birds and mammals. *Nature* 313: 131–133.

Clutton-Brock, T. H., S. D. Albon, and F. E. Guinness. 1986. Great expectations: Maternal dominance, sex ratios and offspring reproductive success in red deer. *Anim. Behav.* 34: 460–471.

Clutton-Brock, T. H., S. D. Albon, and F. E. Guinness. 1988. Reproductive success in red deer. In T. H. Clutton-Brock, ed., *Reproductive Success*, pp. 325–343. University of Chicago Press, Chicago.

Clutton-Brock, T. H., S. D. Albon, and F. E. Guinness. 1989. Fitness costs of gestation and lactation in wild mammals. *Nature* 337: 260–262.

Clutton-Brock, T. H., F. E. Guinness, and S. D. Albon. 1982. *Red Deer: The Behaviour and Ecology of Two Sexes.* University of Chicago Press, Chicago.

Clutton-Brock, T. H., F. E. Guinness, and S. D. Albon. 1983. The costs of reproduction to red deer hinds. *J. Anim. Ecol.* 52: 367–383.

Cockburn, A. 1989. Adaptive patterns in marsupial reproduction. *Trends in Ecology and Evolution* 4: 126–130.

Cockburn, A. 1988. *Social Behaviour in Fluctuating Populations.* Croom Helm, London.

Cockburn, A., M. P. Scott, and C. R. Dickman. 1985. Sex ratio and intrasexual kin competition in mammals. *Oecologia* 66: 427–429.

Cockburn, A., M. P. Scott, and D. J. Scotts. 1985. Inbreeding avoidance and male-biased natal dispersal in *Antechinus* spp. (Marsupiala: Dasyuridae). *Anim. Behav.* 33: 908–915.

Cole, L. J., and W. F. Kirkpatrick. 1915. Sex ratios in pigeons. *Rhode Island Agr. Ex. Sta. Bull.*, no. 162.

Coleman, R. M., and M. R. Gross. In press. Understanding parental investment theory: Can animals commit the Concorde fallacy? *Anim. Behav.*

Coleman, R. M., and R. D. Whittall. 1988. Clutch size and the cost of incubation in the Bengalese finch (*Lonchura striata* var. *domestica*). *Behav. Ecol. Sociobiol.* 23: 367–372.

Coleman, R. M., M. R. Gross, and R. C. Sargent. 1985. Parental investment decision rules: A test in bluegill sunfish. *Behav. Ecol. Sociobiol.* 18: 59–66.

Colgan, P. W., and M. R. Gross. 1977. Dynamics of aggression in male pumpkinseed sunfish (*Lepomis gibbosus*) over the reproductive phase. *Z. Tierpsychol.* 43: 139–151.

Collins, D. A. 1986. Interactions between adult male and infant yellow baboons (*Papio c. cynocephalus*) in Tanzania. *Anim. Behav.* 34: 430–443.

Collopy, M. W. 1986. Food consumption and growth energetics of nestling golden eagles. *Wilson Bull.* 98: 445–458.

Congdon, J. D., and J. W. Gibbons. 1987. Morphological constraint on egg size: A challenge to optimal egg size theory? *Proc. Nat. Acad. Sci.* 84: 4145–4147.

Congdon, J. D., A. E. Dunham, and D. W. Tinkle. 1982. Energy budgets and life histories of reptiles. In C. Gans and F. Billett, eds., *Biology of the Reptilia.* Academic Press, New York.

Congreve, W. M., and S.E.P. Frame. 1930. Seven weeks in eastern and northern Iceland. *Ibis* 6: 192–228.

Constanz, G. D. 1985. Allopaternal care in the tessellated darter *Etheostoma olmstedi* (Pisces: Peradae). *Environ. Biol. Fishes* 14: 175–183.

Corben, C. J., G. J. Ingram, and M. J. Tyler. 1974. Gastric brooding: Unique form of parental care in an Australian frog. *Science* 186: 946–947.

Costa, D. P., and R. L. Gentry. 1986. The ranging and reproductive energetics of the northern fur seal. In R. L. Gentry and G. L. Kooyman, eds., *Fur Seals: Maternal Strategies on Land and at Sea.* Princeton University Press, Princeton, N.J.

Costa, D. P., B. J. Le Boeuf, A. C. Huntley, and C. Ortiz. 1986. The energetics of lactation in the northern elephant seal, *Mirounga angustirostris*. *J. Zool.* 209: 21–33.

Costa, D. P., F. Trillmich, and J. P. Croxall. 1988. Intra-specific allometry of neonatal size in the Antarctic fur seal (*Arctocephalus gazella*). *Behav. Ecol. Sociobiol.* 22: 361–364.

Coulson, J. C., and G. Hickling. 1961. Variation in the secondary sex-ratio of the grey seal *Halichoerus grypus* during the breeding season. *Nature* 190: 28.

Cowan, I. McT., and A. J. Wood. 1955. The growth rate of black-tailed deer. *J. Wildl. Mgmt.* 19: 331–336.

Cowgill, V. M., and G. E. Hutchinson. 1963a. Sex ratio in childhood. *Human Biol.* 35: 1–5.

Cowgill, V. M., and G. E. Hutchinson. 1963b. Differential mortality among the sexes in childhood and its possible evolutionary significance. *Proc. Nat. Acad. Sci.* 49: 425–429.

Craig, R. 1980. Sex investment ratios in social Hymenoptera. *Science* 20: 163–165.

Crespi, B. J. 1988. Sex ratio selection in a bivoltine thrips. I. Conditional sex-ratio manipulation and fitness variation. *Evolution* 42: 1199–1211.

Crespi, B. J. 1989. Facultative viviparity in a thrips. *Nature* 337: 357–358.

Crespi, B. J. In press. Sex ratio selection in Thysanoptera. In D. L. Wrensch and D. A. Krainacker, eds., *Evolution and Diversity of Sex Ratio in Insects and Mites*. Chapman and Hall, London.

Crockett, C. M., and R. Rudran. 1984. Interbirth intervals in red howler monkeys: Parental investment and ecological constraints. Paper presented at Animal Behavior Society Meeting, Washington, D.C. 1984.

Crome, F.H.J. 1976. Some observations on the biology of the cassowary in north Queensland. *Emu* 76: 8–14.

Crompton, A. W., and F. A. Jenkins. 1979. Origin of mammals. In J. A. Lillegraven, Z. Kielan-Jaworowska, and W. A. Clemens, eds., *Mesozoic Mammals: The First Two-Thirds of Mammalian History*, pp. 74–90. University of California Press, Berkeley.

Crompton, A. W., C. R. Taylor, and J. A. Jagger. 1978. Evolution of homeothermy in mammals. *Nature* 272: 333–336.

Cronmiller, J. R., and C. F. Thomson. 1980. Experimental manipulation of brood size in red winged blackbirds. *Auk* 97: 559–565.

Cronmiller, J. R., and C. F. Thomson. 1981. Sex ratio adjustment in malnourished red winged blackbird broods. *J. Field Ornithol.* 56: 65–67.

Crook, J. H. 1962. The adaptive significance of pair formation types in weaver birds. *Symp. Zool. Soc. Lond.* 8: 57–70.

Crook, J. H. 1964. The evolution of social organisation and visual communication in the weaver birds Ploceinae. *Behaviour Supplement* 10: 1–178.

Croxall, J. P. 1982. Energy costs of incubation and moult in petrels and penguins. *J. Anim. Ecol.* 51: 171–194.

Crump, M. L. 1981. Variation in propagule size as a function of environmental uncertainty for tree frogs. *Am. Nat.* 117: 724–737.

Crump, M. L. 1984. Intraclutch egg size variability in *Hyla crucifer* (Anura: Hylidae). *Copeia* 1984: 302–308.

Crump, M. L., and R. H. Kaplan. 1979. Clutch energy partitioning of tropical tree frogs (Hylidae). *Copeia* 1979: 626–635.

Cullen, E. 1957. Adaptations in the kittiwake to cliff nesting. *Ibis* 99: 275–302.

Cundiff, L. 1972. The role of maternal effects in animal breeding. VIII. Comparative aspects of maternal effects. *J. Anim. Sci.* 35: 1335–1337.

Curio, E. 1975. The functional organization of anti-predator behaviour in the pied fly-catcher: A study of avian visual perception. *Anim. Behav.* 23: 1–115.

Curio, E. 1983. Why do young birds reproduce less well? *Ibis* 125: 400–404.

Curio, E. 1987. Animal decision-making and the Concorde fallacy. *Trends in Evolution and Ecology* 6: 1458–152.

Curio, E. 1988. Relative realized lifespan and delayed cost of parental care. *Am. Nat.* 131: 825–836.

Curio, E., and K. Regelmann. 1987. Do great tit *Parus major* parents gear their brood defence to the quality of their young? *Ibis* 129: 344–352.

Curio, E., K. Regelmann, and U. Zimmermann. 1984. The defence of first and second broods by great tit (*Parus major*) parents: A test of predictive sociobiology. *Z. Tierpsychol.* 60: 101–127.

Daly, M. 1978. The cost of mating. *Am. Nat.* 112: 771–774.

Daly, M. 1979. Why don't male mammals lactate? *J. theor. Biol.* 78: 325–345.

Daly, M., and S. Daly. 1975. Behaviour of *Psammomys obesus* (Rodentia: Gerbillinae) in the Algerian Sahara. *Z. Tierpsychol.* 37: 298–321.

Daly, M., and M. Wilson. 1984. A sociobiological analysis of human infanticide. In G. Hausfater and S. B. Hrdy, eds., *Infanticide: Comparative and Evolutionary Perspectives*, pp. 487–502. Aldine, Chicago.

Dapson, R. W., P. R. Ramsey, M. H. Smith, and D. F. Urbston. 1979. Demographic differences in contiguous populations of white-tailed deer. *J. Wildl. Mgmt.* 43: 889–898.

Darwin, C. 1871. *The Descent of Man and Selection in Relation to Sex.* John Murray, London.

Davies, N. B. 1985. Cooperation and conflict among dunnocks, *Prunella modularis* in a variable mating system. *Anim. Behav.* 33: 628–648.

Davies, N. B. 1990. Mating systems. In J. R. Krebs and N. B. Davies, eds., *Behavioural Ecology III.* Blackwell Scientific Publications, Oxford.

Davies, N. B., and A. I. Houston. 1986. Reproductive success of dunnocks *Prunella modularis* in a variable mating system. II. Conflicts of interest among breeding adults. *J. Anim. Ecol.* 55: 139–154.

Davies, S.J.J.F. 1976. The natural history of the emu in comparison with that of other ratites. *Proc. 16th International Ornithological Congress*, Canberra, Australia, 1974, pp. 109–120.

Dawkins, R. 1976. *The Selfish Gene.* Oxford University Press, Oxford.

Dawkins, R., and H. J. Brockmann. 1980. Do digger wasps commit the Concorde fallacy? *Anim. Behav.* 28: 892–896.

Dawkins, R., and T. R. Carlisle. 1976. Parental investment, mate desertion and a fallacy. *Nature* 262: 131–133.

Dawkins, R., and J. R. Krebs. 1979. Arms races between and within species. *Proc. Roy. Soc. B.* 205: 489–511.

Day, C.S.D., and B. G. Galef. 1977. Pup cannibalism: One aspect of maternal behavior in golden hamsters. *J. Comp. Physiol. Psychol.* 91: 1179–1189.

Deag, J. 1980. Interactions between males and unweaned Barbary macaques: Testing the agonistic buffering hypothesis. *Behaviour* 56: 81–92.

Deag, J., and J. H. Crook. 1971. Social behaviour and "agonistic buffering" in the wild Barbary macaque, *Macaca sylvana. Folia Primatol.* 15: 183–280.

Dean, J. K. 1981. The relationship between lifespan and reproduction in the grasshopper *Melanophus. Oecologia* 48: 365–368.

Deecaraman, M., and T. Subramoniam. 1983. Mating and its effect on female reproductive physiology with special reference to the fate of male accessory sex gland secretion in the stomatopod, *Squilla holoschista. Marine Biol.* 77: 161–170.

Demaree, S. R. 1975. Observations on roof-nesting killdeers. *Condor* 77: 487–488.

De Martini, E. 1976. The adaptive significance of territoriality and egg cannibalism in the painted greenling, *Oxylebius pictus*, a northeastern Pacific marine fish. Ph.D. diss., University of Washington, Seattle.

De Martini, E. 1987. Paternal defence, cannibalism and polygyny: Factors affecting the reproductive success of painted greenling (Pisces, Hexagrammidae). *Anim. Behav.* 35: 1145–1158.

Dempster, E. R., and D. Lowry. 1952. Continuous selection for egg production in poultry. *Genetics* 37: 693–708.

De Steven, D. 1978. Clutch size, breeding success and parental survival in the tree sparrow (*Iridoprocne bicolor*). *Evolution* 34: 278–291.

Dhillon, J. S., R. M. Acharya, M. S. Tiwana, and S. C. Aggarwal. 1970. Factors affecting the interval between calving and conception in Hariana cattle. *Anim. Prod.* 12: 81–87.

Dhondt, A. A. 1970. The sex ratio of nestling great tits. *Bird Study* 17: 282–286.

Dhondt, A. A. 1987. Reproduction and survival of polygynous and monogamous blue tits *Parus caeruleus. Ibis* 129: 327–334.

Diesel, R. 1989. Parental care in an unusual environment: *Metopaulias depressus* (Decapoda: Grapsidae), a crab that lives in epiphytic bromeliads. *Anim. Behav.* 38: 561–575.

Digiacomo, R. F., and P. W. Shaughnessy. 1979. Fetal sex ratio in the rhesus monkey (*Macaca mulatta*). *Folia Primatol.* 31: 246–250.

Dittus, W.P.J. 1977. The social regulation of population density and age-sex distribution in the toque monkey. *Behaviour* 63: 281–311.

Dittus, W.P.J. 1979. The evolution of behaviours regulating population density and age-specific sex ratios in a primate population. *Behaviour* 69: 265–301.

Dittus, W.P.J. 1986. Sex differences in fitness following a group take-over among toque macaques: Testing models of social evolution. *Behav. Ecol. Sociobiol.* 19: 257–266.

Dixon, A.F.G. 1985. *Aphid Ecology*. Blackie, Glasgow and London.

Dobson, F. S. 1982. Competition for mates and predominant juvenile male dispersal in mammals. *Anim. Behav.* 30: 1183–1192.

Dobzhansky, T. L., and H. Levine. 1955. Developmental nemeostasis in natural populations of *Drosophila pseudo-obscura. Genetics* 40: 797–808.

Doidge, D. W., J. P. Croxall, and J. R. Baker. 1984. Density dependent pup mortality in the Antarctic fur seal *Arctocephalus gazella* at South Georgia. *J. Zool.* 202: 449–460.

Dolhinow, P., J. J. McKenna, and J.V.H. Laws. 1979. Rank and reproduction among female langur monkeys (They're not just getting older, they're getting better). *Aggressive Behavior* 5: 19–30.

Dominey, W. J. 1981. Anti-predator function of bluegill sunfish nesting colonies. *Nature* 290: 586–588.

Downhower, J. F., and L. Brown. 1981. The timing of reproduction and its behavioural consequences for mottled sculpins *Cottus bairdi*. In R. D. Alexander and D. W. Tinkle, eds., *Natural Selection and Social Behavior*, pp. 78–95. Chiron Press, New York.

Drent, R. H., and S. Daan. 1980. The prudent parent: Energetic adjustments in avian breeding. *Ardea* 68: 225–252.

Drent, R. H., and S. Daan. 1981. The prudent parent: Energetic adjustment in avian breeding. In H. Klomp and J. W. Woldendorp, eds., *The Integrated Study of Bird Populations*, pp. 225–252. North Holland Publishing Co., Amsterdam.

Drewry, G. 1970. The role of amphibians in the ecology of Puerto Rican rain forests. In *Puerto Rico Nuclear Center Rain Forest Project Annual Report*, pp. 16–85. Puerto Rico Nuclear Center, San Juan.

Drewry, G. 1974. Ecology of *Eleutherodactylus coqui* Thomas in montane rain forest of eastern Puerto Rico (abstract). *54th Annual Meeting Amer. Soc. Ichthyol. Herpetol.*, p. 5.

Drummond, H., and C. G. Chavelas. 1989. Food shortage influences sibling aggression in the blue-footed booby. *Anim. Behav.* 38: 806–819.

Drummond, H., E. Gonzalez, and J. L. Osorno. 1986. Parent-offspring cooperation in the blue-footed booby (*Sula nebouxii*): Social roles in infanticidal brood reduction. *Behav. Ecol. Sociobiol.* 19: 365–372.

Drummond, H., J. L. Osorno, R. Torres, and C. Garcia. In press. Sexual size dimorphism and sibling competition: Implications for avian sex ratios. *Anim. Behav.*

Dunbar, R.I.M. 1988. *Primate Social Systems.* Croom Helm, London.

Dunbrack, R. L., and M. A. Ramsay. 1989. The evolution of viviparity in amniote vertebrates: Egg retention versus egg size reduction. *Am. Nat.* 133: 138–148.

Duncan, P., P. H. Harvey, and S. M. Wells. 1984. On lactation and associated behaviour in a natural herd of horses. *Anim. Behav.* 32: 255–263.

East, M. 1981. Alarm calling and parental investment in the robin, *Erithacus rubecula*. *Ibis* 123: 223–230.

Eberhard, W. G. 1975. The ecology and behavior of a subsocial pentatomid bug and two scelionid wasps: Strategy and counterstrategy in a host and its parasites. *Smithson. Contr. Zool.* 205: 1–39.

Edwards, J. S. 1962. Observations on the development and predatory habit of two reduviid Heteroptera, *Rhinocoris carmelita* Stål and *Platymeris rhadmanthus* Gerst. *Proc. Roy. Ent. Soc. Lond. A* 37: 89–98.

Edwards, T. C., and M. W. Collopy. 1983. Obligate and facultative brood reduction in eagles: An examination of the factors that influence fratricide. *Auk* 100: 630–635.

Eickwort, A. C. 1981. Presocial insects. In H. R. Hermann, ed., *Social Insects*, pp. 199–280. Academic Press, New York.

Eickwort, K. R. 1973. Cannibalism and kin selection in *Labidomera clivicollis* (Coleoptera: Chrysomelidae). *Am. Nat.* 107: 452–453.

Eisenberg, J. 1981. *The Mammalian Radiations*. University of Chicago Press, Chicago.

Ekman, J., and C. Askenmo. 1986. Reproductive cost, age specific survival and a comparison of the reproductive strategy in two European tits (Genus *Parus*). *Evolution* 40: 159–168.

Elgar, M. A., and H.C.J. Godfray. 1987. Sociality and sex ratios in spiders. *Trends in Ecology and Evolution* 2: 6–7.

Elgar, M., and L. I. Heaphy. In press. Clutch size and egg weight in turtles. *Am. Nat.*

Emlen, J. T. 1970. Age specificity and ecological theory. *Ecology* 51: 588–601.

Emlen, S. T. 1984. Cooperative breeding in birds and mammals. In J. R. Krebs and N. B. Davies, eds., *Behavioural Ecology: An Evolutionary Approach*, pp. 305–309. Blackwell Scientific Publications, Oxford.

Emlen, S. T., and L. W. Oring. 1977. Ecology, sexual selection and the evolution of mating systems. *Science* 197: 215–223.

Emlen, S. T., and S. L. Vehrencamp. 1983. Cooperative breeding strategies among birds. In A. H. Brush and G. A. Clark, eds., *Perspectives in Ornithology*, pp. 93–120. Cambridge University Press, Cambridge.

Emlen, S. T., J. H. Emlen, and S. A. Levin. 1986. Sex ratio selection in species with helpers at the nest. *Am. Nat.* 127: 1–8.

Erasmus, J. E. 1962. Part-period selection for egg production. *Proc. 12th World Poultry Congr. Sydney 1962* 17–18.

Erckmann, W. J. 1981. The evolution of sex role reversal and monogamy in shore birds. Ph.D. diss, University of Washington, Seattle.

Erckmann, W. J. 1983. The evolution of polyandry in shorebirds: An evaluation of hypotheses. In S. K. Wasser, ed., *Social Behavior of Female Vertebrates*, pp. 113–168. Academic Press, New York.

Evans, L. T. 1959. A motion picture study of maternal behavior of the lizard, *Eumeces obsoletus* Baird and Girard. *Copeia* 1959: 103–110.

Evans, L. T. 1961. Studies as related to behavior in the organization of populations of reptiles. In F. Blair, ed., *Vertebrate Speciation*, pp. 148–178. University of Texas Press, Austin.

Evans, R. M. 1990. The relationship between parental input and parental investment. *Anim. Behav.* 39: 797–813.

Ewbank, R. 1964. Observations on the suckling habits of twin lambs. *Anim. Behav.* 12: 34–37.

Ewer, R. F. 1973. *The Carnivores*. Weidenfeld and Nicolson, London.

Ezaki, Y. 1988. Mate desertion by male great reed warblers *Acrocephalus arundinaceus* at the end of the breeding season. *Ibis* 130: 427–437.

Faaborg, J., and C. B. Patterson. 1981. The characteristics and occurrence of cooperative polyandry. *Ibis* 123: 477–484.

Faaborg, J. F., T. de Vries, C. B. Patterson, and C. R. Griffin. 1980. Preliminary observations on the occurrence and evolution of polyandry in the Galapagos hawk (*Buteo galapagoensis*). *Auk* 97: 581–590.

Fagen, R.B.L. 1972. An optimal life history strategy in which reproductive effort decreases with age. *Am. Nat.* 106: 258–261.

Fairbanks, L. A., and M. T. McGuire. 1986. Age, reproductive value and dominance related behaviour in vervet monkey females: Cross-generational social influences on social relationships and reproduction. *Anim. Behav.* 34: 1718–1721.

Fedigan, L. M. 1983. Dominance and reproductive success in primates. *Yearbook of Physical Anthropology* 26: 9–129.

Feifarek, B. P., C. A. Wyngaard, and J.D.G. Allan. 1983. The cost of reproduction in a freshwater copepod. *Oecologia* 56: 166–168.

Feldman, M. W., and I. Eshel. 1982. On the theory of parent–offspring conflict: A two-locus genetic model. *Am. Nat.* 119: 285–292.

Fernald, R. D., and N. R. Hirata. 1977. Field study of *Haplochromis burtoni*: Habitats and co-habitats. *Env. Biol. Fish.* 2: 299–308.

Festa-Bianchet, M. 1989. Individual differences, parasites and the costs of reproduction for bighorn ewes (*Ovis canadensis*). *J. Anim. Ecol.* 58: 785–795.

Fiala, K. L. 1981a. Sex ratio constancy in the red-winged blackbird. *Evolution* 35: 898–910.

Fiala, K. L. 1981b. Reproductive costs and the sex ratio in red winged blackbirds. In: R. D. Alexander and D. W. Tinkle, eds., *Natural Selection and Social Behavior*, pp. 198–214. Chiron Press, New York.

Fiala, K. L., and J. D. Congdon. 1983. Energetic consequences of sexual size dimorphism in red-wing blackbirds. *Ecology* 64: 642–647.

Fiedler, K. 1954. Vergleichende Verhaltensstudien an Seenadeln, Schlangennadeln und Seepferdchen (Syngnathidae). *Z. Tierpsychol.* 11: 358–416.

Finke, M. A., D. J. Milinkovich, and C. F. Thompson. 1987. Evolution of clutch size: An experimental test in the house wren (*Troglodytes aedon*). *J. Anim. Ecol.* 56: 99–114.

Fisher, R. A. 1930. *The Genetical Theory of Natural Selection.* Oxford University Press, Oxford.

Fleming, T. H., and R. J. Rauscher. 1978. On the evolution of litter size in *Peromyseus leucopus*. *Evolution* 32: 45–55.

Flook, D. R. 1970. *A Study of Sex Differential in the Survival of Wapiti.* Canadian Wildlife Service Report Series No. 11. Department of Indian Affairs and Northern Development, Ottawa.

Forester, D. C. 1979. The adaptiveness of parental care in *Desmognathus ochrophaeus* (Urodela: Plethodontidae). *Copeia* 1979: 332–341.

Forsyth, A. 1981. Sex ratio and parental investment in an ant population. *Evolution* 35: 1252–1253.

Fowler, C. W. 1987. A review of density dependence in populations of large mammals. In H. Genoways, ed., *Current Mammalogy*, vol. 1, pp. 401–441. Plenum Press, New York.

Fowler, K., and L. Partridge. 1989. A cost of mating to female fruitflies. *Nature* 338: 760–761.

Fowler, L. G. 1972. Growth and mortality of fingerling chinook salmon as affected by egg size. *Prog. Fish. Cult.* 34: 66–69.

Frame, L. H., J. R. Malcolm, G. W. Frame, and H. van Lawick. 1979. Social organisation of African wild dogs (*Lycaon pictus*) on the Serengeti Plains, Tanzania, 1967–1978. *Z. Tierpsychol.* 50: 225–249.

Frank, S. A. 1987. Individual and population sex allocation patterns. *Theoret. Popul. Biol.* 31: 47–74.

Frank, S. A., and B. Crespi. 1989. Synergism between sib-rearing and sex ratio in Hymenoptera. *Behav. Ecol. Sociobiol.* 24: 1555–162.

Frank, S. A., and I. R. Swingland. 1988. Sex ratio under conditional sex expression. *J. theoret. Biol.* 135: 415–418.

Fraser, D. F. 1980. On the environmental control of oocyte maturation in a plethodontid salamander. *Oecologia* 46: 302–307.

Fraser, D., and R. Morley Jones. 1975. The "teat order" of suckling pigs. I. Relation to birth weight and subsequent growth. *J. Agric. Sci. Camb.* 84: 387–391.

Fredga, K., A. Gropp, I. Winkling, and F. Frank. 1977. A hypothesis explaining the exceptional sex ratio of the wood lemming (*Myopus schesticolor*). *Hereditas* 85: 101–104.

Freed, L. A. 1987. Prospective infanticide and protection of genetic paternity in tropical house wrens. *Am. Nat.* 130: 948–954.

Fricke, X. 1974. Öko-ethologie des monogamen Anemonefisches *Amphiprion bicinctus* (Freiswasseruntersuchung aus dem Roten Meer). *Z. Tierpsychol.* 36: 429–512.

Fricke, H. 1980. Mating systems, maternal and biparental care in triggerfish Balistidae. *Z. Tierpsychol.* 53: 105–122.

Frisch, R. E. 1984. Body fat, puberty and fertility. *Biol. Rev.* 59: 161–188.

Frith, H. J. 1959. Breeding of the Mallee Fowl, *Leipoa ocellata* Gould (Megapodiidae). *CSIRO Wildl. Res.* 4: 31–65.

Frith, H. J. 1962. *The Mallee-Fowl.* Angus and Robertson, Sydney.

Fryer, G., and T. D. Iles. 1972. *The Cichlid Fishes of the Great Lakes of Africa.* T.F.H. Publ., Neptune City, N.J.

Fuchs, J. 1982. Optimality of parental investment: The influence of nursing on reproductive success of mother and female young in house mice. *Behav. Ecol. Sociobiol.* 10: 39–51.

Fujioka, M. 1985a. Sibling competition and siblicide in asynchronously hatching broods of the cattle egret *Bubulcus ibis. Anim. Behav.* 33: 1218–1242.

Fujioka, H. 1985b. Food delivery and sibling competition in experimentally even-aged broods of the cattle egret. *Behav. Ecol. Sociobiol.* 17: 67–74.

Gadgil, M., and W. H. Bossert. 1970. Life historical consequences of natural selection. *Am. Nat.* 104: 1–24.

Galbraith, H. 1988. Effects of egg size and composition on the size, quality and survival of lapwing *Vanellus vanellus. J. Zool. Lond.* 21: 383–398.

Gale, W. F., and W. G. Deutsch. 1985. Fecundity and spawning frequency of captive tessellated darters—fractional spawners. *Trans. Amer. Fish. Soc.* 114: 220–229.

Gandelman and Simon. 1978. Spontaneous pup-killing by mice in response to large litters. *Dev. Psychobiol.* 11: 235–241.

Garnaud, J. 1950. La reproduction et l'incubation branchiale chez *Apogon imberbis* G. et G. *Bull. Inst. Oceanogr.* (Monaco) 977: 1–10.

Garnaud, J. 1962. Monographie de l'Apogon mediterranéan, *Apogon imberbis* (Linne) 1758. *Bull. Inst. Oceanogr.* (Monaco) 1248: 1–183.

Garnett, M. C. 1981. Body size: Its heritability and influence on juvenile survival among great tits *Parus major*. *Ibis* 123: 31–41.

Geist, V. 1971. *Mountain Sheep*. University of Chicago Press, Chicago.

Gentry, R. C., and G. L. Kooyman. 1986. *Fur Seals: Maternal Strategies on Land and at Sea*. Princeton University Press, Princeton, N.J.

Getz, L. L., and C. S. Carter. 1980. Social organisation in *Microtus ochrogaster* populations. *The Biologist* 62: 56–69.

Getz, L. L., C. S. Carter, and L. Gavish. 1981. The mating system of the prairie vole, *Microtus ochrogaster*: Field and laboratory evidence for pair bonding. *Behav. Ecol. Sociobiol.* 8: 189–194.

Gibbons, D. 1987. Hatching asynchrony reduces parental investment in the jackdaw. *J. Anim. Ecol.* 56: 403–414.

Giesel, J. T. 1979. Genetic co-variation of survivorship and other fitness indices in *Drosophila melanogaster*. *Exp. Gerontol.* 14: 323–328.

Giesel, J. T., and E. E. Zettler. 1980. Genetic correlations of life-historical parameters and certain fitness indices in *Drosophila melanogaster*: r_m, r_s and diet breadth. *Oecologia* 47: 299–302.

Giesel, J. T., P. A. Murphy, and M. N. Manlove. 1982. The influence of temperature on genetic interrelationships of life history traits in a population of *Drosophila melanogaster*: What tangled data sets we weave. *Am. Nat.* 119: 464–479.

Gill, D. E. 1974. Intrinsic rates of increase, saturation density and competitive ability. II. The evolution of competitive ability. *Am. Nat.* 198: 103–116.

Gittleman, J. L. 1981. The phylogeny of parental care in fishes. *Anim. Behav.* 29: 936–941.

Gittleman, J. L. 1984. The behavioural ecology of carnivores. Ph.D. diss., University of Sussex, England.

Gittleman, J. L. 1985. Functions of communal care in mammals. In P. J. Greenwood, P. H. Harvey, and M. Slatkin, eds., *Evolution: Essays in Honour of John Maynard Smith*, pp. 181–205. Cambridge University Press, Cambridge.

Gittleman, J. L. 1986. Carnivore life history patterns: Allometric, phylogenetic and ecological allocations. *Am. Nat.* 127: 744–771.

Gittleman, J. L. 1988. Carnivore group living: Comparative trends. In J. L. Gittleman, ed., *Carnivore Behavior, Ecology and Evolution*, pp. 183–207. Cornell University Press, Ithaca, New York.

Gittleman, J. L., and O. T. Oftedal. 1987. Comparative growth and lactation energetics in carnivores. *Symp. Zool. Soc. Lond.* 57: 41–77.

Gittleman, J. L., and S. D. Thompson. In press. Energy allocation in mammalian reproduction. In *Energetics and Animal Behavior. Am. Zool.*

Gladstone, D. E. 1979. Promiscuity in monogamous colonial birds. *Am. Nat.* 114: 545–557.

Glucksman, A. 1974. Sexual dimorphism in mammals. *Biol. Rev.* 49: 423–475.

Gochfeld, M. 1984. Antipredator behavior. In J. Burger and B. L. Olla, eds., *Shorebirds: Breeding Behavior and Populations*, pp. 289–377. Plenum Press, New York.

Godfray, H.C.J. 1986. The evolution of clutch size in parasitic wasps. *Am. Nat.* 129: 221–233.

Godfray, H.J.C. 1987. Genetic models of clutch size evolution in parasitic wasps. *Am. Nat.* 129: 221–233.

Goldizen, A. W. 1987. Tamarins and marmosets: Communal care of offspring. In B. B. Smuts et al., eds., *Primate Societies*, pp. 34–43. University of Chicago Press, Chicago.

Gomendio, M. 1988. Mother-offspring relationships and consequences for fertility in rhesus macaques. Ph.D. diss., University of Cambridge.

Gomendio, M. 1989. Suckling behaviour and fertility in rhesus macaques (*Macaca mulatta*). *J. Zool.* 217: 449–467.

Gomendio, M., T. H. Clutton-Brock, S. D. Albon, F. E. Guinness, and M. J. Simpson. 1990. Contrasting costs of sons and daughters and the evolution of mammalian sex ratios. *Nature* 343: 261–263.

Goodall, J. 1986. *The Chimpanzees of Gombe*. Belknap Press of Harvard University Press, Cambridge, Mass.

Goodman, D. 1974. Natural selection and a cost ceiling in reproductive effort. *Am. Nat.* 108: 247–268.

Goodman, D. 1979. Regulating reproductive effort in a changing environment. *Am. Nat.* 113: 735–748.

Gordon, I. J. 1989. The interspecific allometry of reproduction: Do larger species invest relatively less in their offspring? *Functional Ecology* 3: 285–288.

Gosling, L. M. 1986. Selective abortion of entire litters in the coypu: Adaptive control of offspring production in relation to quality and sex. *Am. Nat.* 127: 772–795.

Gosling, M. M., S. J. Baker, and K.M.H. Wright. 1984. Differential investment by female coypus (*Myocastor coypus*) during lactation. *Symp. Zool. Soc. Lond.* 51: 273–300.

Gottfried, B. M. 1979. Anti-predator aggression in birds nesting in old field habitats: An experimental analysis. *Condor* 81: 251–257.

Göttlander, K. 1987. Parental feeding behaviour and sibling rivalry. *Ornis. Scand.* 18: 269–276.

Gotto, R. V. 1962. Egg-number and ecology in commensal and parasitic copepods. *Ann. Mag. Nat. Hist.* 13: 97–107.

Gould, S. J. 1984. Only his wings remained. *Natural History* 9/84: 10–15.

Goulden, C. H., and L. L. Henry. 1984. Lipid energy reserves and their role in *Cladocera* spp. In D. G. Meyers and J. R. Strickler, eds., *Trophic Interactions within Aquatic Systems*, pp. 167–185. Westview, Boulder, Colorado.

Goulden, C. H., L. Henry, and D. Berrigan. 1987. Egg size, post-embryonic yolk and survival ability. *Oecologia* 72: 28–31.

Gowaty, P. 1983. Male parental care and apparent monogamy among eastern bluebirds (*Sialia sialis*). *Am. Nat.* 12: 144–157.

Gowaty, P. A., and M. R. Lennartz. 1985. Sex ratios of nestling and fledgling red-cockaded woodpeckers (*Picoides borealis*) favor males. *Am. Nat.* 126: 347–353.

Gowaty, P. A., J. H. Plissner, and T. G. Williams. 1989. Behavioural correlates of uncertain parentage: Mate guarding and nest guarding by eastern bluebirds. *Anim. Behav.* 38: 272–284.

Gowen, W., and L. E. Johnson. 1946. On the mechanism of heterosis. I. Metabolic capacity of different races of *Drosophila melanogaster* for egg production. *Am. Nat.* 80: 149–179.

Grafen, A. 1980. Opportunity cost, benefit and degree of relatedness. *Anim. Behav.* 28: 967–968.

Grafen, A. 1984. Natural selection, kin selection and group selection. In J. R. Krebs and N. B. Davies, eds., *Behavioural Ecology: An Evolutionary Approach*, 2d ed., pp. 62–84. Blackwell Scientific Publications, Oxford.

Grafen, A., and R. Sibly. 1978. A model of mate desertion. *Anim. Behav.* 26: 645–652.

Grahame, J., and G. M. Branch. 1985. Reproductive patterns of marine invertebrates. *Oceanogr. Mar. Biol. Ann. Rev.* 23: 373–398.

Grant, W. D. 1974. Adaptive aspects of the mountain plover social system. *Living Bird* 12: 69–74.

Graul, W. D., S. R. Derrickson, and D. W. Mock. 1977. The evolution of avian polyandry. *Am. Nat.* 111: 812–816.

Green, B. 1984. Composition of milk and energetics of growth in marsupials. *Symp. Zool. Soc. Lond.* 51: 369–387.

Greenlaw, J. S., and W. Post. 1985. Evolution of monogamy in seaside sparrows, *Ammodramus maritimus*: Tests of hypotheses. *Anim. Behav.* 33: 373–383.

Greenwood, P. J. 1980. Mating systems, philopatry and dispersal in birds and mammals. *Anim. Behav.* 28: 1140–1162.

Greer, A. E. 1971. Crocodilian nesting habits and evolution. *Fauna* 2: 20–28.

Greer, K. R., and R. E. Howe. 1964. Winter weights of Northern Yellowstone elk, 1961–1962. *Transactions of Int. 29th North American Wildlife and Natural Resources Conference*, pp. 237–248. Wildlife Management Institute, Washington, D.C.

Greig-Smith, P. W. 1980. Parental investment in nest defence by stonechats (*Saxicola torquata*). *Anim. Behav.* 28: 604–619.

Griffiths, M. 1968. *Echidnas*. Pergamon Press, New York.

Griffiths, M. 1978. *The Biology of Monotremes*. Academic Press, New York.

Gronell, A. M. 1984. Courtship, spawning and social organisation of the pipefish, *Corythoichthys intestinalis* (Pisces: Syngnathidae) with notes on two congeneric species. *Z. Tierpsychol.* 65: 1–24.

Gross, M. R. 1984. Sunfish, salmon and the evolution of alternative reproductive tactics in fishes. In R. J. Wootton and G. Potts, eds., *Fish Reproduction: Strategies and Tactics*, pp. 55–75. Academic Press, New York.

Gross, M. R., and R. C. Sargent. 1985. The evolution of male and female parental care in fishes. *Am. Zool.* 25: 807–822.

Gross, M. R., and R. Shine. 1981. Parental care and mode of fertilization in ectothermic vertebrates. *Evolution* 35: 775–793.

Grubb, P. 1974. Population dynamics of the Soay sheep. In P. A. Jewell, C. Milner, and J. M. Boyd, eds., *Island Survivors: The Ecology of Soay Sheep on St. Kilda*, pp. 242–272. Athlone Press, London.

Gubernick, D. J., and J. R. Alberts. 1989. Postpartum maintenance of paternal behaviour in the biparental California mouse, *Peromyscus californicus*. *Anim. Behav.* 37: 656–664.

Guerrero, R. 1970. Sex ratio: A statistical association with type and time of insemination in the menstrual cycle. *Int. J. Fert.* 15: 221–225.

Guerrero, R. 1974. Association of the type and time of insemination within the menstrual cycle with the human sex ratio at birth. *New Engl. J. Med.* 291: 1056–1059.

Guinness, F. E., T. H. Clutton-Brock, and S. D. Albon. 1978. Factors affecting calf mortality in red deer. *J. Anim. Ecol.* 47: 817–832.

Gunn, R. G. 1964a. Levels of first winter feeding in relation to performance of Cheviot hill ewes. I. Body growth and development during treatment period. *J. Agric. Sci. Camb.* 62: 99–122.

Gunn, R. G. 1964b. Levels of first winter feeding in relation to performance of Cheviot hill ewes. II. Body growth and development during the summer after treatment, 12–18 months. *J. Agric. Sci. Camb.* 62: 123–149.

Gustaffson, L., and W. J. Sutherland. 1988. The costs of reproduction in the collared flycatcher *Ficedula albicollis*. *Nature* 335: 813–865.

Gwynn, A. M. 1953. The egg-laying and incubation periods of Rockhopper, Macaroni and Gentoo penguins. *A.N.A.R.E. Rept. Series B. Zool.* 1: 1–29.

Gwynne, D. T. 1983. Male nutritional investment and the evolution of sexual differences in the Tettigoniidae and other Orthoptera. In D. T. Gwynne and G. K. Morris, eds., *Orthoptera Mating Systems: Sexual Competition in a Diverse Group of Insects*, pp. 337–366. Westview, Boulder, Colorado.

Gwynne, D. T. 1984a. Male mating effort, confidence of paternity, and insect sperm composition. In R. L. Smith, ed., *Sperm Competition and the Evolution of Mating Systems*, pp. 117–149. Academic Press, New York.

Gwynne, D. T. 1984b. Courtship feeding increases female reproductive success in bush crickets. *Nature* 307: 361–363.

Gwynne, D. T. 1988. Courtship feeding and the fitness of female katydids (Orthoptera: Tettigoniidae). *Evolution* 42: 545–555.

Gwynne, D. T., B. J. Bowen, and C. G. Codd. 1984. The function of the katydid spermatophore and its role in fecundity and insemination (Orthoptera: Tettigoniidae). *Aust. J. Zool.* 32: 15–22.

Hagan, H. R. 1951. *Embryology of the Viviparous Insects*. Ronald Press, New York.

Hahn, D. C. 1981. Asynchronous hatching in the laughing gull: Cutting losses and reducing rivalry. *Anim. Behav.* 29: 421–427.

Haig, D. 1989. Seed size and adaptation. *Trends in Evolution and Ecology* 4: 145.

Haig Thomas, R., and J. S. Huxley. 1927. Sex ratios in pheasant species crosses. *J. Genetic.* 18: 233–246.

Hairston, N. G. 1983. Growth, survival and reproduction of *Plethodon jordani*: Trade-offs between selection pressures. *Copeia* 1983: 1024–1035.

Hall, C. O., and D. R. Marble. 1931. The relationship between the first year egg production and the egg production of later years. *Poultry Sci.* 10: 194–203.

Hall, R. D., J. P. Leakey, and W. M. Robertson. 1979. The effects of protein malnutrition on the behavior of rats during the suckling period. *Devel. Psychobiol.* 12: 455–466.

Hamilton, J. B., and G. E. Mestler. 1969. Mortality and survival: Comparison of eunuchs with intact men and women in a mentally retarded population. *J. Gerontol.* 24: 395–411.

Hamilton, J. B., R. J. Hamilton, and G. E. Mestler. 1969. Duration of life and causes of death in domestic cats: Influence of sex, gonadectomy and inbreeding. *J. Gerontol.* 24: 427–437.

Hamilton, W. D. 1961. Geometry for the selfish herd. *J. theoret. Biol.* 31: 295–311.

Hamilton, W. D. 1964. The genetical evolution of social behaviour. *J. theoret. Biol.* 7: 1–52.

Hamilton, W. D. 1967. Extraordinary sex ratios. *Science* 156: 477–488.

Handford, P., and M. A. Mares. 1985. The mating systems of ratites and tinamous: An evolutionary perspective. *Biol. J. Linn. Soc.* 25: 77–104.

Hannon, J. J. 1984. Factors limiting polygyny in the willow ptarmigan. *Anim. Behav.* 32: 153–161.

Hanwell, A., and M. Peaker. 1977. Physiological effects of lactation on the mother. *Symp. Zool. Soc. Lond.* 41: 297–312.

Harcourt, A. H. 1987. Dominance and fertility among primates. *J. Zool.* 213: 471–487.

Harcourt, A. H. 1989a. Environment, competition and reproductive performance of female monkeys. *Trends in Evolution and Ecology* 4: 101–105.

Harcourt, A. H. 1989b. Social influences on competitive ability: Alliances and their consequences. In V. Standen and R. A. Foley, eds., *Comparative Socioecology of Mammals and Humans*, pp. 223–242. Blackwell Scientific Publications, Oxford.

Harcourt, A. H., and K. J. Stewart. 1987. The influence of help in contests on dominance rank in primates: Hints from gorillas. *Anim. Behav.* 35: 182–190.

Haresign, T. W., and S. E. Shumway. 1981. Permeability of the marsupium of the pipe fish *Syngnathus fuscus* to [^{14}C]-alpha amino isobutyric acid. *Comp. Biochem. Physiol.* 69A: 603–604.

Harms, J. 1946. Die Fortpflanzung von *Salamandra atra. Biol. Zentralbl.* 65: 254–272.

Harmsen, R., and F. Cooke. 1983. Binomial sex ratio distribution in the lesser snow goose: A theoretical enigma. *Am. Nat.* 121: 1–8.

Harper, A. B. 1986. The evolution of begging: Sibling competition and parent-offspring conflict. *Am. Nat.* 128: 99–114.

Harper, J. L., P. H. Lovell, and K. G. Moore. 1970. The shapes and sizes of seeds. *Ann. Rev. Ecol. Syst.* 1: 327–356.

Harris, J. A., and H. R. Lewis. 1922. The correlation between first and second year egg production in the domestic fowl. *Genetics* 7: 274–318.

Harris, V. A. 1964. *The Life of the Rainbow Lizard.* Hutchinson, London.

Harten, A. von. 1936. Pflege und Zucht des *Poecilobryon unifasciatus. Wochenschr. Aquar.-Terrarienk.* 33: 353–355.

Hartsock, T. G., and H. B. Graves. 1976. Neonatal behavior and nutrition-related mortality in domestic swine. *J. Anim. Sci.* 42: 235–241.

Hartsock, T. G., H. B. Graves, and B. R. Baumgardt. 1977. Agonistic behavior and the nursing order in suckling piglets: Relationships with survival, growth and body composition. *J. Anim. Sci.* 44: 320–330.

Hartung, J. J. 1977. An implication about human mating systems. *J. theoret. Biol.* 66: 737–745.

Hartung, J. 1980. Parent-offspring conflict—a retraction. *J. theoret. Biol.* 87: 815–817.

Harvey, I. F., and P. S. Corbet. 1985. Territorial behaviour of larvae enhances mating success of male dragonflies. *Anim. Behav.* 33: 561–565.

Harvey, P. H., and T. H. Clutton-Brock. 1985. Life history variation in primates. *Evolution* 39: 559–581.

Harvey, P. H., and T. H. Clutton-Brock. 1986. Life history variation in primates. *Evolution* 39: 559–581.

Harvey, P. H., and P. J. Greenwood. 1978. Anti-predator defence strategies: Some evolutionary problems. In J. R. Krebs and N. B. Davies, eds., *Behavioural Ecology*, pp. 129–154. Blackwell Scientific Publications, Oxford.

Harvey, P. H., and G. M. Mace. 1982. Comparisons between taxa and adaptive trends: Problems of methodology. In King's College Sociobiology Group, eds., *Current Problems in Sociobiology*, pp. 343–362. Cambridge University Press, Cambridge.

Harvey, P. H., and M. D. Pagel. 1988. The allometric approach to species differences in brain size. *Human Evolution* 3: 461–472.

Harvey, P. H., and A. F. Read. 1988. How and why do mammalian life histories vary? In M. J. Boyce, ed., *Evolution of Life Histories of Mammals*, pp. 213–231. Yale University Press, New Haven, Conn.

Harvey, P. H., and R. M. Zammuto. 1985. Patterns of mortality and age at first reproduction in natural populations of mammals. *Nature* 315: 319–320.

Harvey, P. H., M. D. Pagel, and J. A. Rees. In press. Mammalian metabolism and life histories. *Am. Nat.*

Harvey, P. H., D.E.L. Promislow, and A. F. Read. 1989. Causes and correlates of life history differences among mammals. In V. Standen and R. Foley, eds., *Comparative Socioecology*, pp. 305–318. Blackwell Scientific Publications, Oxford.

Harvey, P. H., A. F. Read, and D.E.L. Promislow. 1989. Life history variation in placental mammals: Unifying data with theory. *Oxford Surveys in Evolutionary Biology* 6: 13–31.

Harvey, P. H., M. J. Stenning, and B. Campbell. 1985. Individual variation in seasonal breeding success of pied flycatchers (*Ficedula hypoleuca*). *J. Anim. Ecol.* 54: 391–398.

Hatziolos, M. E., and R. Caldwell. 1983. Role reversal in the stomatopod *Pseudosquilla ciliata* (Crustacea). *Anim. Behav.* 31: 1077–1087.

Haukioja, E., and R. Salovaara. 1978. Summer weight of reindeer (*Rangifer tarandus*) calves and its importance for their future survival. *Rep. Kevo Subarctic Res. Stat.* 14: 1–4.

Hauser, M. 1988. Variation in maternal responsiveness in free-ranging vervet monkeys: A response to infant mortality risk. *Am. Nat.* 131: 573–587.

Hauser, M. D., and L. A. Fairbanks. 1988. Mother-offspring conflict in vervet monkeys: Variation in response to ecological conditions. *Anim. Behav.* 36: 802–813.

Hays, H. 1972. Polyandry in the spotted sandpiper. *Living Bird* 11: 43–57.

Heath, D. J. 1977. Simultaneous hermaproditism: Cost and benefit. *J. theoret. Biol.* 81: 151–155.

Hébert, P. N., and R.M.R. Barclay. 1986. Asynchronous and synchronous hatching: Effect on early growth and survivorship of herring gull *Larus argentatus* broods. *Can. J. Zool.* 64: 2357–2362.

Hector, A.C.K., R. M. Seyfarth, and M. J. Raleigh. 1989. Male parental care, female choice and an effect of an audience in vervet monkeys. *Anim. Behav.* 38: 262–271.

Hegmann, J. P., and H. Dingle. 1982. Phenotypic and genetic covariance structure in milkweed bug life-history traits. In H. Dingle and J. P. Hegmann, eds., *Evolution and Genetics of Life Histories*, pp. 177–188. Springer-Verlag, New York.

Heinrich, B., and G. A. Bartholomew. 1979. The ecology of the African dung beetle. *Sci. Am.* 241: 118–127.

Hempel, G. 1965. On the importance of larval survival for the population dynamics of marine food fish. *Rep. Conf. Oceanic Fish. Invest.* 10: 13–23.

Henderson, B. 1975. Role of the chick's begging behavior in the regulation of parental feeding behavior of *Larus glaucescens. Condor* 77: 488–492.

Herre, E. A. 1985. Sex ratio adjustment in fig wasps. *Science* 228: 896–898.

Hildén, O. 1975. Breeding biology of Temminck's stint *Calidris temminckii. Orn. Fenn.* 52: 117–146.

Hildén, O., and S. Vuolanto. 1972. Breeding biology of the red-necked phalarope *Phalaropus lobatus* in Finland. *Orn. Fenn.* 49: 72–85.

Hinton, H. E. 1981. *Biology of Insect Eggs.* Pergamon Press, Oxford.

Hirschfield, M. F., and D. W. Tinkle. 1975. Natural selection and the evolution of reproductive effort. *Proc. Nat. Acad. Sci.* 72: 2227–2231.

Hislop, J.R.G. 1975. The breeding and growth of whiting *Merlangius merlangus* in captivity. *J. Cons. Perm. Int. Explor. Mer.* 36: 119–127.

Hitchcock, R. R., and R. E. Mirarchi. 1984. Comparisons between single parent and normal mourning dove nestlings during the fledging period. *Wilson Bull.* 96: 494–495.

Ho, F. W., and P. S. Dawson. 1966. Egg cannibalism by *Tribolium* larvae. *Ecology* 47: 318–322.

Hoar, W. S. 1969. Reproduction. In W. S. Hoar and D. J. Randall, eds., *Fish Physiology.* Academic Press, New York.

Hobcraft, J., J. W. McDonald, and C. Rutstein. 1983. Child spacing effects on infant and early child mortality. *Population Index* 49: 585–618.

Hoesch, W. 1959. Zur Biologie des südafrikanischen Laufhünchens *Turnix sylvatica lepurana. J. Orn.* 100: 341–349.

Hoesch, W. 1960. Zum Bruterverhalten des Laufhünchens *Turnix sylvatica lepurana. J. Orn.* 101: 265–275.

Hofman, M. A. 1983. Evolution of the brain in neonatal and adult placental mammals: A theoretical approach. *J. theoret. Biol.* 105: 317–322.

Hogarth, P. J. 1976. *Viviparity.* Edward Arnold, London.

Hogstedt, G. 1981. Should there be positive or negative correlation between survival of adults in a bird population and their clutch size? *Am. Nat.* 118: 568–571.

Holby, L. B., and M. C. Healey. 1986. Selection for adult size in Coho salmon. *Can. J. Fish. Aquat. Sci.* 43: 949–1057.

Holcomb, L. C., and G. Twiest. 1970. Growth rates and sex ratios of red-winged blackbird nestlings. *Wilson Bull.* 82: 294–303.

Holmes, W. G., and P. W. Sherman. 1983. Kin recognition in animals. *Amer. Zool.* 71: 46–55.

Hoogland, J. L. 1981. Sex ratio and local resource competition. *Am. Nat.* 117: 796–797.

Hoogland, J. L., and P. W. Sherman. 1976. Advantages and disadvantages of bank swallow (*Riparia riparia*) coloniality. *Ecol. Monogr.* 46: 33–58.

Hopson, J. A. 1973. Endothermy, small size and the origin of mammalian reproduction. *Am. Nat.* 107: 446–452.

Horrocks, J., and W. Hunte. 1983a. Maternal rank and offspring rank in vervet monkeys: An appraisal of the mechanisms of rank acquisition. *Anim. Behav.* 31: 772–782.

Horrocks, J. A., and W. Hunte. 1983b. Rank relations in vervet sisters: A critique of the role of reproductive value. *Am. Nat.* 122: 417–421.

Horsfall, J. A. 1984. Brood reduction and brood division in coots. *Anim. Behav.* 32: 216–225.

Houston, A. I., and N. B. Davies. 1985. The evolution of cooperation and life history in the dunnock *Prunella modularis*. In R. M. Sibley and R. H. Smith, eds., *Behavioural Ecology*, pp. 471–487. Blackwell Scientific Publications, Oxford.

Howe, H. F. 1976. Egg size, hatching asynchrony, sex, and brood reduction in the common grackle. *Evolution* 57: 1195–1207.

Howe, H. F. 1977. Sex ratio adjustment in the common grackle. *Science* 198: 744–747.

Howe, H. F. 1979. Evolutionary aspects of parental care in the common grackle, *Quiscalus quiscula* L. *Evolution* 33: 41–51.

Hubbs, C. L. 1958. Geographic variations in egg complement of *Percina caprode* and *Etheostoma spectabile*. *Copeia* 1958: 102–105.

Huck, U. W. 1984. Infanticide and the evolution of pregnancy block in rodents. In G. Hausfater and S. B. Hrdy, eds., *Infanticide: Comparative and Evolutionary Perspectives*, pp. 349–365. Aldine, Chicago.

Huck, U. W., J. D. Labov, and R. D. Lisk. 1986. Food restricting young hamsters (*Mesocricetus auratus*) affects sex ratio and growth of subsequent offspring. *Biol. Reprod.* 35: 592–598.

Huck, U. W., J. D. Labov, and R. D. Lisk. 1987. Food-restricting first generation juvenile female hamsters (*Mesocricetus auratus*) affects sex ratio and growth of third generation offspring. *Biol. Reprod.* 37: 612–617.

Huck, U. W., N. C. Pratt, J. D. Labov and R. D. Lisk. 1988. Effects of age and parity on litter size and offspring sex ratios in golden hamsters (*Mesocricetus auratus*). *J. Reprod. Fert.* 83: 209–214.

Hudson, P. J. 1979. The parent-chick feeding relationship of the puffin, *Fratercula arctica*. *J. Anim. Ecol.* 48: 889–898.

Huggins, C., and T.L.-Y. Dao. 1954. Lactation induced by luteotrophin in women with mammary cancer. Growth of the breast of the human male following estrogenic treatment. *Canc. Res.* 14: 303–306.

Hurxthal, L. M. 1979. Breeding behaviour of the ostrich (*Struthio camelus massaicus* Newman) in Nairobi National Park. Ph.D. diss., University of Nairobi, Kenya.

Husby, M. 1986. On the adaptive value of brood reduction in birds: Experiments with the magpie *Pica pica*. *J. Anim. Ecol.* 55. 75–83.

Hussell, D.J.T. 1972. Factors affecting clutch size in Arctic passerines. *Ecol. Monogr.* 42: 317–364.

Imms, A. D. 1970. *A General Textbook of Entomology*. Methuen & Co., London.

Inger, R. F., H. K. Voris, and P. Walker. 1986. Larval transport in a Bornean ranid frog. *Copeia* 1986: 523–525.

Ingram, G. J., M. Anstis, and C. J. Corben. 1975. Observations on the Australian leptodactylid frong, *Assa darlingtoni*. *Herpetologica* 31. 425–429.

Ito, Y. 1980. *Comparative Ecology*. Cambridge University Press, Cambridge.

Ito, Y., and Y. Iwasa. 1981. Evolution of litter size. I. Conceptual re-examination. *Res. Popul. Ecol.* (Kyoto) 23: 344–359.

Ives, A. R. 1987. Testing parent-offspring conflicts in insect parasitoids. *Trends in Evolution and Ecology* 2: 231–233.

James, W. H. 1971. Timing of fertilization and sex ratio of offspring. A review. *Ann. Hum. Biol.* 3: 549–556.

James, W. H. 1987. Hormone levels of parents and sex ratios of offspring. *J. theoret. Biol.* 129: 139–140.

Jameson, E. W. Jr. 1981. *Patterns of Vertebrate Biology.* Springer-Verlag, New York.

Jamieson, I. G., and P. W. Colgan. 1989. Eggs in the nests of males and their effect on mate choice in the three-spined stickleback. *Anim. Behav.* 38: 859–865.

Janzen, D. H. 1977a. Promising direction of study in tropical plant-animal interactions. *Ann. Mo. Bot. Gard.* 61: 706–736.

Janzen, D. H. 1977b. Variation in seed size within a crop of Costa Rican *Micuna andreana* (Leguminosae). *Am. J. Bot.* 64: 347–349.

Jarvis, J. H., and P. E. King. 1972. Reproduction and development in the pycnogonid *Pycnogonum littorale. Marine Biol.* 13: 146–154.

Jarvis, J. H., and P. E. King. 1975. Egg development and the reproductive cycle in the pycnogonid *Pycnogonum littorale. Marine Biol.* 13: 146–154.

Jarvis, J.U.M. 1981. Eusociability in a mammal: Cooperative breeding in naked mole rat colonies. *Science* 212: 571–573.

Jenness, R., and R. E. Sloan. 1970. The composition of milks of various species: A review. *Dairy Sci. Abstr.* 32: 599–612.

Jenni, D. A. 1974. Evolution of polyandry in birds. *Amer. Zool.* 14: 129–144.

Jenni, D. A., and G. Collier. 1972. Polyandry in the American jacana (*Jacana spinosa*). *Auk* 89: 743–765.

Jewell, P. A., C. Milner, and J. M. Boyd, eds. 1974. *Island Survivors: The Ecology of the Soay Sheep of St. Kilda.* Athlone Press, London.

Johannes, R. E. 1978. Reproductive strategies of coastal marine fishes in the tropics. *Env. Biol. Fish.* 3: 65–84.

Johnson, C. N. 1985. Ecology, social behaviour and reproductive success in a population of red-necked wallabies. Ph.D. diss., University of New England, Armidale, N.S.W., Australia.

Johnson, C. N. 1986. Philopatry, reproductive success of females and maternal investment in the red-necked wallaby. *Behav. Ecol. Sociobiol.* 19: 143–50.

Johnson, C. N. 1987. Relationship between mother and infant red-necked wallabies (*Macropus rufogriseus banksianus*). *Ethology* 74: 1–20.

Johnson, C. N. 1988. Dispersal and the sex ratio at birth in primates. *Nature* 332: 726–728.

Johnson, C. N., and P. Jarman. 1983. Geographical variation in offspring sex ratios in kangaroos. *Search* 14: 152–154.

Johnson, E. J., and C. B. Best. 1982. Factors affecting feeding and brooding of gray catbird nestlings. *Auk* 99: 148–156.

Jones, W. T. 1982. Sex ratio and host size in a parasitoid wasp. *Behav. Ecol. Sociobiol.* 10: 207–210.

Jull, M. A. 1928. Second year egg production in relation to first year egg production in the domestic fowl. *Poultry Science* 7: 276–286.

Kaestner, A. 1969. *Invertebrate Zoology*, vol. 2. *Arthropod Relatives, Chelicerata, Myriapoda.* Translated by H. W. Levi and L. R. Levi. Wiley, New York.

Kalela, O., and T. Oksala. 1966. Sex ratio in the wood lemming *Myopus schisticolor* (Hilljeb.) in nature and captivity. *Ann. Univ. Turknensis*, Ser. AII, 37: 1–24.

Kaplan, R. H. 1980a. The implications of ovum size variability for offspring fitness and clutch size within several populations of salamanders (*Ambystoma*). *Evolution* 34: 51–64.

Kaplan, R. H. 1980b. Ontogenetic energetics in *Ambystoma. Physiol. Zool.* 53: 43–56.

Kaplan, R. H., and W. S. Cooper. 1984. The evolution of developmental plasticity in reproductive characteristics: An application of the "adaptive coin flipping" principle. *Am. Nat.* 123: 393–410.

Kaplan, R. H., and W. S. Cooper. 1988. On the evolution of coin-flipping plasticity: A response to McGinley, Temme and Geber. *Am. Nat.* 132: 753–755.

Karlsson, B., and C. Wiklund. 1984. Egg weight variation and lack of correlation between egg weight and offspring fitness in the small brown butterfly *Lasiommata megera. Oikos* 43: 376–385.

Karlsson, B., and G. Wiklund. 1985. Egg weight variation in relation to egg mortality and starvation evidence of newly hatched larvae in some Satyrid butterflies. *Ecol. Entomol.* 10: 205–211.

Karlsson, J. 1983. Breeding of the starling (*Sturnus vulgaris*). Ph.D. diss., University of Lund, Sweden.

Kaston, B. J. 1965. Some little known aspects of spider behavior. *Amer. Midl. Naturalist* 73: 336–356.

Kaston, B. J. 1978. *How to Know the Spiders*, 3d ed. Brown, Dubuque, Iowa.

Kawai, M. 1958. On the system of social ranks in a natural troop of Japanese monkeys. I. Basic rank and dependant rank. *Primates* 1: 111–130.

Kear, J. 1970. The adaptive radiation of parental care in waterfowl. In J. H. Crook, ed., *Social Behaviour in Birds and Mammals*, pp. 357–392. Academic Press, London.

Keenleyside, M.H.A. 1978. Parental behavior in fishes and birds. In E. S. Reese and F. J. Lighter, eds., *Contrasts in Behavior*, pp. 3–30. Wiley, New York.

Keenleyside, M.H.A. 1980. Parental care patterns of fishes. *Am. Nat.* 117: 1019–1022.

Keenleyside, M.H.A. 1983. Mate desertion in relation to adult sex ratio in the biparental cichlid fish *Herotilapia multispinosa. Anim. Behav.* 31: 683–688.

Kendeigh, S. C. 1952. Parental care and its evolution in birds. *Illinois Biological Monograph* 22: 1–358.

Kessel, B. 1957. A study in the breeding biology of the European starling (*Sturnus vulgaris* L.) in North America. *Amer. Midl. Nat.* 58: 251–331.

Kidwell, J. F., and L. Malick. 1967. The effect of genotype, mating status, weight and egg production on longevity in *Drosophila melanogaster. J. Heredity* 58: 169–172.

King, B. H. 1987. Offspring sex ratios in parasitoid wasps. *Q. Rev. Biol.* 62: 367–396.

King, B. H. 1988. Sex ratio manipulation in response to host size by the parasitoid wasp *Spalangia cameroni*: A laboratory study. *Evolution* 42: 1190–1198.

King, P. E. 1973. *Pycnogonids*. Hutchinson, London.

King, P. E. 1974. *British Sea Spiders*. Linnean Society Synopsis of British Fauna, no. 5. Academic Press, London.

King, P. E., and J. H. Jarvis. 1970. Egg development in a littoral pycnogonid, *Nymphon gracile. Marine Biol.* 7: 294–304.

Kleiman, D. G. 1977. Monogamy in mammals. *Q. Rev. Biol.* 52: 39–69.

Kleiman, D. G. 1985. Paternal care in New World primates. *Amer. Zool.* 857–859.

Kleiman, D. G., and J. F. Eisenberg. 1973. Comparisons of canid and felid social systems from an evolutionary perspective. *Anim. Behav.* 21: 637–659.

Kleiman, D. G., and J. R. Malcolm. 1981. The evolution of male parental investment in mammals. In D. J. Gubernick and P. H. Klopfer, eds., *Parental Care in Mammals*, pp. 347–387. Plenum Press, New York.

Klein, D. R. 1968. The introduction, increase and crash of reindeer on St. Matthew Island. *J. Wildl. Mgmt.* 32: 350–367.

Klopfer, P., and M. Klopfer. 1977. Compensatory responses of goat mothers to their impaired young. *Anim. Behav.* 25: 286–291.

Kluge, A. G. 1978. Sexual selection, territoriality and the evolution of parental care in *Hyla rosenbergi.* Abstract, Joint Annual Meeting, ASIH-SSAR-Herp. League, Tempe, Arizona.

Knapton, R. W. 1984. Parental investment: The problem of currency. *Can. J. Zool.* 62: 2673–2674.

Knight, R. L., and S. A. Temple. 1986a. Why does intensity of avian nest defence increase during the nesting cycle? *Auk* 103: 318–327.

Knight, R. L., and S. A. Temple. 1986b. Methodological problems in studies of avian nest defence. *Anim. Behav.* 34: 561–566.

Knight, R. L., and S. A. Temple. 1986c. Nest defence in the American goldfinch. *Anim. Behav.* 34: 887–897.

Knowlton, N. 1979. Reproductive synchrony, parental investment and the evolutionary dynamics of sexual selection. *Anim. Behav.* 27: 1022–1033.

Knowlton, N. 1982. Parental care and sex role reversal. In King's College Sociobiology Group, eds., *Current Problems in Sociobiology*, pp. 203–222. Cambridge University Press, Cambridge.

Kojola, I. 1989. Mother's dominance and differential investment in reindeer calves. *Anim. Behav.* 38: 177–185.

Kojola, I., and E. Eloranta. 1989. Influences of maternal body weight, age and parity on sex ratio in semidomesticated reindeer (*Rangifer t. tarandus*). *Evolution* 43: 1331–1336.

Kolding, S., and T. M. Fenchel. 1981. Patterns of reproduction in different populations of five species of the amphipod genus *Gammarus. Oikos* 37: 167–172.

König, B., J. Riesler, and H. Markl. 1988. Maternal care in house mice (*Mus musculus*). II. The energy cost of lactation as a function of litter size. *J. Zool.* 216: 195–210.

Koufopanou, V., and G. Dell. 1984. Measuring the cost of reproduction. IV. Predation experiments with *Daphnia pulex. Oecologia* 64: 81–86.

Kovacs, K., and D. M. Lavigne. 1986. Maternal investment and neonatal growth in phocid seals. *J. Anim. Ecol.* 55: 1035–1051.

Kramer, D. L. 1973. Parental behaviour in the blue gourami *Trichogaster tricopterus* (Pisces, Belontiidae) and its induction during exposure to varying numbers of conspecific eggs. *Behaviour* 47: 14–32.

Kronester-Frei, A. 1975. Licht- und elektronenmikroskopische Untersuchungen am Brutepithel des Männchens von *Nerophis lumbriciformis* (Pennant, 1776). Syngnathidae, unter spezieller Berücksichtigung der strukturellen Veränderung der Eihülle. *Forma und Funtio* 8: 419–462.

Kruse, K. C. 1990. Male backspace availability in the giant waterbug (*Belostoma flumineum* Say.). *Behav. Ecol. Sociobiol.* 26: 281–289.

Kruuk, H. 1972. *The Spotted Hyena.* University of Chicago Press, Chicago.

Kuester, J., and A. Paul. 1986. Male-infant relationships in free-ranging Barbary macaques (*Macaca sylvanus*) of Aftenberg Salem/FRG: Testing the "male care" hypothesis. *Amer. J. Primatol.* 10: 315–327.

Labov, J. B., U. W. Huck, P. Vaswani, and R. D. Lisk. 1986. Sex ratio manipulation and decreased growth of male offspring of prenatally and postnatally undernourished golden hamsters. *Behav. Ecol. Sociobiol.* 18: 241–249.

Lacey, E. P., L. Real, J. Antonovics, and D. G. Haeckel. 1983. Variance models in the study of life histories. *Am. Nat.* 122: 114–131.

Lack, D. 1947. The significance of clutch size. *Ibis* 89: 302–352.

Lack, D. 1954. *The Natural Regulation of Animal Numbers.* Oxford University Press, Oxford.

Lack, D. 1966. *Population Studies of Birds.* Clarendon Press, Oxford.

Lack, D. 1968. *Ecological Adaptations for Breeding in Birds.* Methuen & Co., London.

Lancaster, D. A. 1964a. Biology of the brushland tinamou, *Nothoprocta cinerascens. Bull. Amer. Mus. Nat. Hist.* 127: 269–314.

Lancaster, D. A. 1964b. Life history of the Boucard tinamou in British Honduras. III. Breeding biology. *Condor:* 253–276.

Lane, E. A., and T. S. Hyde. 1973. The effect of maternal stress on fertility and sex ratio: A pilot study with rats. *J. Abnormal Psychol.* 82: 73–80.

Lang, E. M. 1963. Flamingoes raise their young on a liquid containing blood. *Experientia* 19: 532–533.

Lank, D. B., L. W. Oring, and S. J. Maxson. 1985. Mate and nutrient limitation of egg laying in the spotted sandpiper *Actitis macularia*, a polyandrous shorebird. *Ecology* 66: 1513–1524.

Latham, R. M. 1947. Differential ability of male and female game birds to withstand starvation and climatic extremes. *J. Wildl. Mgmt.* 11: 139–149.

Laws, R. M. 1956. Growth and sexual maturity in aquatic mammals. *Nature* 178: 193–194.

Laws, R. M., I.S.C. Parker, and R.C.B. Johnstone. 1975. *Elephants and Their Habitats.* Oxford University Press, Oxford.

Lazarus, J. 1987. The concepts of sociobiology. In H. Beloff and A. M. Colman, eds., *Psychology Survey 6*, pp. 192–216. British Psychological Society, Leicester, England.

Lazarus, J. 1989. The logic of mate desertion. *Anim. Behav.* 39: 657–671.

Lazarus, J., and I. R. Inglis. 1978. The breeding behaviour of the pink-footed goose: Parental care and vigilant behaviour during the fledging period. *Behaviour* 65: 62–88.

Lazarus, J., and I. R. Inglis. 1986. Shared and unshared parental investment, parent-offspring conflict and brood size. *Anim. Behav.* 34: 1791–1804.

Leader-Williams, N. 1988. *Reindeer on South Georgia*. Cambridge University Press, Cambridge.

Leader-Williams, N., and C. Ricketts. 1982. Seasonal and sexual patterns of growth and condition in introduced reindeer on South Georgia. *Oikos* 38: 27–39.

Le Boeuf, B. J., and K. T. Briggs. 1977. The cost of living in a seal harem. *Mammalia* 41: 167–195.

Le Boeuf, B. J., R. Condit, and J. Reiter. 1989. Parental investment and the secondary sex ratio in northern elephant seals. *Behav. Ecol. Sociobiol.* 25: 109–117.

Lee, A. K., and A. Cockburn. 1985. *Evolutionary Ecology of Marsupials*. Cambridge University Press, Cambridge.

Lee, P. C. 1987. Nutrition, fertility and maternal investment in primates. *J. Zool.* 213: 409–422.

Lee, P. C., and C. J. Moss. 1986. Early maternal investment in male and female African elephant calves. *Behav. Ecol. Sociobiol.* 18: 353–361.

Legates, J. 1972. The role of maternal effects in animal breeding. IV. Maternal effects in laboratory species. *J. Anim. Sci.* 35: 1294–1302.

Leigh, E. G. 1970. Sex ratio and differential mortality between the sexes. *Am. Nat.* 104: 205–210.

Leitch, J., F. E. Hytten, and W. Z. Billewicz. 1959. Maternal and neonatal weights of some Mammmalia. *Proc. Zool. Soc. Lond.* 133: 11–28.

Le Mesurier, A. D. 1987. A comparative study of the relationship between host size and brood size in *Apanteles* sp. (Hymenoptera: Braconidae). *Ecol. Entomol.* 12: 383–393.

Lengerken, H. von. 1939. Die Brutforsorge- und Brutpflegeinstinkte der Käfer. Akademische Verlagsgesellschaft MBH, Leipzig, G.D.R.

Lenninington, S. 1980. Bi-parental care in killdeer: An adaptive hypothesis. *Wilson Bull.* 92: 8–20.

Lennington, S., and T. Mace. 1975. Mate fidelity and nesting site tenacity in the killdeer. *Auk* 92: 192–251.

Lerner, I. M. 1958. *The Genetic Basis of Selection*. Wiley, New York.

Lessells, C. M., and M. I. Avery. 1987. Sex ratio selection in species with helpers at the nest: Some extensions of the repayment model. *Am. Nat.* 129: 610–620.

Lessells, C. M., and M. I. Avery. 1989. Hatching asynchrony in European bee-eaters *Merops apiaster*. *J. Anim. Ecol.* 58: 815–835.

Leutenegger, W. 1973. Maternal-fetal weight relationships in primates. *Folia primatol.* 20: 280–293.

Leutenegger, W. 1976. Allometry of neonatal size in eutherian mammals. *Nature* 263: 229–230.

Leutenegger, W. 1977. Neonatal-maternal weight relationship in macaques: An example of intrageneric scaling. *Folia Primatol.* 27: 152–159.

Leutenegger, W. 1980. Monogamy in callitrichids: A consequence of phyletic dwarfism? *Int. J. Primatol.* 1: 95–98.

Lewontin, R. C. 1978. Adaptation. *Scientific American* 239: 156–165.

Liberg, O., and T. von Schantz. 1985. Sex-biased philopatry and dispersal in birds and mammals: The Oedipus hypothesis. *Am. Nat.* 126: 129–135.

Licht, P., and W. R. Moberly. 1965. Thermal requirements for embryonic development in the tropical lizard. *Copeia* 1965: 515–517.

Lidicker, W. Z. 1976. Social behaviour and density regulation in confined populations of four species of rodent. *Res. Pop. Ecol.* 7: 57–72.

Lill, A. 1974. The evolution of clutch size and male "chauvinism" in the white-bearded manakin. *Living Bird* 13: 211–232.

Lill, A. 1986. Time energy budgets during reproduction in the evolution of single parenting in the superb lyrebird. *Aust. J. Zool.* 34: 351–371.

Lillegraven, J. A. 1974. Biogeographical considerations of the marsupial-placental dichotomy. *Ann. Rev. Ecol. Syst.* 5: 74–94.

Lillegraven, J. A. 1975. Biological considerations of the marsupial-placental dichotomy. *Evolution* 29: 707–722.

Lillegraven, J. A., S. D. Thompson, B. K. McNab, and J. L. Patton. 1987. The origin of eutherian mammals. *Biol. J. Linn. Soc.* 32: 281–336.

Linden, H. 1981. Estimation of juvenile mortality in the Capercaillie *Tetrao urogallus* and the black grouse, *Tetrao tetrix* from indirect evidence. *Finnish Game Research* 39: 35–51.

Linden, M., and A. Pape Møller. 1989. Cost of reproduction and covariation of life history traits in birds. *Trends in Ecology and Evolution* 4: 367–370.

Linsenmair, K. E., and C. Linsenmair. 1971 Paarbildung und Paarzusammenhaltung bei der monogamen Wüstenassel *Hemilepistus reamuri* (Crustacea, Isopoda, Oniscoidea). *Z. Tierpsychol.* 29: 134–155.

Linzell, J. L. 1972. Milk yield, energy loss in milk and mammary gland weight in different species. *Dairy Sci. Abstr.* 34: 351–360.

Liske, E., and W. J. Davis. 1987. Courtship and mating behaviour of the Chinese praying mantis, *Tenodera aridifolia sinensis. Anim. Behav.* 35: 1524–1537.

Llewellyn, L. C. 1974. Spawning, development and distribution of the Southern Pigmy Perch, *Nannoperca australis australis* Guenther from inland waters in Eastern Australia. *Aust. J. Mar. Freshwat. Res.* 25: 121–149.

Lloyd, D. G. 1984. Variation strategies of plants in heterogeneous environments. *Biol. J. Linn. Soc.* 21: 357–385.

Lloyd, D. G. 1987. Selection of offspring size at independence and other size-versus-number strategies. *Am. Nat.* 129: 800–817.

Lockie, J. D. 1955. The breeding and feeding of jackdaws and rooks with notes on carrion crows and other corvidae. *Ibis* 97: 31–369.

Lohrl, L. B. 1968. Das Nesthäkchen als biologisches Problem. *J. Ornithol.* 109: 383–395.

Loiselle, P. V. 1978. Prevalence of male brood care in teleosts. *Nature* 276: 98–99.

Loiselle, P. V., and G. W. Barlow. 1978. Do fishes lek like birds? In E. S. Reese and F. J. Lighter, eds., *Contrasts in Behavior*, pp. 31–76. Wiley, New York.

Long, C. A. 1972. Two hypotheses on the origin of lactation. *Am. Nat.* 106: 141–144.

Loudon, A.S.I. 1985. Neonatal mortality and lactation in mammals. *Symp. Zool. Soc. Lond.* 54: 183–207.

Loudon, A.S.I., and P. A. Racey, eds. 1987. *Reproductive Energetics in Mammals.* Symp. Zool. Soc. Lond. 57.

Loudon, A.S.I., A. Darroch, and J. A. Milne. 1984. The lactation performance of red deer on hill and improved species pastures. *J. Agric. Sci. Camb.* 102: 149–158.

Low, B. S. 1978. Environmental uncertainty and the parental strategies of marsupials and placentals. *Am. Nat.* 112: 197–213.

Luckinbill, L. S., R. Arking, and M. J. Clare. 1985. Selection for delayed senescence in *Drosophila melanogaster*. *Evolution* 38: 996–1003.

Lundberg, C. A., and R. A. Vaisanen. 1979. Selective correlation of egg size with chick mortality in the black headed gull (*Larus ridibundus*). *Condor* 81: 141–156.

Lynch, M. J. 1984. The limits to life history evolution in *Daphnia*. *Evolution* 38: 465–482.

Lyon, B., R. D. Montgomerie, and L. D. Hamilton. 1987. Male parental care and monogamy in snow buntings. *Behav. Ecol. Biol.* 20: 377–382.

Lyon, M., L. Goldman and R. Hoage. 1985. Parent-offspring conflict following a birth in the primate, *Callimico goeldii*. *Anim. Behav.* 4: 1364–1365.

MacArthur, J. W., and W.H.T. Baillie. 1932. Sex differences in mortality in abraxas-type species. *Q. Rev. Biol.* 7: 313–325.

McCance, R. A., and E. M. Widdowson. 1962. Nutrition and growth. *Proc. R. Soc. B* 156: 326–337.

McCann, T. S., M. A. Fedak, and J. Harwood. 1989. Parental investment in southern elephant seals *Mirounga leonina*. *Behav. Ecol. Sociobiol.* 25: 81–87.

McClure, P. A. 1981. Sex biased litter reduction in food-restricted wood rats (*Neotama floridana*). *Science* 211: 1058–1060.

McClure, P. A. 1987. The energetics of reproduction and life histories of cricetine rodents (*Neotoma floridana* and *Sigmodon hispidus*). *Symp. Zool. Soc. Lond.* 57: 241–258.

McCullough, D. R. 1979. *The George River Deer Herd: Population Ecology of a k-Selected Species*. University of Michigan Press, Ann Arbor.

McDiarmid, R. W. 1978. Evolution of parental care in frogs. In G. M. Burghardt and M. Bekoff, eds., *The Development of Behavior: Comparative and Evolutionary Aspects*, pp. 127–147. Garland, New York.

MacDowell, E. C., and E. M. Lord. 1926. Relative viability of male and female mouse embryos. *Am. J. Anat.* 37: 127–140.

Mace, G. M. 1979. The evolutionary ecology of small mammals. Ph.D. diss., University of Sussex, England.

McEdward, L. R., and L. K. Coulter. 1987. Egg volume and energetic content are not correlated among sibling offspring of starfish: Implications for life history theory. *Evolution* 41: 914–917.

McEwan, E. H. 1968. Growth and development of barren ground caribou. II. Postnatal growth rates. *Can. J. Zool.* 46: 1023–1029.

McGillivray, W. B. 1983. Intraseasonal reproductive costs of the house sparrow (*Passer domesticus*). *Auk* 100: 25–32.

McGinley, M. A. 1984. The adaptive value of male-biased sex ratios among stressed animals. *Am. Nat.* 124: 597–599.

McGinley, M. A. 1989. The influence of a positive correlation between clutch size and offspring fitness on the optimal offspring size. *Evol. Ecol.* 3: 150–156.

McGinley, M. A., and E. L. Charnov. 1988. Multiple resources and the optimal balance between size and number of offspring. *Evol. Ecol.* 2: 77–84.

McGinley, M. A., D. H. Temme, and M. A. Geber. 1987. Parental investment in offspring in variable environments: Theoretical and empirical considerations. *Am. Nat.* 130: 370–398.

McKaye, K. R. 1981. Natural selection and the evolution of interspecific brood care in fishes. In R. D. Alexander and D. W. Tinkle, eds., *Natural Selection and Social Behavior*, pp. 173–183. Chiron Press, New York.

McKaye, K. R., and N. M. McKaye. 1977. Communal care and kidnapping of young by parental cichlids. *Evolution* 31: 674–681.

McMillen, M. M. 1979. Differential mortality by sex in fetal and neonatal deaths. *Science* 204: 89–91.

McNab, B. K. 1978. The comparative energetics of neotropical marsupials. *J. Comp. Physiol.* 125: 115–128.

McNab, B. K. 1980. Food habits, energetics and population biology of mammals. *Am. Nat.* 116: 106–124.

McNab, B. K. 1987. The reproduction of marsupial and eutherian mammals in relation to energy expenditure. *Symp. Zool. Soc. Lond.* 57: 29–39.

McNab, B. K. 1988. Complications inherent in scaling the basal rate of metabolism of mammals. *Q. Rev. Biol.* 63: 25–53.

Macnair, M. R., and G. A. Parker. 1978. Models of parent-offspring conflict. II. Promiscuity. *Anim. Behav.* 26: 111–122.

Macnair, M. R., and G. A. Parker. 1979. Models of parent-offspring conflict. III. Intrabrood conflict. *Anim. Behav.* 27: 1202–1209.

McNamara, J. M., and A. I. Houston. 1986. The common currency for behavioral decisions. *Am. Nat.* 127: 358–378.

McNeilly, A. S. 1987. Prolactin and the control of gonadotrophin secretion. *J. Endocr.* 115: 1–5.

McNeilly, A. S. 1988. Suckling and the control of gonadotropin secretion. In E. Knobil and J. Neill, eds., *The Physiology of Reproduction*, pp. 2323–2349. Raven Press, New York.

McShea, W. J., and D. M. Madison. 1986. Sex ratio shifts within litters of meadow voles (*Microtus pennsylvanicus*). *Behav. Ecol. Scoiobiol.* 18: 431–436.

McVey, M. E. 1988. The opportunity for sexual selection in a territorial dragonfly, *Erythemis simplicicollis*. In T. H. Clutton-Brock, ed., *Reproductive Success*, pp. 44–58. University of Chicago Press, Chicago.

Magrath, R. D. 1988. Hatching asynchrony in altricial birds: Nest failure and adult survival. *Am. Nat.* 131: 893–950.

Magrath, R. D. 1989. Hatching asynchrony and reproductive success in the blackbird. *Nature* 339: 536–538.

Malcolm, J. R. 1985. Paternal care in Canids. *Amer. Zool.* 25: 853–859.

Malcolm, J. R., and K. Marten. 1982. Natural selection and the communal rearing of pups in African wild dogs (*Lycaon pictus*). *Behav. Ecol. Sociobiol.* 10: 1–13.

Mallory, F. F., and R. J. Brooks. 1978. Infanticide and other reproductive strategies in the collared lemming, *Dicrostonyx groenlandicus*. *Nature* 273: 144–146.

Mangel. M., and C. W. Clark. 1986. Towards a unified foraging theory. *Ecology* 67: 1127–1138.

Mangel, M., and C. W. Clark. 1988. *Dynamic Modeling in Behavioral Ecology*. Princeton University Press, Princeton, N.J.

Mann, R.H.K., and C. A. Mills. 1979. Demographic aspects of fish fecundity. *Symp. Zool. Soc. Lond.* 44: 161–177.

Martin, K. 1984. Reproductive defence priorities of male willow ptarmigan (*Lagopus lagopus*): Enhancing mate survival or extending paternity options? *Behav. Ecol. Sociobiol.* 16: 57–63.

Martin, K., F. G. Cooch, R. F. Rockwell, and F. Cooke. 1985. Reproductive performance in lesser snow geese: Are two parents essential? *Behav. Ecol. Sociobiol.* 17: 257–263.

Martin, K., and F. Cooke. 1987. Bi-parental care in willow ptarmigan: A luxury? *Anim. Behav.* 35: 369–379.

Martin, R. D. 1981. Relative brain size and metabolic rate in terrestrial vertebrates. *Nature* 293: 57–60.

Martin, R. D. 1983. Human brain evolution in an ecological context. *52nd James Arthur Lecture on the Evolution of the Human Brain.* Amer. Mus. Nat. Hist., New York.

Martin, R. D. 1984. Scaling effects and adaptive strategies in mammalian lactation. *Symp. Zool. Soc. Lond.* 51: 81–117.

Martin, R. D., and A. M. MacLarnon. 1985. Gestation period, neonatal size and maternal investment in placental mammals. *Nature* 313: 220–223.

Martin, S. G. 1974. Adaptations for polygynous breeding in the bobolink, *Dolichonyx oryzivorous. Am. Zool.* 14: 109–119.

Masters, J. C., M. R. Centner, and W. Caithness. 1982. Sex ratios in galagos revisited. *Sth. Afr. J. Sci.* 78: 198–202.

Maxson, S. J., and L. W. Oring. 1980. Breeding season time and energy budgets of the polyandrous spotted sandpiper. *Behaviour* 74: 200–263.

May, R. M., and D. I. Rubenstein. 1984. Reproductive strategies. In C. R. Austin and R. V. Short, eds., *Reproductive Fitness*, pp. 1–23. Cambridge University Press, Cambridge.

Maynard Smith, J. 1977. Parental investment: A prospective analysis. *Anim. Behav.* 25: 1–9.

Maynard Smith, J. 1978. *The Evolution of Sex.* Cambridge University Press, Cambridge.

Maynard Smith, J. 1980. A new theory of sexual investment. *Behav. Ecol. Sociobiol.* 7: 247–251.

Maynard Smith, J. 1982a. *Evolution and the Theory of Games.* Cambridge University Press, Cambridge.

Maynard Smith, J. 1982b. The evolution of social behaviour—a classification of models. In King's College Sociobiology Group, eds., *Current Problems in Sociobiology*, pp. 29–44. Cambridge University Press, Cambridge.

Maynard Smith, J., and M. G. Ridpath. 1972. Wife sharing in the Tasmanian native hen *Tribonyx mortierri*: A case of kin selection? *Am. Nat.* 106: 447–452.

Maynard Smith, J., R. Burian, S. Kauffman, P. J. Campbell, B. Goodwin, R. Lande, D. Raup, and L. Wolpert. 1985. Developmental constraints and evolution. *Q. Rev. Biol.* 60: 265–287.

Meikle, D. B., and L. C. Drickamer. 1986. Food availability and secondary sex ratio variation in wild and laboratory house mice *Mus musculus. J. Reprod. Fert.* 78: 587–591.

Meikle, D. B., B. L. Tilford, and S. H. Vesey. 1984. Dominance rank, secondary sex ratio, and reproduction of offspring in polygynous primates. *Am. Nat.* 124: 173–188.

Mell, R. 1929. *Beiträge zur Fauna Sinica IV. Grundzüge einer Ökologie der chinesischen Reptilien und einer perpetologischen Tiergeographie Chinas.* Walter de Gruyter, Berlin.

Mendl, M. 1988. The effects of litter size on variation in mother-offspring relationships and behavioural and physical development in several mammalian species (principally rodents). *J. Zool.* 215: 15–34.

Mendl, M., and E. S. Paul. 1989. Observation of nursing and sucking behaviour as an indicator of milk transfer and parental investment. *Anim. Behav.* 37: 513–514.

Mendoza, S. P., and W. A. Mason. 1986. Parental division of labor and differentiation of attachments in a monogamous primate (*Callicebus moloch*). *Anim. Behav.* 34: 1336–1347.

Merritt, E. S. 1962. Selection for egg production in geese. *Proc. 12th World Poultry Cong. Sydney*, pp. 83–87.

Mertz, D. B. 1975. Senescent decline in flour beetle strains selected for early adult fitness. *Physiolog. Zool.* 48: 1–23.

Mertz, D. B., and J. R. Robertson. 1970. Some developmental consequences of handling egg-eating and population density for flour beetle larvae. *Ecology* 51: 989–998.

Messieh, S. 1976. Fecundity studies on Atlantic herring from the Southern Gulf of St. Lawrence and along the Nova Scotia Wash. *Trans. Am. Fish. Soc.* 105: 384–394.

Metcalf, R. A. 1980. Sex ratios, parent offspring conflict and local competition for males in the social wasps *Polistes metricus* and *Polistes variatus*. *Am. Nat.* 116: 648–654.

Metcalf, R. A., J. A. Stamps, and V. V. Krishnan. 1979. Parent-offspring conflict which is not limited by degree of kinship. *J. theoret. Biol.* 76: 99–107.

Metzen, W. D. 1977. Nesting ecology of alligators on the Ofkefenokee National Wildlife Refuge. *Proc. Ann. Southeast Assn. Game and Fish Comm.* 31: 29–32.

Michelson, P. G. 1979. Avian community ecology at two sites on Espenberg Peninsula in Kotzebue Sound, Alaska. In *Environmental Assessment of the Alaska Continental Shelf*, pp. 289–607. Final Reports of Principal Investigators, 5. U.S. Dept. of Commerce, OCSEAP.

Michod, R. E. 1979. Evolution of life histories in response to age-specific mortality factors. *Am. Nat.* 113: 531–550.

Milinski, M. 1978. Kin selection and reproductive value. *Z. Tierpsychol.* 47: 328–329.

Millar, J. S. 1977. Adaptive features of mammalian reproduction. *Evolution* 31: 370–386.

Millar, J. S. 1978. Energetics of reproduction in *Peromyscus leucopus*: The cost of lactation. *Ecology* 59: 1055–1061.

Millar, J. S. 1981. Post-partum characteristics of eutherian mammals. *Evolution* 35: 1149–1163.

Millar, J. S. 1984. The role of design constraints on the evolution of mammalian reproductive rates. *Acta Zool. Fenn.* 171: 133–136.

Miller, R. S. 1969. Competition and species diversity. *Brookhaven Sympos. Biol.*, no. 22: 63–70.

Milne, J. A. 1987. The effect of litter and maternal size on reproductive performance of grazing ruminants. *Symp. Zool. Soc. Lond.* 57: 189–201.

Milne, L. J., and M. Milne, 1976. The social behavior of burying beetles. *Sci. Amer.* 235: 84–89.

Missakian, E. A. 1972. Genealogical and cross-genealogical dominance relations in a group of free-ranging rhesus monkeys (*Macaca mulatta*) on Cayo Santiago. *Primates* 13: 169–180.

Mitchell, B., B. W. Staines, and D. Welch. 1977. *Ecology of Red Deer: A Research Review Relevant to Their Management.* Institute of Terrestrial Ecology, Cambridge, England.

Mitchell, G. D. 1968. Paternalistic behavior in primates. *Psychol. Bull.* 71: 399–417.

Mock, D. W. 1983. On the study of avian mating systems. In A. H. Brush and G. A. Clark, eds., *Perspectives in Ornithology*, pp. 55–84. Cambridge University Press, Cambridge.

Mock, D. W. 1984a. Infanticide, siblicide and avian nesting mortality. In G. Hausfater and S. B. Hrdy, eds., *Infanticide: Comparative and Evolutionary Perspectives*, pp. 3–30. Aldine, New York.

Mock, D. W. 1984b. Siblicidal aggression and resource monopolization in birds. *Science* 225: 731–733.

Mock, D. W., and M. Fujioka. 1990. Monogamy and long-term pair bonding in vertebrates. *Trends in Ecology and Evolution* 5: 39–43.

Mock, D. W., and G. A. Parker. 1986. Advantages and disadvantages of egret and heron brood reduction. *Evolution* 40: 459–470.

Mock, D. W., and B. J. Ploger. 1987. Parental manipulation of optimal hatch asynchrony in cattle egrets: An experimental study. *Anim. Behav.* 35: 150–160.

Moehlmann, P. D. 1986. Ecology of cooperation in canids. In D. I. Rubenstein and R. W. Wrangham, eds., *Ecological Aspects of Social Evolution*, pp. 64–86. Princeton University Press, Princeton, N.J.

Moehlmann, P. 1988. Intraspecific variation in canid mating systems. In J. L. Gittleman, ed., *Carnivore Behavior, Ecology and Evolution*, pp. 143–163. Cornell University Press, Ithaca, N.Y.

Moen, A. M. 1973. *Wildlife Ecology.* W. H. Freeman & Co., San Francisco.

Mohler, L. L., J. H. Wampole, and E. Fichter. 1951. Mule deer in Nebraska National Forest. *Wildl. Mgmt.* 15: 129–157.

Møller, A. P. 1986. Mating systems among European passerines: A review. *Ibis* 128: 234–250.

Montgomerie, R. D., and P. J. Weatherhead. 1988. Risks and rewards of nest defence by parent birds. *Q. Rev. Biol.* 63: 167–187.

Montgomery, G. G., and M. E. Sunquist. 1978. Habitat selection and use by two-toed and three-toed sloths. In G. G. Montgomery, ed., *The Ecology of Arboreal Folivores*, pp. 329–360. Smithsonian Institution Press, Washington, D.C.

Moore, C. L. 1982. Maternal behavior of rats is affected by hormonal condition of pups. *J. Comp. Physiol. Psychol.* 96: 123–129.

Moore, C. L., and G. A. Morelli. 1979. Mother rats interact differently with male and female offspring. *J. Comp. Physiol. Psychol.* 93: 677–684.

Mori, U. 1979a. Reproductive behavior. In M. Kawai, ed., *Ecological and Sociological Studies of Gelada Baboons*, pp. 183–197. Karger, Basel, Switzerland.

Mori, U. 1979b. Individual relationships within a unit. In M. Kawai, ed., *Ecological and Sociological Studies of Gelada Baboons*, pp. 94–124. Karger, Basel, Switzerland.

Moriya, A., and T. Hiroshige. 1978. Sex ratio of offspring of rats bred at 5°C. *Int. J. Biometeor.* 22: 312–315.

Morris, J. A. 1963. Continuous selection for egg production using short-term records. *Aust. J. Agric. Res.* 14: 909–925.

Morton, S. R., H. F. Richter, S. D. Thompson, and R. W. Braithwaite. 1982. Comments on the relative advantages of marsupial and eutherian reproduction. *Am. Nat.* 120: 128–134.

Moss, R., and J. Oswald. 1985. Population dynamics of Capercaillie in a north-east Scottish glen. *Ornis. Scand.* 16: 229–238.

Mrowka, W. 1987a. Brood adoption in a mouthbrooding cichlid fish: Experiments and a hypothesis. *Anim. Behav.* 35: 922–923.

Mrowka, 1987b. Egg stealing in a mouthbrooding cichlid fish. *Anim. Behav.* 35: 923–925.

Muehleis, P. M., and S. Y. Long. 1976. The effects of altering the pH of seminal fluid on the sex ratio of rabbit off-spring. *Fertility and Senility* 27: 1438–1445.

Muldal, A. M., J. D. Moffatt, and R. J. Robertson. 1986. Parental care of nestlings by male red-winged blackbirds. *Behav. Ecol. Sociobiol.* 19: 105–114.

Mumme, R. L., W. D. Koenig, and F. A. Pitelka. 1983. Reproductive competition in the communal acorn woodpecker: Sisters destroy each others' eggs. *Nature* 306: 583–584.

Murdoch, W. W. 1966. Population stability and life-history phenomena. *Am. Nat.* 150: 5–11.

Murphy, P. A., J. T. Giesel, and M. N. Manlove. 1983. Temperature effects on life history variation in *Drosophila simulans*. *Evolution* 37: 1181–1192.

Myers, J. H. 1978. Sex ratio adjustment under food stress: Maximization of quality or numbers of offspring? *Am. Nat.* 112: 368–381.

Myrberg, A. 1965. A descriptive analysis of the African cichlid fish, *Pelmatochromis guentheri* (Sauvage). *Anim. Behav.* 13: 312–329.

Myrberg, A. 1966. Parental recognition of young in cichlid fishes. *Anim. Behav.* 14: 565–571.

Nafus, D. E., and I. H. Schreiner. 1988. Parental care in a tropical nymphalid butterfly *Hypolimnas anomala*. *Anim. Behav.* 36: 1425–1431.

Neill, W. T. 1964. Viviparity in snakes: Some ecological and zoogeographical considerations. *Am. Nat.* 98: 35–55.

Nethersole-Thompson, D. 1973. *The Dotterel*. William Collins, Glasgow, Scotland.

Nettleship, D. N. 1972. Breeding success of the common puffin (*Fratercula arctica*) on different habitats at Great Island Newfoundland. *Ecol. Monogr.* 42: 239–268.

Newton, I. 1979. *Population Ecology of Raptors*. Poyser, Berkhamstead, England.

Newton, I., and M. Marquiss. 1979. Sex ratio among nestlings of the European sparrow-hawk. *Am. Nat.* 113: 309–315.

Nice, M. M. 1962. Development of behavior in precocial birds. *Trans. Linn. Soc. N.Y.* 8: 1–211.

Nicoll, C. S., and H. A. Bern. 1971. On the actions of prolactin among the vertebrates: Is there a common denominator? In G.E.W. Wolstenholme and J. Knight, eds., *Lactogenic Hormones*, pp. 299–374. Churchill Livingstone, Edinburgh, Scotland.

Nicolson, N. A. 1982. Weaning and development of independence in olive baboons. Ph.D. diss., Harvard University, Cambridge, Mass.

Nikolsky, G. V. 1963. *The Ecology of Fishes*. Academic Press, London.

Nisbet, I.C.T. 1973. Courtship feeding, egg size and breeding success in common terns. *Nature* 241: 141–142.

Noakes, D.L.G. 1979. Parent-touching behaviour by young fishes: Incidence, function and causation. *Env. Biol. Fish.* 4: 389–400.

Noakes, D.L.G., and G. W. Barlow. 1973. Ontogeny of parent-contacting in young *Cichlasoma citrinellum* (Pisces, Cichlidae). *Behaviour* 46: 221–255.

Nol, E. 1986. Incubation period and foraging technique of shorebirds. *Am. Nat.* 128: 115–119.

Nonacs, P. 1986. Ant reproductive strategies and sex allocation theory. *Q. Rev. Biol.* 61: 1–21.

Noonan, K. M. 1981. Sex ratios of parental investment in colonies of the social wasp (*Polistes fuscatus*). *Science* 199: 1354–1356.

Nordskog, A. W. 1977. Success and failure of quantitative genetic theory in poultry. In E. Pollacks, O. Kempthorne and T. B. Bailey, eds., *Proceedings of the International Conference on Quantitative Genetics*, pp. 568–569. Ohio State University, Columbus.

Nordskog, A. W., and M. Festing. 1962. Selection and correlated responses in the fowl. *Proc. 12th World Poultry Cong. Sydney*, pp. 25–29.

Nordskog, A. W., H. S. Tolman, D. W. Casey, and C. Y. Lin. 1974. Selection in small populations of chickens. *Poultry Science* 53: 1188–1219.

Nowosad, R. F. 1975. Reindeer survival in the Mackenzie Delta herd, birth to four months. In *Proc. First International Reindeer and Caribou Symposium*, pp. 199–208. Biological Papers of the University of Alaska, Special Report No. 1. Fairbanks, Alaska.

Nuechterlein, G. L., and A. Johnson. 1981. The downy young of the hooded grebe. *Living Bird* 19: 69–71.

Nur, N. 1983. On parental investment during the breeding season. *Anim. Behav.* 31: 309–311.

Nur, N. 1984a. The consequences of brood size for breeding blue tits. I. Adult survival, weight change and the cost of reproduction. *J. Anim. Ecol.* 53: 479–496.

Nur, N. 1984b. The consequences of brood size for breeding blue tits. II. Nestling weight, offspring survival and optimal brood size. *J. Anim. Ecol.* 53: 497–518.

Nur, N. 1984c. Feeding frequencies of nestling blue tits (*Parus caeruleus*): Costs, benefits and a model of optimal feeding frequency. *Oecologia* 65: 125–137.

Nussbaum, R. A. 1985. The evolution of parental care in salamanders. *Misc. Publ. Mus. Zool. Univ. Michigan* 169: 1–50.

Nussbaum, R. A. 1987. Parental care and egg size in salamanders: An examination of the safe harbor hypothesis. *Res. Popul. Ecol.* (Kyoto) 29: 27–44.

Nussbaum, R. A., and D. L. Schultz. 1989. Coevolution of parental care and egg size. *Am. Nat.* 133: 591–603.

O'Connor, R. J. 1978. Brood reduction in birds: Selection for fratricide, infanticide and suicide. *Anim. Behav.* 26: 79–96.

O'Connor, R. 1984. *The Growth and Development of Birds.* Wiley, Chichester, England.

Odhiambo, T. R. 1959. An account of parental care in *Rhinocoris albopilosus* Signoret (Hemiptera—Heteroptera: Reduviidae) with notes on its life history. *Proc. Roy. Entomol. Soc. London A* 34: 175–185.

Odhiambo, T. R. 1960. Parental care in bugs and non-social insects. *New Scientist* 8: 449–451.

Oftedal, O. T. 1981. Milk, protein and energy intakes of suckling mammalian young: A comparative study. Ph.D. diss., Cornell University, Ithaca, N.Y.

Oftedal, O. T. 1984a. Milk composition, milk yield and energy output at peak lactation: A comparative review. *Symp. Zool. Soc. Lond.* 51: 33–85.

Oftedal, O. T. 1984b. Body size and reproductive strategy as correlates of milk energy yield in lactating mammals. *Acta Zool. Fenn.* 171: 183–186.

Oftedal, O. T. 1985. Pregnancy and lactation. In R. J. Hudson and R. G. White, eds., *Bioenergetics of Wild Herbivores*, pp. 215–238. CRC Press, Florida.

Oftedal, O. T., D. J. Boness, and R. A. Tedman. 1987. The behavior, physiology and anatomy of lactation in the Pinnipedia. *Curr. Mammal.* 1: 175–245.

Okansen, L. 1981. All-female litters as a reproductive strategy: Defense and generalization of the Trivers-Willard hypothesis. *Am. Nat.* 177: 109–111.

Oliver, J. A. 1956. Reproduction in the king cobra, *Ophiophagus hannah* Cantor. *Zoologica* 41: 145–152.

Ollason, J. C., and G. M. Dunnet. 1988. Variation in breeding success in fulmars. In T. H. Clutton-Brock, ed., *Reproductive Success*, pp. 263–278. University of Chicago Press, Chicago.

Oppenheimer, J. R. 1970. Mouth brooding in fishes. *Anim. Behav.* 18: 493–503.

Orians, G. H. 1980. *Some Adaptations of Marsh-nesting Blackbirds.* Princeton University Press, Princeton, N.J.

Oring, L. W. 1982. Avian mating systems. In D. J. Farner and K. Parkes, eds., *Avian Biology*, vol. 6, pp. 1–92. Academic Press, New York.

Oring, L. W., and M. L. Knudson. 1972. Monogamy and polyandry in the spotted sandpiper. *Living Bird* 11: 59–73.

Oring, L. W., and D. B. Lank. 1984. Breeding area fidelity, natal philopatry, and the social systems of sandpipers. In J. Burger and B. L. Olla, eds., *Shorebirds: Breeding Behavior and Populations*, pp. 125–146. Plenum Press, New York.

Oring, L. W., and S. J. Maxson. 1978. Instances of simultaneous polyandry by a spotted sandpiper *Actitis macularia. Ibis* 120: 349–353.

Oster, G., I. Eshel, and D. Cohen. 1977. Worker-queen conflict and the evolution of social insects. *Theor. Popul. Biol.* 12: 49–85.

Otte, D., and K. Stayman. 1979. Beetle horns: Some patterns in functional morphology. In M. S. Blum and N. A. Blum, eds., *Sexual Selection and Reproductive Competition in Insects*, pp. 259–292. Academic Press, New York.

Owen, R. E., F. H. Rodd, and R. C. Plowright. 1980. Sex ratios in bumble bee colonies: Complications due to orphaning? *Behav. Ecol. Sociobiol.* 7: 287–291.

Owiny, A. M. 1974. Some aspects of the breeding biology of the equatorial land snail *Limcolaria martensiana. J. Zool.* 172: 191–206.

Packard, G. C. 1966. The influence of ambient temperature and aridity on modes of reproduction and excretion of amniote vertebrates. *Am. Nat.* 100: 677–682.

Packard, G. C., and M. J. Packard. 1988. The physiological ecology of reptilian eggs and embryos. In C. Gans and R. B. Huey, eds., *Biology of the Reptilia*, vol. 16, pp. 523–605. Alan R. Liss, New York.

Packard, G. C., C. R. Tracy, and J. J. Roth. 1977. The physiological ecology of reptilian eggs and embryos and the evolution of viviparity within the class Reptilia. *Biol. Rev.* 52: 71–105.

Packer, C. R. 1977. Reciprocal altruism in *Papio anubis. Nature* 265: 441–443.

Packer, C. R. 1980. Male care and exploitation of infants in *Papio anubis. Anim. Behav.* 28: 512–520.

Packer, C. R., and A. E. Pusey. 1987. Intrasexual cooperation and the sex ratio in African lions. *Am. Nat.* 130: 636–642.

Page, R. E., and R. A. Metcalf. 1982. Multiple mating, sperm utilization and social evolution. *Am. Nat.* 119: 263–281.

Pagel, M. D., and P. H. Harvey. 1988. How mammals produce large-brained offspring. *Evolution* 42: 948–957.

Pagel, M. D., and P. H. Harvey. 1990. Diversity in the brain sizes of newborn mammals: Allometry, energetics or life history tactics. *Bioscience 1990.*

Pamilo, P. 1982. Genetic evolution of sex ratios in eusocial Hymenoptera: Allele frequency simulations. *Am. Nat.* 119: 638–656.

Parker, G. A. 1985. Models of parent-offspring conflict. V. Effects of the behaviour of two parents. *Anim. Behav.* 33: 519–533.

Parker, G. A., and M. Begon. 1986. Optimal egg size and clutch size: Effects of environment and maternal phenotype. *Am. Nat.* 128: 573–592.

Parker, G. A., and S. P. Courtney. 1984. Models of clutch size in insect oviposition. *Theor. Popul. Biol.* 26: 27–48.

Parker, G. A., and M. R. Macnair. 1978. Models of parent-offspring conflict. I. Monogamy. *Anim. Behav.* 26: 97–110.

Parker, G. A., and M. R. Macnair. 1979. Models of parent-offspring conflict. IV. Suppression: Evolutionary retaliation of the parent. *Anim. Behav.* 27: 1210–1235.

Parker, G. A., and D. W. Mock. 1987. Parent-offspring conflict over clutch size. *Evol. Ecol.* 1: 161–174.

Parker, G. A., D. W. Mock, and T. C. Lamey. In press. How selfish should stronger sibs be? *Am. Nat.*

Parkes, A. S. 1925. Studies in the sex ratio and related phenomena. 7. The foetal sex ratio in the pig. *J. Agric. Sci.* 15: 15–30.

Parkes, A. S. 1926. The mammalian sex ratio. *Biol. Rev.* 2: 1–51.

Parmelee, D. F., and R. B. Payne. 1973. On multiple broods and the breeding strategy of arctic sanderlings. *Ibis* 115: 218–226.

Parry, G. D. 1981. The meanings of r- and K-selection. *Oecologia* 48: 260–264.

Parsons, J. 1970. Relationship between egg size and post-hatching chick mortality in the herring gull (*Larus argentatus*). *Nature* 228: 1221–1222.

Parsons, J. 1975. Asynchronous hatching and chick mortality in the herring gull *Larus argentatus. Ibis* 117: 517–520.

Parsons, J. 1976. Factors determining the number and size of eggs laid in the herring gull. *Condor* 78: 481–492.

Partridge, L. 1988. Reproductive success in *Drosophila*. In T. H. Clutton-Brook, ed., *Reproductive Success*, pp. 11–23. University of Chicago Press, Chicago.

Partridge, L. 1989a. Lifetime reproductive success and life-history evolution. In I. Newton, ed., *Lifetime Reproductive Success In Birds*, pp. 421–440. Academic Press, London.

Partridge, L. 1989b. An experimentalist's approach to the role of cost of reproduction in evolution of life histories. In P. J. Grubb and J. B. Whittaker, eds., *Toward a More Exact Ecology*. Blackwell Scientific Publications, Oxford.

Partridge, L., and P. H. Harvey. 1985. Costs of reproduction. *Nature* 316: 20.

Partridge, L., and P. H. Harvey. 1988. The ecological context of life history evolution. *Science* 241: 1449–1455.

Patterson, C. B. 1979. Relative parental investment in the red-winged blackbird. Ph.D. diss., Indiana University, Bloomington.

Patterson, C. B., and J. M. Emlen. 1980. Variation in nestling sex ratios in the yellow-headed blackbird. *Am. Nat.* 115: 743–747.

Patterson, C. B., W. I. Erckmann, and G. H. Orians. 1980. An experimental study of parental investment and polygyny in male blackbirds. *Am. Nat.* 116: 757–769.

Paul, A., and J. Kuester. 1987. Sex ratio adjustment in a seasonally breeding primate species: Evidence from the Barbary macaque population at Affenberg Salem. *Ethology* 74: 117–132.

Paul, A., and D. Thommen. 1984. Timing of birth, female reproductive success and infant sex ratio in semi-free-ranging Barbary macaques (*Macaca sylvanus*). *Folia Primatol.* 42: 2–16.

Payne, P. R., and E. F. Wheeler. 1968. Comparative nutrition in pregnancy and lactation. *Proc. Nutr. Soc.* 27: 129–138.

Peaker, M., ed. 1977. *Comparative Aspects of Lactation*. Symp. Zool. Soc. Lond. 41.

Pearse, A. S., M. T. Patterson, J. S. Jankin, and G. W. Wharton. 1936. The ecology of *Passalus corhutus* Fabricius, a beetle which lives in rotting logs. *Ecol. Monogr.* 6: 455–490.

Pearson, A. K., and O. Pearson. 1955. Natural history and breeding behavior of the tinamou *Nothoprocta ornata*. *Auk* 72: 113–127.

Pearson, T. H. 1968. The feeding ecology of sea birds breeding on the Faroe Islands, Northumberland. *J. Anim. Ecol.* 37: 521–552.

Pederson, J. C. and K. T. Harper. 1984. Does summer range quality influence sex ratios among mule deer fawns in Utah? *J. Range Mgmt.* 37: 64–66.

Perrigo, G. 1987. Breeding and feeding strategies in deer mice and house mice when females are challenged to work for their food. *Anim. Behav.* 35: 1298–1316.

Perrins, C. M. 1965. Population fluctuations and clutch size in the great tit (*Parus major*). *J. Anim. Ecol.* 34: 601–647.

Perrone, M. 1978a. Mate size and breeding success in a monogamous cichlid fish. *Envir. Biol. Fishes.* 3: 193–201.

Perrone, M. 1978b. The economy of brood defense by parental cichlid fishes *Cichlasoma maculicauda*. *Oikos* 31: 137–141.

Perrone, M., and T. M. Zaret. 1979. Parental care patterns of fishes. *Am. Nat.* 113: 351–361.

Peters, R. H. 1983. *The Ecological Implications of Body Size.* Cambridge University Press, Cambridge.

Petersen, J. J. 1972. Factors affecting sex ratios of a mermithid parasite of mosquitoes. *J. Nemat.* 4: 83–87.

Petersen, J. J., H. C. Chapman, and D. B. Woodward. 1968. The bionomics of a mermithid nematode of larval mosquitoes in southwestern Louisiana. *Mosq. News* 28: 346–352.

Petrie, M. 1983a. Female moorhens compete for small fat males. *Science* 220: 413–415.

Petrie, M. 1983b. Mate choice in role-reversed species. In P.P.G. Bateson, ed., *Mate Choice*, pp. 167–179. Cambridge University Press, Cambridge.

Philipi, T., and J. Seger. 1989. Hedging one's evolutionary bets. *Trends in Evolution and Ecology* 4: 41–44.

Pianka, E. R. 1971. Ecology of the agamid lizard *Amphibolurus isolepis* in Western Australia. *Copeia* 1971: 527–536.

Pianka, E. R. 1972. r and K or b and d selection? *Am. Nat.* 106: 581–588.

Pianka, E. R. 1976. Natural selection of optimal reproduction tactics. *Am. Zool.* 16: 775–784.

Pianka, E. R., and W. S. Parker. 1975. Age-specific reproductive tactics. *Am. Nat.* 109: 453–464.

Pickering, S. P. 1983. Aspects of the behavioural ecology of feral goats (*Capra* domestic). Ph.D. diss., University of Durham, England

Pienkowski, M. W., and J.J.D. Greenwood. 1979. Why change mates? *Biol. J. Linn. Soc.* 12: 85–94.

Pierotti, R., and E. C. Murphy. 1987. Intergenerational conflicts in gulls. *Anim. Behav.* 35: 435–444.

Pietsch, T. W., and D. B. Grobecker. 1980. Parental care as an alternative reproductive mode in an antennariid angler fish. *Copeia* 1980: 551–553.

Pinkowski, B. C. 1977. Breeding adaptations in the Eastern bluebird. *Condor* 79: 289–302.

Pinkowski, B. C. 1978. Feeding of nestling and fledgling Eastern bluebirds. *Wilson Bull.* 90: 84–98.

Pitelka, F. A., R. T. Holmes, and S. F. MacLean. 1974. Ecology and evolution of social organization in arctic sandpipers. *Amer. Zool.* 14: 185–204.

Platt, H. 1978. *A Survey of Perinatal Mortality and Disorders in the Thoroughbred.* Animal Health Trust, Newmarket, England.

Pond, C. M. 1977. The significance of lactation in the evolution of mammals. *Evolution* 31: 177–199.

Pond, C. M. 1983. Parental feeding as a determinant of ecological relationships in Mesozoic terrestrial ecosystems. *Acta Palaeontol. Pol.* 28: 215–224.

Pope, N. S., T. P. Gordon, and M. E. Wilson. 1986. Age, social rank and lactational status influence ovulatory patterns in seasonally breeding rhesus monkeys. *Biol. Reprod.* 35: 353–359.

Pratt, D. M., and V. A. Anderson. 1979. Giraffe cow-calf relationships and social development of the calf in the Serengeti. *Z. Tierpsychol.* 51: 233–251.

Pratt, N. C., U .W. Huck, and R. D. Lisk. In press. Do pregnant hamsters react to stress by producing fewer males? *Anim. Behav.*

Prentice, A. M., and R. G. Whitehead. 1987. The energetics of human reproduction. *Symp. Zool. Soc. Lond.* 57: 275–304.

Pressley, P. H. 1981. Parental effort and the evolution of nest-guarding tactics in the three spine stickleback *Gasterosteus aculeatus* L. *Evolution* 35: 282–295.

Prevost, J. 1961. *Ecologie du Manohot Empereur.* Hermann, Paris.

Prevost, J., and V. Vilter. 1963. Histologie de la secretion oesophagienne du manchot empereur. *Proc. Int. Ornithol. Congr.* 13: 1085–1094.

Price, P. W. 1973. Reproductive strategies in parasitoid wasps. *Am. Nat.* 197: 684–693.

Promislow, D.E.L., and P. H. Harvey. 1990. Living fast and dying young: A comparative analysis of life-history variation among mammals. *J. Zool.* 220: 417–438.

Prout, T., and F. McChesney. 1985. Competition among immatures affects their adult fertility: Population dynamics. *Am. Nat.* 126: 521–558.

Pugesek, B. H. 1981. Increased reproductive effort with age in the California gull (*Larus californicus*). *Science* 212: 822–823.

Pugesek, B. H., and K. L. Diem. 1983. A multivariate study of the relationship of parental age to reproductive success in California gulls. *Ecology* 64: 829–839.

Pukowski, E. 1933. Ökologische Untersuchungen an *Necrophorus* F. *Z. f. Morphol. und Ökol. der Tiere* 27: 518–586.

Pulliainen, E. 1971. *Breeding Behaviour of the Dotterel, Charadrius morinellus.* Vario Subarctic Station Report 24.

Qasim, S. Z. 1956. Time and duration of the spawning season in some marine teleosts in relation to their distribution. *Journal du Conseil* 21: 144–155.

Quiring, D. T., and J. N. McNeill. 1984. Influence of intra-specific larval competition and mating in the longevity and reproductive performance of females of the leaf miner, *Agromyza frontella* (Rondani) (Diptera: Agromyzidae). *Can. J. Zool.* 62: 2197–2200.

Rahn, H., and A. Ar. 1974. The avian egg: Incubation time and water loss. *Condor* 76: 147–152.

Rahn, H., C. V. Pagnelli, and A. Ar. 1975. Relation of avian egg weight to body weight. *Auk* 92: 750–765.

Ralls, K. 1976. Mammals in which females are larger than males. *Quart. Rev. Biol.* 31: 245–276.

Ralls, K. 1977. Sexual dimorphism in mammals: Avian models and unanswered questions. *Am. Nat.* 111: 917–938.

Ralls, K., R. L. Brownwell, and J. Ballou. 1980. Differential mortality by sex and age in mammals, with specific reference to the sperm whale. *Rep. Int. Whal. Comm.* (Special issue) 2: 223–243.

Ralston, J. S. 1977. Egg guarding by male assassin bugs of the genus *Zelus* (Hemiptera: Reduviidae). *Psyche* 84: 103–107.

Ramsay, M. A., and R. L. Dunbrack. 1986. Physiological constraints on life history phenomena: The example of small bear cubs at birth. *Am. Nat.* 127: 735–743.

Randolph, P. A., J. C. Randolph, W. Mattingley, and M. M. Foster. 1977. Energy costs of reproduction in the cotton rat, *Sigmodon hispidus*. *Ecology* 58: 31–45.

Rass, T. S. 1942. Analogous or parallel variations in structure and development of fishes in Northern and Arctic seas. *Jubilee Publ. Moscow Soc. Naturalists* 1805–1940: 1–60.

Ratcliffe, L. M. 1974. Nest defence behavior in the lesser snow goose. B.Sc. thesis, Queens University, Kingston, Ontario.

Raveling, D. 1970. Dominance relationships and agonistic behaviour of Canada geese in winter. *Behaviour* 37: 291–319.

Rawlins, R., and M. Kessler. 1986. Secondary sex ratio variation in the Cayo Santiago macaque population. *Am. J. Primatol.* 10: 9–23.

Read, A. F., and P. H. Harvey. 1989. Life history differences among the eutherian radiations. *J. Zool.* 219: 329–353.

Redondo, T., and J. Carranza. 1989. Offspring reproductive value and nest defense in the magpie (*Pica pica*). *Behav. Ecol. Sociobiol.* 5: 369–378.

Reese, E. 1964. Ethology and marine zoology. In *Oceanogr. Mar. Biol. Ann. Rev.* 1964: 455–488.

Reese, E. S. 1975. A comparative field study of the social behavior and related ecology of reef fishes of the family Chaetodontidae. *Z. Tierpsychol.* 32: 319–324.

Regelmann, K., and E. Curio. 1983. Determinants of brood defence in the great tit. *Behav. Ecol. Sociobiol.* 13: 131–145.

Regelmann, K., and E. Curio. 1986. How do great tit (*Parus major*) pair-mates cooperate in brood defence? *Behaviour* 97: 10–36.

Reid, M. C., and R. D. Montgomerie. 1985. Seasonal patterns of nest defence by Baird's sandpipers. *Can. J. Zool.* 63: 2207–2211.

Reinhardt, V., and A. Reinhardt. 1981. Natural suckling performance and age of weaning in Zebu cattle (*Bos indicus*). *J. Agric. Sci.* 96: 309–312.

Reish, D. J. 1957. The life history of the polychaetous annelid *Neanthes candata* (delle Chiaje), including a summary of development in the family Nereidae. *Pacific Sci.* 11: 216–228.

Reiss, M. J. 1985. The allometry of reproduction: Why larger species invest relatively less in their offspring. *J. theoret. Biol.* 113: 529–544.

Reiss, M. J. 1987a. Evolutionary conflict over the control of offspring sex ratio. *J. theoret. Biol.* 125: 25–39.

Reiss, M. J. 1987b. The intraspecific relationship of parental investment to female body weight. *Functional Ecology* 1: 105–107.

Reiss, M. J. 1989. *The Allometry of Growth and Reproduction.* Cambridge University Press, Cambridge.

Reiter, J., N. L. Stinson, and B. J. Le Boeuf. 1978. Northern elephant seal development: The transition from weaning to nutritional development. *Behav. Ecol. Sociobiol.* 3: 337–367.

Rendel, J. M. 1943. Variation in the weights of hatched and unhatched duck's eggs. *Biometrika* 33: 48–56.

Reyer, H. U. 1984. Investment and relatedness: A cost/benefit analysis of breeding and helping in the pied kingfisher *Ceryle rudis. Anim. Behav.* 32: 1163–1178.

Reynolds. J. D. 1987. Mating system and nesting biology of the red-necked phalarope *Phalaropus lobatus*: What constrains polyandry? *Ibis* 129: 225–242.

Reznick, D. 1983. The structure of guppy life histories: The tradeoff between growth and reproduction. *Ecology* 64: 862–873.

Reznick, D. 1985. Costs of reproduction: An evaluation of the critical evidence. *Oikos* 44: 257–267.

Reznick, D., and J. A. Endler. 1982. The impact of predation on life history evolution in Trinidadian guppies (*Poecilia reticulata*). *Evolution* 36: 160–177.

Reznick, D. N., E. Perry, and J. Travis. 1986. Measuring the costs of reproduction: A comment on papers by Bell. *Evolution* 40: 1338–1344.

Richmond, A. R. 1978. An experimental study of advantages of monogamy in the cardinal. Ph.D. diss., Indiana University, Bloomington.

Richter, W. 1982. Hatching asynchrony: The nest failure hypothesis and brood reduction. *Am. Nat.* 120: 818–832.

Richter, W. 1983. Balanced sex ratios in dimorphic altricial birds: The contribution of sex-specific growth dynamics. *Am. Nat.* 121: 158–171.

Richter, W. 1984. Nestling survival and growth in the yellow-headed blackbird *Xanthocephalus xanthocephalus*. *Ecology* 65: 597–608.

Ricklefs, R. E. 1969. An analysis of nestling mortality in birds. *Smithson. Contr. Knowl. (Zool.)* 9: 1–48.

Ricklefs, R. E. 1974. Energetics of reproduction in birds. In R. A. Paynter, ed., *Avian Energetics*, pp. 152–297. Nuttall Ornithological Club, Cambridge, Mass.

Ricklefs, R. E. 1977a. On the evolution of reproductive strategies in birds: Reproductive effort. *Am. Nat.* 111: 453–478.

Ricklefs, R. E. 1977b. Variation in size and quality of the starling egg. *Auk* 94: 167–168.

Ricklefs, R. E. 1977c. Reactions of some Panamanian birds to human intrusion at the nest. *Condor* 79: 376–379.

Ridley, M. 1978. Paternal care. *Anim. Behav.* 26: 904–932.

Ridley, M. 1983. *The Explanation of Organic Diversity: The Comparative Method and Adaptations for Mating*. Oxford University Press, Oxford.

Ridley, M., and C. Rechten. 1981. Female sticklebacks prefer to spawn with males whose nests contain eggs. *Behaviour* 76: 152–161.

Ridpath, M. G.. 1964. The Tasmanian native hen. *Aust. Nat. Hist.* 14: 346–350.

Riedmann, M. L., and B. J. Le Boeuf. 1982. Mother-pup separation and adoption in northern elephant seals. *Behav. Ecol. Sociobiol.* 11: 203–215.

Rivers, J.P.W., and M. A. Crawford. 1974. Maternal nutrition and the sex ratio at birth. *Nature* 252: 297–298.

Robbins, C. T. 1983. *Wildlife Feeding and Nutrition*. Academic Press, New York.

Robbins, C. T., and B. C. Robbins. 1979. Fetal and neonatal growth patterns and maternal reproductive effort in ungulates and subungulates. *Am. Nat.* 114: 101–116.

Roberts, A. M. 1978. The origins of fluctuations in the human secondary sex ratio. *J. Biosoc. Sci.* 10: 169–182.

Robertson, D. R. 1973. Field observations on the reproductive behaviour of a pomacentrid fish, *Acanthochromis polyacanthus*. *Z. Tierpsychol.* 32: 319–324.

Robertson, D.R., N.V.C. Polunin, and K. Leighton. 1979. The behavioural ecology of three Indian Ocean surgeonfishes (*Acanthurus lineatus, A. leucosternon* and *Ze-*

brasoma scopas), their feeding strategies and social and mating systems. *Env. Biol. Fishes* 4: 125–170.

Robertson, R. J., and G. C. Biermann. 1979. Parental investment strategies determined by expected benefits. *Z. Tierpsychol.* 50: 124–128.

Robinette W. L., J. S. Gashwiler, J. B. Low, and D. A. Jones. 1957. Differential mortality by sex and age among mule deer. *J. Wildl. Mgmt.* 21: 1–16.

Robinson, J. V. 1980. A necessary modification of Trivers' parent-offspring conflict model. *J. theoret. Biol.* 83: 533–536.

Roderick, B. J., and J. D. Storer. 1961. Correlation between mean litter size and mean life span among 12 inbred strains of mice. *Science* 134: 48–49.

Rodriguez, C. A., and S. Guerrero. 1976. La historia natural y el comportamiento de *Zygopachylus albomarginis* (Chamberlain) (Arachnida, Opiliones: Gonyleptidae). *Biotropica* 8: 242–247.

Rood, J. P. 1986. Ecology and social evolution in the mongooses. In D. I. Rubenstein and R. W. Wrangham, eds., *Ecological Aspects of Social Evolution*, pp. 131–152. Princeton University Press, Princeton, NJ.

Rose, M. 1984a. Genetic covariation in *Drosophila* life history: Untangling the data. *Am. Nat.* 123: 565–569.

Rose, M. 1984b. Laboratory evolution of postponed senescence in *Drosophila melanogaster. Evolution* 38: 1004–1010.

Rose, M. R., and B. Charlesworth. 1981a. Genetics of life history in *Drosophila melanogaster.* I. Sib analysis of adult females. *Genetics* 97: 173–186.

Rose, M. R., and B. Charlesworth. 1981b. Genetics of life history in *Drosophila melanogaster.* II. Exploratory selection experiments. *Genetics* 97: 187–196.

Røskaft, E. 1985. The effect of enlarged brood size on the future reproductive potential of the rook. *J. Anim. Ecol.* 54: 255–260.

Røskaft, E., and T. Slagsvold, 1985. Differential mortality of male and female offspring in experimentally manipulated broods of the rook. *J. Anim. Ecol.* 54: 261–266.

Røskaft, E., Y. Espmark, and T. Järvi. 1983. Reproductive effort and breeding success in relation to age by the rook *Corvus frugilegus. Ornis. Scand.* 14: 169–174.

Ross, C. 1988. The intrinsic rate of natural increase and reproductive effort in primates. *J. Zool.* 214: 199–220.

Ross, H. A. 1979. Multiple clutches and shorebird egg and body weight. *Am. Nat.* 113: 618–622.

Roth, L. M., and E. R. Willis. 1957. An analysis of oviparity and viviparity in the Blattaria. *Trans. Amer. Ent. Soc.* 83: 221–240.

Rowell, T. E., and J. Chism. 1986. The ontogeny of sex differences in the behavior of patas monkeys. *Int. J. Primatol.* 7: 83–106.

Royama, T. 1966. Factors governing feeding rate, food requirements and brood size of nestling great tits *Parus major. Ibis* 108: 313–347.

Rubenstein, D. 1982. Reproductive value and behavioral strategies: Coming of age in monkeys and horses. In P.P.G. Bateson and P. H. Klopfer, eds., *Perspectives in Ethology*, Vol. 5, *Ontogeny*, pp. 469–487. Plenum Press, New York.

Rubenstein, D. I. 1986. Ecology and sociality in horses and zebras. In D. I. Rubenstein and R. W. Wrangham, eds., *Ecological Aspects Of Social Evolution*, pp. 282–302. Princeton University Press, Princeton, N.J.

Russell, E. M. 1982a. Patterns of parental care and parental investment in marsupials. *Biol. Rev. Camb. Phil. Soc.* 57: 423–486.

Russell, E. M. 1982b. Parental investment and desertion of young in marsupials. *Am. Nat.* 119: 744–748.

Rutberg, A. T. 1986. Lactation and fetal sex ratios in American bison. *Am. Nat.* 127: 89–94.

Rutberg, A. T., and S. Rohwer. 1980. Breeding strategies of male yellowheaded blackbirds: Results of a removal experiment. *Auk* 97: 619–622.

Rutowski, R. L. 1982. Mate choice and lepidopteran mating behavior. *Fla. Ent.* 65: 71–82.

Rutowski, R. L., M. Newton, and J. Schaefer. 1983. Interspecific variation in the size of the nutrient investment made by male butterflies during copulation. *Evolution* 37: 708–713.

Ryan, M. R., and J. J. Dinsmore. 1986. The behavioral ecology of breeding American coots in relation to age. *Condor* 82: 320–327.

Ryden, O. O., and G. Bengtsson. 1980. Differential begging and locomotory behaviour by early and late hatched nestlings affecting the distribution of food in asynchronously hatched broods of altricial birds. *Z. Tierpsychol.* 53: 209–224.

Ryder, J. P. 1983. Sex ratio and egg sequence in ring-billed gulls. *Auk* 100: 726–729.

Sacher, G. A., and E. F. Staffeldt. 1974. Relationship of gestation time to brain weight for placental mammals. *Am. Nat.* 108: 593–616.

Sackett, G. P. 1981. Receiving severe aggression correlates with fetal gender in pregnant pigtail monkeys. *Develop. Psychobiol.* 14: 267–272.

Sackett, G. P., R. A. Holm, A. E. Davis, and C. E. Fahrenbuch. 1975. Prematurity and low birth weight in pigtail macaques: Incidence, prediction and effects of infant development. In S. Kondo, M. Kawai, A. Ehara, and S. Kawamura, eds., *Proc. Fifth Congr. Int. Primat. Soc.*, pp. 189–206. Japan Science Press, Tokyo.

Sacks, M. 1964. Life history of an aquatic gastrotrich. *Trans. Amer. Microscop. Soc.* 83: 358–362.

Sade, D. S. 1972. A longitudinal study of social behavior in rhesus monkeys. In R. Tuttle, ed., *The Functional and Evolutionary Biology of Primates*, pp. 378–398. Aldine, Chicago.

Saito, Y. 1986. Biparental defence in a spider mite (Acari: Tetranychidae) infesting *sasa* bamboo. *Behav. Ecol. Sociobiol.* 18: 377–86.

Sadleir, R.M.F.S. 1969. *The Ecology of Reproduction in Wild and Domestic Mammals.* Methuen & Co., London.

Salthe, S. N. 1969. Reproductive modes and number and size of ova in the Urodeles. *Amer. Midl. Natur.* 81: 467–490.

Salthe, S. N., and W. E. Duellman. 1973. Quantitative constraints associated with reproductive mode in anurans. In J. L. Vial, ed., *Evolutionary Biology of the Anurans*, pp. 229–249. University of Missouri Press, Columbia.

Salthe, S. N., and J. S. Mecham. 1974. Reproductive and courtship patterns. In B. Lofts, ed., *Physiology of the Amphibia*, pp. 209–521. Academic Press, New York.

Sargent, R. C. 1981. Sexual selection and reproductive effort in the three spine stickleback, *Gasterosteus aculeatus.* Ph.D. diss., State University of New York, Stonybrook, New York.

Sargent, R. C. 1988. Paternal care and egg survival both increase with clutch size in the fathead minnow, *Pimephales promelas. Behav. Ecol. Sociobiol.* 23: 33–37.

Sargent, R. C. 1989. Allopaternal care in the fathead minnow *Pimephales promelas*: Stepfathers discriminate against their adopted eggs. *Behav. Ecol. Sociobiol.* 25: 379–385

Sargent, R. C., and M. R. Gross. 1985. Parental investment decision rules and the Concorde fallacy. *Behav. Ecol. Sociobiol.* 17: 43–45.

Sargent, R. C., and M. R. Gross. 1986. Williams' principle: An explanation of parental care in teleost fishes. In T. J. Pitcher, ed., *The Behaviour of Teleost Fishes*, pp. 275–293. Croom Helm, London.

Sargent, R. C., P. D. Taylor, and M. R. Gross. 1987. Parental care and the evolution of egg size in fishes. *Am. Nat.* 129: 32–46.

Sasvari, L. 1986. Reproductive effort of widowed birds. *J. Anim. Ecol.* 55: 553–564.

Sato, T. 1986. A brood parasitic catfish of mouthbreeding cichlid fishes in Lake Tanganyika. *Nature* 323: 58–59.

Sauer, E.G.F., and E. M. Sauer. 1966. The behavior and ecology of the South African ostrich. *Living Bird* 5: 45–75.

Schachak, M. E., A. Chapman, and Y. Steinberger. 1976. Feeding, energy flow and soil turnover in the desert isopod, *Hemilepistus reamuri. Oecologia* 24: 57–69.

Schafer, E. 1954. Zur Biologie des Steisshuhnes, *Nothocercus bonapartei. J. Ornithol.* 59: 219–232.

Schaffer, N. M. 1974a. Selection for optimal life histories: The effects of age structure. *Ecology* 55: 291–303.

Schaffer, N. M. 1974b. Optimal reproductive effort in fluctuating environments. *Am. Nat.* 108: 783–798.

Schaller, G. B. 1971. *The Serengeti Lion*. University of Chicago Press, Chicago.

Schamel, D., and D. Tracy. 1977. Polyandry, replacement clutches and site tenacity in the red phalarope (*Phalaropus fulicarius*) at Barrow, Alaska. *Bird Banding* 48: 314–324.

Scheel, J. J. 1970. Notes on the biology of the African tree-toad, *Nectophryne afra* Buchholz and Peters, 1875 (Bufonidae, Anura) from Fernando Po. *Rev. Zool. Bot. Afr.* 81 : 225–236.

Schifferli, L. 1973. The effect of egg weight on the subsequent growth of nestling great tits *Parus major. Ibis* 115: 549–558.

Schifferli, L. 1976. Factors affecting weight and condition in the house sparrow, particularly when breeding. D.Phil. thesis, University of Oxford.

Schinkel, P. G., and B. F. Short. 1961. The influence of nutritional level during prenatal and early post-natal life on adult fleece and body characteristics. *Aust. J. Agric. Res.* 12: 176–202.

Schreiner, J. 1906. Die Lebensweise und Metamorphose des Rebenschneiders oder grosskopfungen Zwiebelhornkäfers (*Lethrus apterus* Laxm.) (Coleoptera, Scarabaeidae). *Horae Societatis Entomologicae Rossicae* 37: 197–208.

Schulman, S. R., and B. Chapais. 1980. Reproductive value and rank relations among macaque sisters. *Am. Nat.* 115: 580–593.

Schwagmeyer, P. 1988. Scramble competition polygyny in an asocial mammal: Mate mobility and mating success. *Am. Nat.* 131: 885–892.

Schwagmeyer, P. 1989. Sperm competition in the thirteen-lived ground squirrels: Differential fertilization success under field conditions. *Am. Nat.* 133: 257–265.

Scott, D. K. 1980. Functional aspects of prolonged parental care in Bewicks swans. *Anim. Behav.* 28: 938–952.

Scott, D. K. 1984. Parent-offspring association in mute swans (*Cygnus olor*). *Z. Tierpsychol.* 64: 74–86.

Scott, M. P. 1990. Brood guarding and the evolution of male parental care in burying beetles. *Behav. Ecol. Sociobiol.* 26: 31–39.

Scott, M. P., and J.F.A. Traniello. 1990. Behavioural and ecological correlates of male and female parental care and reproductive success in burying beetles (*Nicrophorus* spp.). *Anim. Behav.* 39: 274–283.

Seigel, R. A., M. M. Huggins, and N. B. Ford. 1987. Reduction in locomotor ability as a cost of reproduction in gravid snakes. *Oecologia* 73 (4): 481–485.

Sergeer, A. M. 1940. Researches on the viviparity of reptiles. *Moscow Soc. of Naturalists* 1: 1–34.

Service, P. M., and M. R. Rose. 1985. Genetic covariation among life-history components: The effect of novel components. *Evolution* 39: 943–945.

Shachak, M. 1980. Energy allocation and life history strategy of the desert isopod *H. reamuri. Oecologia* 45: 404–413.

Shann, E. W. 1923. The embryonic development of the porbeagle shark, *Lamna cornubica. Proc. Zool. Soc. Lond.* 1923: 161–171.

Shapiro, D. Y., D. A. Hensley, and R. S. Appledoorn. 1988. Pelagic spawning and egg transport in coral-reef fish. *Env. Ecol. Fishes* 22: 3–14.

Sharman, G. B. 1965. Marsupials and the evolution of viviparity. In J. D. Butterworth and C. L. Duddington, eds., *Viewpoints in Biology*, pp. 1–28. Butterworth, London.

Shine, R. 1978. Propagule size and parental care: The "safe harbor" hypothesis. *J. theoret. Biol.* 75: 417–424.

Shine, R. 1980. "Costs" of reproduction in reptiles. *Oecologia* 46: 92–100.

Shine, R. 1981. Venomous snakes in cold climates: Ecology of the Australian genus *Drysdalia* (Serpentes: Elapidae). *Copeia* 1981: 14–25.

Shine, R. 1983a. Reptilian reproductive modes: The oviparity-viviparity continuum. *Herpetol.* 39: 1–8.

Shine, R. 1983b. Reptilian viviparity in cold climates: Testing assumptions of an evolutionary hypothesis. *Oecologia* 57: 397–405.

Shine, R. 1984. Physiological and ecological questions on the evolution of reptilian viviparity. In R. S. Seymour, ed., *Respiration and Metabolism of Embryonic Vertebrates*, pp. 147–154. Dr W. Junk, Dordrecht.

Shine, R. 1985. The evolution of viviparity in reptiles: An ecological analysis. In C. Gans and F. Billett, eds., *Biology of the Reptilia*, vol. 15, pp. 605–694. Wiley, New York.

Shine, 1987a. Reproductive mode may determine geographic distribution in Australian venomous snakes (*Pseudechis*, Elapidae). *Oecologia* 71: 608–612.

Shine, R. 1987b. The evolution of viviparity: Ecological correlates of reproductive mode within a genus of Australian snakes (*Pseudechis*: Elapidae). *Copeia* 1987: 551–563.

Shine, R. 1988a. Constraints on reproductive investment: A comparison between aquatic and terrestrial snakes. *Evolution* 42: 17–27.

Shine, R. 1988b. The evolution of large body size in females: A critique of Darwin's "fecundity" advantage model. *Am. Nat.* 131: 124–131.

Shine, R. 1988c. Parental care in reptiles. In C. Gans, ed., *Biology of the Reptilia*, vol. 16, pp. 276–329. Alan R. Liss, New York.

Shine, R. 1989a. Alternative models for the evolution of offspring size. *Am. Nat.* 134: 311–317.

Shine, R. 1989b. Ecological influences on the evolution of vertebrate viviparity. In D. B. Wake and G. Roth, eds., *Complex Organismal Functions: Integration and Evolution in Vertebrates*, pp. 263–278. Wiley, New York.

Shine, R., and J. F. Berry. 1978. Climatic correlates of live-bearing in squamate reptiles. *Oecologia* 33: 261–268.

Shine, R., and J. J. Bull. 1979. The evolution of live bearing in lizards and snakes. *Am. Nat.* 113: 905–923.

Shine, R., and L. J. Guillette. 1988. The evolution of viviparity: A physiological model and its ecological consequences. *J. theoret. Biol.* 132: 43–50.

Short, R. V. 1976a. Lactation—the central control of reproduction. In *Breast Feeding and the Mother*. Ciba Foundation Symposium 45, pp. 73–86. Elsevier, Holland.

Short, R. V. 1976b. The evolution of human reproduction. *Proc. Roy. Soc. B.* 195: 3–24.

Short, R. V. 1983. The biological bases for the contraceptive effects of breast feeding. In D. B. Jellife and E. F. B. Jellife, eds., *Advances in International Maternal and Child Health*, pp. 27–39. Oxford University Press, Oxford.

Shrode, J. B., and S. D. Gerking. 1977. Effects of constant and fluctuating temperatures on reproductive performance of a desert pupfish, *Cyprinodon n. nevadensis*. *Physiol. Zool.* 50: 1–10.

Sibly, R., and P. Calow. 1983. An integrated approach to life-cycle evolution using selective landscapes. *J. theoret. Biol.* 102: 527–547.

Sibly, R. M., and P. Calow. 1986. *Physiological Ecology of Animals*. Blackwell Scientific Publications, Oxford.

Silk, J. B. 1983. Local resource competition and facultative adjustment of sex ratios in relation to competitive activities. *Am. Nat.* 12: 56–66.

Silk, J. B. 1986. Maternal investment in captive bonnet macaques *Macaca radiata*. *Am. Nat.* 132: 1–19.

Silk, J. B. 1988. Maternal investment in captive bonnet macaques *Macaca radiata*. *Am. Nat.* 132: 1–19.

Silk, J. B., C. B. Clark-Wheatley, P. S. Rodman, and A. Samuels. 1981. Differential reproductive success and facultative adjustment of sex ratios among captive female bonnet macaques (*Macaca radiata*). *Anim. Behav.* 29: 1106–1120.

Silver, R., H. Andrews, and G. F. Ball. 1985. Parental care in an ecological perspective: A quantitative analysis of avian subfamilies. *Amer. Zool.* 25: 823–840.

Silverstone, M. P. 1976. A revision of the poison-arrow frogs of the genus *Phyllobates* Bibron in Sagra (Family Dendrobatidae). *Nat. Hist. Mus., Los Angeles Co., Sci. Bull.* 27: 1–53.

Silvertown, J. 1989. The paradox of seed size and adaptation. *Trends in Ecology and Evolution* 4: 24–26.

Simon, M. P. 1983. The ecology of parental care in a terrestrial breeding frog from New Guinea. *Behav. Ecol. Sociobiol.* 14: 61–67.

Simpson, A. E., and M.J.A. Simpson. 1985. Short-term consequences of different breeding histories for captive rhesus macaque mothers and young. *Behav. Ecol. Sociobiol.* 18: 83–89.

Simpson, M.J.A., and A. E. Simpson. 1982. Birth sex ratios and social rank in rhesus monkey mothers. *Nature* 300: 440–441.

Simpson, M.J.A., A. E. Simpson, J. Hooley, and M. Zunz. 1981. Infant-related influences on birth intervals in rhesus monkeys. *Nature* 290: 49–51.

Sinervo, B. 1990. The evolution of maternal investment in lizards: An experimental and comparative analysis of egg size and its effect on offspring performance. *Evolution* 44: 279–294.

Singh, O. N., R. N. Singh, and R.R.P. Srivastava. 1965. Study in post-partum interval to first service in Tharparkar cattle. *Indian J. Vet. Sci.* 35: 245–248.

Skogland, T. 1985. The effects of density-dependent resource limitation on size of wild reindeer. *Oecologia* 60: 156–168.

Skogland, T. 1986. Sex ratio variation in relation to maternal condition and parental investment in wild reindeer (*Rangifer tarandus*). *Oikos* 46: 417–419.

Skutch, A. F. 1949. Do tropical birds rear as many young as they can nourish? *Ibis* 91: 430–455.

Skutch, A. F. 1976. *Parent Birds and Their Young.* University of Texas Press, Austin.

Slagsvold, T. 1982. Sex, size and natural selection in the hooded crow *Corvus corone cornix. Ornis. Scand.* 13: 165–175.

Slagsvold, T. 1985. Asynchronous hatching in passerine birds: Influence of hatching failure and brood reduction. *Ornis Scand.* 16: 81–87.

Slagsvold, T. 1986a. Hatching asynchrony: Interspecific comparisons of altricial birds. *Am. Nat.* 128: 120–125.

Slagsvold, T. 1986b. Asynchronous versus synchronous hatching in birds: experiments with the pied flycatcher. *J. Anim. Ecol.* 55: 1115–1134.

Slagsvold, T., and J. Lifjeld. 1988. Ultimate adjustment of clutch size to parental feeding capacity in a passerine bird. *Ecology* 69: 1918–1922.

Slagsvold, T., E. Røskaft, and S. Engen. 1986. Sex ratio, differential cost of rearing young and differential mortality during the period of parental care: Fisher's theory applied to birds. *Ornis. Scand.* 17: 117–125.

Slagsvold, T., J. Sandvik, G. Rofstad, O. Lorentsen, and M. Husby. 1984. On the adaptive value of intraclutch egg-size variation in birds. *Auk* 101: 685–697.

Slee, J. 1970. Resistance to body cooling in male and female sheep and the effects of previous exposure to chronic cold, acute cold and repeated cold shocks. *Anim. Prod.* 12: 13–21.

Slee, J. 1972. Habituation and acclimatization of sheep to cold following exposure of varying length and severity. *J. Physiol.* 227: 51–70.

Small, M. F., and S. B. Hrdy. 1986. Secondary sex ratios by maternal rank, parity and age in captive rhesus macaques (*Macaca mulatta*). *Am. J. Primatol.* 11: 359–365.

Small, M. F., and D. G. Smith. 1984. Sex differences in maternal investment by *Macaca mulatta. Behav. Ecol. Sociobiol.* 14: 313–314.

Small, M. F., and D. G. Smith. 1985. Sex ratio of infants produced by male rhesus macaques. *Am. Nat.* 126: 354–361.

Smith, A. T., and B. L. Ivins. 1987. Temporal separation between philopatric juvenile pikas and their parents limits behavioural conflict. *Anim. Behav.* 35: 1210–1214.

Smith, C. C., and S. D. Fretwell. 1974. The optimal balance between size and number of offspring. *Am. Nat.* 108: 499–506.

Smith, H. G., H. Kallander, K. Fontell, and M. Ljingstrom. 1988. Feeding frequency and parental division of labour in the double-brooded great tit *Parus major*. *Behav. Ecol. Sociobiol.* 22: 447–453.

Smith, J. M. 1981. Does high fecundity reduce survival in song sparrows? *Evolution* 35: 1142–1148.

Smith, J.N.M., Y. Yom-Tov, and R. Moses. 1982. Polygyny, male parental care and sex ratio in song sparrows: An experimental study. *Auk* 99: 555–564.

Smith, M. A. 1973. *The British Amphibians and Reptiles*, 5th ed. Collins, London.

Smith, R. L. 1976. Male brooding of the waterbug *Abedus herberti* (Hemiptera: Belostomatidae). *Ann. Entomol. Soc. Am.* 69: 740–747.

Smith, R. L. 1979a. Paternity assurance and altered roles in the mating behaviour of a giant water bug, *Abedus herberti* (Heteroptera: Belostomatidae). *Anim. Behav.* 27: 716–725.

Smith, R. L. 1979b. Repeated copulation and sperm precedence: Paternity assurance for a male brooding waterbug. *Science* 205: 1029–1031.

Smith, R. L. 1980. Evolution of exclusive postcopulatory paternal care in the insects. *Florida Entomologist* 63: 65–78.

Smith, W. P. 1987. Maternal defense in Columbian white-tailed deer: When is it worth it? *Am. Nat.* 130: 310–316.

Smuts, B. 1985. *Sex and Friendship in Baboons*. Aldine, New York.

Snell, T. W., and C. E. King. 1977. Lifespan and fecundity patterns in rotifers: The cost of reproduction. *Evolution* 31: 882–890.

Snow, B. K. 1961. Notes on the behavior of three Cotingidae. *Auk* 78: 150–161.

Snow, B. K., and D. W. Snow. 1979. The ochre-bellied flycatcher and the evolution of lek behaviour. *Condor* 81: 286–292.

Snow, D. W. 1961. The displays of the manakins *Pipra pipra* and *Tyrannentes virescens*. *Ibis* 103: 110–113.

Snow, D. W. 1962. A field study of the black and white manakin, *Manacus manacus*, in Trinidad. *Zoologica* 47: 65–104.

Snow, D. W. 1971. Evolutionary aspects of fruit-eating by birds. *Ibis* 113: 194–202.

Snow, D. 1980. Regional differences between tropical floras and the evolution of frugivory. In R. Nohring, ed., *Acta 17th Congr. Intern. Ornithol.*, pp. 1192–1198. Deutsche Gesellschaft, Berlin.

Sokal, R. R. 1970. Senescence and genetic load: Evidence from *Tribolium*. *Science* 167: 1733–1734.

Sotherland, P. R., and H. Rahn. 1987. On the composition of bird eggs. *Condor* 89: 48–65.

Spellerberg, I. F. 1971. Breeding behaviour of the McCormick skua *Catharacta maccormicki* in Antarctica. *Ardea* 59: 189–230.

Spencer-Booth, Y. 1969. The behaviour of twin rhesus monkeys and comparisons with the behaviour of single infants. *Primates* 9: 75–84.

Spight, T. M. 1975. Factors extending gastropod embryonic development and their selective cost. *Oecologia* 21: 1–16.

Stamps, J. A. 1980. Parent-offspring conflict. In G. W. Barlow and J. Silverberg, eds., *Sociobiology: Beyond Nature/Nurture*, pp. 589–618. Reports, Definition and Debate, AAAS Selected Symposium 35. Westview, Boulder, Colorado.

Stamps, J. 1987. The effects of parent and offspring on food allocation in budgies. *Behaviour* 101: 177–199.

Stamps, J. A. In press. When should avian parents differentially provision sons and daughters? *Am. Nat.*

Stamps, J. A., A. Clark, P. Arrowood, and B. Kus. 1985. Parent-offspring conflict in budgerigars. *Behaviour* 94: 1–39.

Stamps, J. A., A. Clark, P. Arrowood, and B. Kus. 1989. Begging behavior in budgerigars. *Ethology* 81: 177–192.

Stamps, J., B. Kus, and P. Arrowood. 1987. The effects of parent and offspring gender on food allocation in budgerigars. *Behaviour* 101: 177–199.

Stamps, J. A., R. A. Metcalf, and V. V. Krishnan. 1978. A genetic analysis of parent-offspring conflict. *Behav. Ecol. Sociobiol.* 3: 369–392.

Stanley, B. V. 1983. Effect of food supply on reproductive behaviour of male *Gasterosteus aculeatus*. Ph.D. diss., University of Wales.

Stearns, S. C. 1976. Life-history tactics: A review of the ideas. *Q. Rev. Biol.* 51: 3–48.

Stearns, S. C. 1983a. A natural experiment in life-history evolution: Field data on the introduction of mosquitofish (*Gambusia affinis*) to Hawaii. *Evolution* 37: 601–617.

Stearns, S. C. 1983b. The influence of size and phylogeny on patterns of covariation among life-history traits in the mammals. *Oikos* 41: 173–187.

Stearns, S. C., and J. C. Koella. 1986. The evolution of phenotypic plasticity in life history norms: Predictions of reaction norms for age and size at maturity. *Evolution* 40: 893–913.

Steele, D. H. 1977. Correlation between egg size and development period. *Am. Nat.* 111: 371–372.

Steele, R. H. 1986. Courtship feeding in *Drosophila subobscura* I. The nutritional significance of courtship feeding. *Anim. Behav.* 34: 1087–1098.

Steinker, P. J. Unpublished. Scapegoating behaviour in the rhesus macaque.

Steinwascher, K. 1984. Egg size variation in *Aedes aegypti*: Relationship to body size and other variables. *Am. Midl. Nat.* 112: 76–84.

Stenseth, N. C. 1978. Is the female-biased sex ratio in wood lemmings *Myopus schesticolor* maintained by cyclic in-breeding? *Oikos* 30: 83–89.

Stephens, D. W., and J. R. Krebs. 1987. *Foraging Theory*. Princeton University Press, Princeton, N.J.

Stephenson, A. G., and J. A. Winsor. 1986. *Lotus corniculatus* regulates offspring quality through selective fruit abortion. *Evolution* 40: 453–458.

Stewart, K. J. 1988. Suckling and lactational anoestrus in wild gorillas (*Gorilla gorilla*). *J. Reprod. Fert.* 83: 627–634.

Stokland, J. N., and T. Amundsen. 1988. Initial size hierarchy in broods of the shag: Significance of egg size and hatching asynchrony. *Auk* 105: 308–315.

Strathmann, R. R. 1977. Egg size, larval development, and juvenile size in benthic marine invertebrates. *Am. Nat.* 111: 373–376.

Strathmann, R. R., and M. F. Strathmann. 1982. The relationship between adult size and brooding in marine invertebrates. *Am. Nat.* 119: 91–101.

Strum, S. C. 1983. Why males use infants. In D. M. Taub, ed., *Primate Paternalism*, pp. 146–185. Van Nostrand-Reinhold Co., New York.

Stuart-Dick, R. I., and K. B. Higginbottom. In press. In G. Grigg, P. Jarman, and I. Hume, eds., *Kangaroos, Wallabies and Rat-Kangaroos*. Surrey Beatty & Sons, Australia.

Sussman, R. W., and P. A. Garber. 1987. A new interpretation of the social organization and mating system of the Callitrichidae. *Int. J. Primatol.* 8: 73–92.

Sutherland, W. J., A. Grafen, and P. H. Harvey. 1986. Life history correlations and demography. *Nature* 320: 88.

Svardson, G. 1949. Natural selection and egg number in fish. *Rep. Inst. Freshwat. Res. Drottingholm*, no. 29: 115–122.

Svensson, I. 1988. Reproductive costs in two sex-role reversed pipefish species (Syngnathidae). *J. Anim. Ecol.* 57: 929–942.

Symington, M. 1987. Sex ratio and maternal rank in wild spider monkeys: When daughters disperse. *Behav. Ecol. Sociobiol.* 20: 421–425.

Taber, R. O., and R. F. Dasmann. 1954. A sex difference in mortality in young Columbian black-tailed deer. *J. Wildl. Mgmt.* 18: 309–315.

Tait, D.E.W. 1980. Abandonment as a reproductive tactic in grizzly bears. *Am. Nat.* 115: 800–808.

Tallamy, D. W. 1982. Age specific maternal defence in *Gargaphia solani* (Hemiptera: Tingidae). *Behav. Ecol. Sociobiol.* 11: 7–11.

Tallamy, D. W. 1984. Insect parental care. *Bio Science* 34: 20–24.

Tallamy, D. W., and R. F. Denmo. 1982. Life-history tradeoffs in *Gargaphia solani* (Hemiptera: Tingidae): The cost of reproduction. *Ecology* 63: 616–620.

Tallamy, D. W., and L. A. Horton. 1990. Costs and benefits of the egg-dumping alternative in *Gargaphia* lace bugs (Hemiptera: Tingidae). *Anim. Behav.* 39: 352–359.

Tallamy, D. W., and T. K. Wood. 1986. Convergence patterns in subsocial insects. *Ann. Rev. Entomol.* 31: 369–390.

Tantawy, A. O., and M. R. El-Helw. 1966. Studies on natural populations of *Drosophila*. V. Correlated response to selection in *Drosophila melanogaster*. *Genetics* 53: 97–110.

Tantawy, A. O. and F. A. Rakha. 1964. Studies on natural populations of *Drosophila*. IV. Genetic variances of and correlation between four characters in *Drosophila melanogaster* and *D. simulans*. *Genetics* 50: 1349–1355.

Taub, D. M. 1985. Male-infant interactions in baboons and macaques: A critique and re-evaluation. *Amer. Zool.* 25: 861–871.

Taub, S. H. 1969. Fecundity of the white perch. *Progr. Fish. Cult.* 31: 166–168.

Taylor, P. D. 1981. Intra-sex and inter-sex sibling interactions as sex ratio determinants. *Nature* 291: 64–66.

Taylor, P. D., and G. C. Williams. 1984. Demographic parameters in evolutionary equilibrium. *Can. J. Zool.* 62: 2264–2271.

Teather, K. L., and P. Weatherhead. 1988. Sex specific requirements of great-tailed grackle (*Quiscalus mexicanus*) nestlings. *J. Anim. Ecol.* 57: 658–659.

Teather, K. L., and P. Weatherhead. 1989. Sex specific mortality in great-tailed grackles. *Ecology* 70: 1485–1493.

Temin, R. G. 1966. Homozygous viability and fertility loads in *Drosophila melanogaster*. *Genetics* 53: 27–56.

Temme, D. H. 1986. Seed size variability: A consequence of variable genetic quality among offspring? *Evolution* 40: 414–417.

Tener, J. S. 1954. A preliminary study of the musk-oxen of Fosheim Peninsula, Ellesmere Island, N.W.T. Canada Dept. Natl. Parks Board. Canadian Wildl. Service, Ottawa. *Wildl. Mgmt. Bull.*, ser. 1, 9: 1–34.

Terner, C. 1968. Studies of metabolism in embryonic development. I. The oxidative metabolism of unfertilized and embryonated eggs of the rainbow trout. *Comp. Biochem. Physiol.* 24: 933–940.

Tessier, A., L. L. Henry, C. E. Goulden, and M. W. Durand. 1983. Starvation in *Daphnia*: Energy reserves and reproductive allocation. *Limnol. Oceanogr.* 28: 667–676.

Thibault, R. E., and R. J. Schultz. 1978. Reproductive adaptations among viviparous fishes (Cyprinodontiformes: Poeciliidae). *Evolution* 32: 320–333.

Thierry, B., and J. R. Anderson. 1986. Adoption in anthropoid primates. *Int. J. Primatol.* 7: 191–216.

Thomas, J. A., and E. C. Birney. 1979. Parental care and mating system of the prairie vole *Microtus ochrogaster*. *Behav. Ecol. Sociobiol.* 5: 171–186.

Thorne, E. T., R. E. Dean, and W. G. Hepworth. 1976. Nutrition during gestation in relation to successful reproduction in elk. *J. Wildl. Mgmt.* 40: 330–335.

Thornhill, R. 1976. Sexual selection and paternal investment in insects. *Am. Nat.* 110: 153–163.

Thornhill, R. 1981. *Panorpa* (Mecoptera: Panorpidae) scorpionflies. Systems for understanding resource-defence polygyny and alternative male reproductive effort. *Ann. Rev. Ecol.* 12: 355–386.

Thornhill, R. 1986. Relative parental contributions of the sexes to offspring and the operation of sexual selection. In M. Nitecki and J. Kitchell, eds., *The Evolution of Behaviour*, pp. 10–35. Oxford University Press, Oxford.

Thornhill, R., and J. Alcock. 1983. *The Evolution of Insect Mating Systems*. Harvard University Press, Cambridge, Mass.

Thornhill, R., and D. T. Gwynne. 1986. The evolution of sex differences in insects. *Amer. Sci.* 74: 382–389.

Thorson, G. 1950. Reproductive and larval ecology of marine bottom invertebrates. *Biol. Rev.* 25: 1–45.

Thouless, C. R. 1987. Feeding competition in red deer hinds. Ph.D. diss., University of Cambridge.

Thouless, C. R., and F. E. Guinness. 1986. Conflict between red deer hinds: The winner always wins. *Anim. Behav.* 34: 1166–1171.

Thresher, R. E. 1984. *Reproduction in Reef Fishes*. T.F.H. Publications, Neptune City, N.J.

Thresher, R. 1985. Brood-directed parental aggression and early brood loss in the coral reef fish, *Acanthochromis polyacanthus* (Pomacentridae). *Anim. Behav.* 33: 897–907.

Thresher, R. E. 1988. Latitudinal variation in egg sizes of tropical and subtropical North Atlantic shore fishes. *Env. Biol. Fishes* 21: 17–25.

Tilley, S. G. 1972. Aspects of parental care and embryonic development in *Desmognathus achrophaeus*. *Copeia* 1972: 532–540.

Tilson, R. L. 1981. Family formation strategies of Kloss's gibbons. *Folia Primatol.* 35: 259–281.

Tinkle, D. W. 1969. The concept of reproductive effort and its relation to the life histories of lizards. *Am. Nat.* 103: 501–506.

Tinkle, D. W., and J. W. Gibbons. 1977. The distribution and evolution of viviparity in reptiles. *Misc. Publ. Mus. Zool. Univ. Michigan* 154: 1–55.

Tinkle, D. W., and N. F. Hadley. 1973. Reproductive effort and winter activity in the viviparous montane lizard, *Sceloporus jarrovi*. *Copeia* 1973: 272–277.

Tinkle, D. W., and N. F. Hadley. 1975. Lizard reproductive effort: Calorific estimates and comments on its evolution. *Ecology* 56: 427–434.

Tinkle, D. W., H. M. Wilbur, and S. G. Tilley. 1970. Evolutionary strategies in lizard reproduction. *Evolution* 24: 55–74.

Todd, C. D., and R. W. Doyle. 1981. Reproductive strategies of marine benthic invertebrates: A settlement timing hypothesis. *Mar. Ecol. Prog.*, ser. 4: 75–83.

Torchio, P. F., and V. J. Tepedino. 1980. Sex ratio, body size and seasonality in a solitary bee, *Osmia lignia propinqua* Cresson (Hymenoptera: Megachilidae). *Evolution* 34: 993–1003.

Townsend, D. S. 1986. The costs of male parental care and its evolution in a neotropical frog. *Behav. Ecol. Sociobiol.* 19: 187–195.

Townsend, D. S. 1989. Sexual selection, natural selection and a fitness trade-off in a tropical frog with male parental care. *Am. Nat.* 133: 266–272.

Townshend, T. J., and R. J. Wootton. 1984. Effects of food supply on the reproduction of the convict cichlid, *Cichlasoma nigrofasciatum*. *J. Fish. Biol.* 23: 91–104.

Townshend, T. J., and R. J. Wootton. 1985. Adjusting parental investment to changing environmental conditions: The effect of food ration on parental behaviour of the convict cichlid, *Cichlasoma nigrofasciatum*. *Anim. Behav.* 33: 494–501.

Travis, J. 1983. Variation in development patterns of larval anurans in temporary ponds. I. Persistent variation within a *Hyla gratiosa* population. *Evolution* 37: 496–512.

Trevelyan, R., P. H. Harvey, and M. D. Pagel. 1990. Metabolic rates and life histories in birds. *Functional Ecology*.

Trillmich, F. 1986. Maternal investment and sex-allocation in the Galapagos fur seal, *Arctocephalus galapagoensis*. *Behav. Ecol. Sociobiol.* 19: 157–164.

Trillmich, F., and E. Lechner. 1986. Milk of the Galapagos fur seal and sea lion, with a comparison of the milk of eared seals (Otariidae). *J. Zool.* 209: 271–237.

Trivers, R. L. 1972. Parental investment and sexual selection. In B. Campbell, ed., *Sexual Selection and the Descent of Man*, pp. 136–179. Aldine, Chicago.

Trivers, R. L. 1974. Parent-offspring conflict. *Amer. Zool.* 11: 249–264.

Trivers, R. L. 1985. *Social Evolution*. Benjamin Cummings, Menlo Park, Calif.

Trivers, R. L., and H. Hare. 1976. Haplodiploidy and the evolution of social insects. *Science* 191: 249–263.

Trivers, R. L., and D. E. Willard. 1973. Natural selection of parental ability to vary the sex ratio of offspring. *Science* 179: 90–92.

Trollope, J. 1970. Notes on the Barred and Andalusian Hemipodes (*Turnix suscitator* and *Turnix sylvatica*). *Avic. Mag.* 76: 219–227.

Tuomi, J., T. Hakala, and E. Haukioja. 1983. Alternative concepts of reproductive efforts, costs of reproduction and selection in life history evolution. *Am. Zool.* 23: 25–34.

Turner, C .L. 1947. Viviparity in teleost fishes. *Sci. Month.* 65: 508–518.

Tyndale-Biscoe, H., and M. Renfree. 1987. *Reproductive Physiology of Marsupials*. Cambridge University Press, Cambridge.

Underwood, A. J. 1974. On models for reproductive strategy in marine benthic invertebrates. *Am. Nat.* 108: 874–878.

Underwood, A. J. 1979. The ecology of intertidal gastropods. *Adv. Mar. Biol.* 16: 111–120.

Unger, L. M., and R. C. Sargent. 1988. Allopaternal care in the fathead minnow, *Pimephales promelas*: Females prefer males with eggs. *Behav. Ecol. Sociobiol.* 23: 27–32.

Usaki, H. 1977. Underwater observations and experiments on pair formation and related behaviors in the apogonid fish, *Apogon notatus* (Houttuyn). *Publ. Seto Mar. Biol. Lab.* 24: 223–243.

Uyenoyama, M. K., and B. O. Bengtsson. 1981. Towards a genetic theory for the evolution of the sex ratio. II. Haplodiploid and diploid models with sibling and parental control of the brood sex ratio and brood size. *Theor. Popul. Biol.* 20: 57–59.

Vance, R. R. 1973a. On reproductive strategy in marine benthic invertebrates. *Am. Nat.* 107: 339–352.

Vance, R. R. 1973b. More on reproductive strategies in marine benthic invertebrates. *Am. Nat.* 107: 353–361.

Vance, R. R. 1974. Reply to Underwood. *Am. Nat.* 108: 879–880.

van den Berghe, E. P. 1984. Natural selection and reproductive success of female coho salmon (*Oncorhynchus kisutsch*): A study in female competiton. M.Sc. thesis, Simon Fraser University, Burnaby, British Columbia.

van den Berghe, E. P., and M. R. Gross. 1989. Natural selection resulting from female breeding competition in a pacific salmon (Coho: *Oncorhynchus kisutch*). *Evolution* 43: 125–140.

Vandeputte-Poma, J. 1980. Feeding, growth and metabolism of the pigeon, *Columba livia domestica*: Duration and role of crop milk feeding. *J. Comp. Physiol.* 135: 97–99.

van der Have, T. M., J. J. Boomsma, and S. B. Menken. 1988. Sex-investment ratios and relatedness in the monogynous ant *Lasius niger*. *Evolution* 42: 160–172.

van Dijk, T. S. 1979. On the relationship between reproduction, age and survival in two carabid beetles: *Calathus melancephalus* L. and *Pterostichus caerulescens* L. (Coleoptera, Carabidae). *Oecologia* 40: 63–80.

van Jaarsveld, A. S., J. D. Skinner, and M. Lindeque. 1988. Growth, development and parental investment in the spotted hyena *Crocuta crocuta*. *J. Zool.* 216: 45–54.

van Noordwijk, A. J., and G. de Jong. 1986. Acquisition and allocation of resources: Their influence on variation in life history tactics. *Am. Nat.* 128: 137–142.

van Noordwijk, N., and J. H. van Balen. 1988. The great tit, *Parus major*. In T. H. Clutton-Brock, ed., *Reproductive Success*, pp. 119–135. University of Chicago Press, Chicago.

van Rhijn, J. 1984. Phylogenetical constraints in the evolution of parental care strategies by birds. *Netherlands J. Zool.* 34: 103–122.

van Schaik, C. P. 1983. Why are diurnal primates living in groups? *Behaviour* 87: 120–144.

van Schaik, C. P., and M. A. van Noordwijk. 1983. Social stress and the sex ratio of neonates and infants among non-human primates. *Netherlands J. Zool.* 33: 249–265.

van Schaik, C. P., and R.I.M. Dunbar. In press. The evolution of monogamy in large primates: A new hypothesis. *Behaviour*.

Vaz-Ferreira, R., and A. Gehrau. 1975. Comportamiento epimeletico de la rana commun, *Leptodactylus ocellatus* (L.) (Amphibia, Leplodactylidae). 1. Attencion de la cria y activades alimentarias y agresivas relacionades. *Physis* 34: 14.

Vehrencamp, S. L. 1977. Relative fecundity and parental effort in communally nesting anis *Crotophoga sulcirostris*. *Science* 197: 403–405.

Vehrencamp, S. L., and J. W. Bradbury. 1984. Mating systems and ecology. In J. R. Krebs and N. B. Davies, eds., *Behavioural Ecology: An Evolutionary Approach*, pp. 251–278. Blackwell Scientific Publications, Oxford.

Vehrencamp, S., R. R. Koford, and S. Bowen. 1988. The effect of breeding-unit size on fitness components in groove-billed anis. In T. H. Clutton-Brock, ed., *Reproductive Success*, pp. 291–304. University of Chicago Press, Chicago.

Verbeek, M. 1988. Differential predation of eggs in clutches of glaucous-winged gulls *Larus glaucescens*. *Ibis* 130: 512–518.

Verme, L. J. 1969. Reproductive patterns of white-tailed deer related to nutritional plane. *J. Wildl. Mgmt.* 33: 881–887.

Verme, L. J. 1983. Sex ratio variation in *Odocoileus*: A critical review. *J. Wildl. Mgmt.* 47: 573–582.

Verme, L. J., and J. J. Ozoga. 1981. Sex ratio of white-tailed deer and the estrous cycle. *J. Wildl. Mgmt.* 45: 710–715.

Vincent, A.C.J. 1990. Reproductive ecology of seahorses. Ph.D. diss., University of Cambridge.

Vitt, L. J., and D. G. Blackburn. 1983. Reproduction in the lizard *Mabuya heathi* (Scincidae): Commentary on viviparity in New World *Mabuya*. *Can. J. Zool.* 61: 2798–2806.

Voland, E. 1 984. Human sex ratio manipulation: Historical data from a German parish. *J. Hum. Evol.* 13: 99–107.

Waage, J. K., and S. M. Ng. 1984. The reproductive strategy of a parasitic wasp. I. Optimal progeny and sex allocation in *Trichrogramma evanescens*. *J. Anim. Ecol.* 53: 401–416.

Wake, M. H. 1977. Fetal maintenance and its evolutionary significance in the Amphibia: Gymnophiona. *J. Herpetol.* 11: 379–386.

Wall, R., and M. Begon. 1987. Individual variation and the effects of population density in the grasshopper *Chorthippus brunneus*. *Oikos* 49: 15–27.

Wallin, K. 1987. Defence as parental care in tawny owls (*Strix aluco*). *Behaviour* 102: 213–230.

Wallinga, J. H., and H. Bakker. 1978. Effects of long-term selection for litter size in mice on lifetime reproduction. *J. Anim. Sci.* 46: 1563–1571.

Walsberg, G. E. 1978. Brood size and the use of time and energy by *Phainopepla*. *Ecology* 59: 147–153.

Walsberg, G. E. 1983. Ecological energetics: What are the questions? In A. H. Bush and G. A. Clark, eds., *Perspectives in Ornithology*, pp. 135–158. Cambridge University Press, Cambridge.

Walters, J. R. 1982. Parental behavior in lapwings (Charadriidae) and its relationship with clutch sizes and mating systems. *Evolution* 36: 1030–1040.

Wanless, S., M. P. Harris, and J. A. Morris. 1988. The effect of radio transmitters on the behavior of common murres and razorbills during chick rearing. *Condor* 90: 816–824.

Ward, J. A., and G. W. Barlow. 1967. The maturation and regulation of glancing-off the parents by young orange chromides (*Etroplus maculatus*: Pisces—Cichlidae). *Behaviour* 29: 1–56.

Ward, J. A., and J. I. Samarakoon. 1981. Reproductive tactics of the Asian cichlids of the genus *Etroplus* in Sri Lanka. *Env. Biol. Fish.* 6: 95–103.

Ward, P. S. 1983. Genetic relatedness and colony organization in a species complex of pomerine ants. II. Patterns of sex ratio investment. *Behav. Ecol. Sociobiol.* 12: 301–307.

Ware, D. M. 1975. Relation between egg size, growth and natural mortality of larval fish. *J. Fish. Res. Bd. Can.* 32: 2503–2512.

Warham, J. 1975. The crested penguins. In B. Stonehouse, ed., *The Biology of Penguins*, pp. 189–269. Macmillan, London.

Wasser, S. K., and D. P. Barash. 1982. Reproductive suppression among female mammals: Implications for biomedecine and sexual selection theory. *Q. Rev. Biol.* 58: 513–538.

Wattiaux, J. M. 1968. Cumulative parental age effects in *Drosophila subobscura*. *Evolution* 22: 406–421.

Weatherhead, P. J. 1979a. Ecological correlates of monogamy in tundra-breeding savannah sparrows. *Auk* 96: 391–401.

Weatherhead, P. J. 1979b. Do savannah sparrows commit the Concorde fallacy? *Behav. Ecol. Sociobiol.* 5: 373–381.

Weatherhead, P. J. 1982. Risk taking by red-winged blackbirds and the Concorde fallacy. *Z. Tierpsychol.* 60: 199–208.

Weatherhead, P. J. 1983. Secondary sex ratio adjustment in red-winged blackbirds. *Behav. Ecol. Sociobiol.* 12: 57–61.

Wegge, P. 1980. Distorted sex ratios and small broods in a declining capercaillie population. *Ornis. Scand.* 11: 106–109.

Wellington, W. G. 1957. Individual differences as a factor in population dynamics: The development of a problem. *Can. J. Zool.* 35: 293–323.

Wellington, W. G. 1965. Some maternal influences on progeny quality in the western tent caterpillar, *Malacosoma pluviale* (Dyar). *Can. Entomol.* 97: 1–14.

Wells, K. D. 1977a. The social behaviour of anuran amphibians. *Anim. Behav.* 25: 666–693.

Wells, K. D. 1977b. The courtship of frogs. In D. H. Taylor and S. I. Guttman, eds., *The Reproductive Biology of Amphibians*, pp. 233–262. Plenum Press, New York.

Wells, K. D. 1978. Courtship and parental behavior in a Panamanian poison-arrow frog (*Dendrobates auratus*). *Herpetologica* 34: 148–155.

Wells, K. D. 1981. Parental behavior of male and female frogs. In R. D. Alexander and D. W. Tinkle, eds., *Natural Selection and Social Behavior*, pp. 184–197. Chiron Press, New York.

Welty, J. C. 1982. *The Life of Birds*. W. B. Saunders Company. Philadelphia.

Werren, J. H. 1980. Sex ratio adaptation to local mate competition in a parasitic wasp. *Science* 208: 1157–1159.

Werren, J. H., and E. L. Charnov. 1978. Facultative sex ratios and population dynamics. *Nature* 272: 349–350.

Werren, J. H., and P. D. Taylor. 1984. The effects of population recruitment on sex ratio selection. *Am. Nat.* 124: 143–148.

Werren, J. H., M. R. Gross, and R. Shine. 1980. Paternity and the evolution of male parental care. *J. theoret. Biol.* 82: 619–631.

Werschkul, D. F. 1982. Parental investment: Influence of nest guarding by male little blue herons *Florida caerulea*. *Ibis* 124: 343–347.

West, K. J., and R. D. Alexander. 1963. Sub-social behavior in a burrowing cricket *Anurogryllus muticus* (De Geer). Orthoptera: Gryllidae. *Ohio J. Sci.* 63: 19–24.

West-Eberhardt, M. J. 1969. The social biology of polistine wasps. *Misc. Publ. Mus. Zool. Univ. Michigan* 140: 1–101.

West-Eberhardt, M. J. 1975. The evolution of social behavior by kin selection. *Quart. Rev. Biol.* 50: 1–33.

Western, D. 1979. Size, life history and ecology in mammals. *Afr. J. Ecol.* 17: 185–204.

Western, D., and J. Ssemakula. 1982. Life history patterns in birds and mammals and their evolutionary interpretation. *Oecologia* 54: 281–290.

Westmoreland, D. 1989. Offspring age and nest defence in mourning doves: A test of two hypotheses. *Anim. Behav.* 38: 1062–1066.

Weygoldt, P. 1980. Complex brood care and reproductive behavior in captive poison-arrow frogs *Dendrobates pumilio* O. Schmidt. *Behav. Ecol. Sociobiol.* 7: 329–332.

Weygoldt, P. 1984. Beobachtungen zur Fortpflanzungsbiologie von *Dendrobates pumilio* Schmidt 1857 im Terrarium (Salientia: Dendrobatidae). *Salamandra* 20: 111–120.

Weygoldt, P. 1987. Evolution of parental care in dart poison frogs (Amphibia: Anura: Dendrobatidae). *Z. Zool. Systematik und Evolutionsforschung* 1: 51–67.

Whittingham, L. A. 1989. An experimental study of paternal behavior in red-winged blackbirds. *Behav. Ecol. Sociobiol.* 251: 73–80.

Wickler, W. 1965. Neue varianten des Fortpflanzungsverhaltens afrikanischer Cichliden (Pisces: Perciformes). *Naturwissenschaften* 9: 219–227.

Widdowson, E. M. 1976a. Changes in the body and its organs during lactation: Nutritional implications. In *Breast Feeding and the Mother*. Ciba Foundation Symposium 45. Elsevier, Holland.

Widdowson, E. M. 1976b. The response of the sexes to nutritional stress. *Proc. Nutr. Soc.* 35: 1175–1180.

Wiklund, C. G., and J. Stigh. 1983. Nest defence and evolution of reversed sexual dimorphism in snowy owls *Nyctea scandiaca*. *Ornis. Scand.* 14: 58–62.

Wiklund, C., B. Karlsson, and J. Forsberg. 1987. Adaptive versus constraint explanations for egg-to-body relationships in two butterfly families. *Am. Nat.* 130: 828–838.

Wilbur, H. M. 1976. Density-dependent aspects of metamorphosis in *Ambystoma* and *Rana sylvatica*. *Ecology* 57: 1289–1296.

Wilbur, H. M. 1977a. Density-dependent aspects of metamorphosis in *Bufo americanus*. *Ecology* 58: 196–200.

Wilbur, H. M. 1977b. Propagule size, number, and dispersion pattern in *Ambystoma* and *Asclepias*. *Am. Nat.* 111: 43–68.

Wiley, R. H. 1974. Evolution of social organisation and life-history patterns among grouse. *Q. Rev. Biol.* 49: 201–227.

Williams, G. C. 1966a. Natural selection, the costs of reproduction and a refinement of Lack's principle. *Am. Nat.* 100: 687–690.

Williams, G. C. 1966b. *Adaptation and Natural Selection*. Princeton University Press, Princeton, N.J.

Williams, G. C. 1975. *Sex and Evolution*. Princeton University Press, Princeton, N.J.

Williams, G. C. 1979. The question of adaptive sex ratio in out-crossed vertebrates. *Proc. Roy. Soc.* 205: 567–580.

Willson, M. F. 1966. Breeding ecology of the yellow-headed blackbird. *Ecol. Monogr.* 76: 51–77.

Willson, M. F., and E. F. Pianka. 1963. Sexual selection, sex ratio and mating system. *Am. Nat.* 97: 405–407.

Wilson, E. O. 1971. *Insect Societies*. Harvard University Press, Cambridge, Mass.

Wilson, E. O. 1975. *Sociobiology, the New Synthesis*. Belknap Press of Harvard University Press, Cambridge, Mass.

Windt, W., and E. Curio. 1986. Clutch defence in great tit (*Parus major*) pairs and the Concorde fallacy. *Ethology* 72: 236–242.

Winkler, D. W. 1987. A general model for parental care. *Am. Nat.* 130: 526–543.

Winkler, D. W., and K. Wallin. 1987. Offspring size and number: A life history model linking effort per offspring and total effort. *Am. Nat.* 129: 708–720.

Winkler, D. W., and G. S. Wilkinson. In press. Parental Care in Birds and Mammals: An Evolutionary Perspective. *Oxford Reviews in Evolutionary Biology*.

Winn, H. E. 1958. Comparative reproductive behavior and ecology of fourteen species of darters (Pisces: Percidae). *Ecol. Monogr.* 28: 15191.

Wintle, C. C. 1975. Notes on the breeding habits of the Kurrichane button-quail. *Honeyguide* 82: 27–30.

Withey, D. S. 1983. Intrapopulation variation in growth of sessile organisms: Natural populations of the intertidal barnacle *Balanus balanoides*. *Oikos* 40: 14–23.

Witt, R., G. Schmidt, and J. Schmidt. 1981. Social rank and Darwinian fitness in a multimale group of Barbary macaques (*Macaca sylvana* L). *Folia Primatol.* 36: 201–211.

Wittenberger, J. F. 1979a. The evolution of mating systems in birds and mammals. In P. Master and J. Vandenburgh, eds., *Handbook of Behavioral Neurobiology: Social Behavior and Communication*, pp. 271–349. Plenum Press, New York.

Wittenberger, J. F. 1979b. A model for delayed reproduction in iteroparous animals. *Am. Nat.* 114: 439–446.

Wittenberger, J. F. 1981. *Animal Social Behavior*. Duxbury Press, Boston.

Wittenberger, J. F. 1982. Factors affecting how male and female bobolinks apportion parental investment. *Condor* 84: 22–39.

Wolf, J. O. 1988. Maternal investment and sex ratio adjustment in American bison calves. *Behav. Ecol. Sociobiol.* 23: 127–133.

Wolf, L., E. D. Ketterson, and V. Nolan, Jr. 1988. Paternal influence on growth and survival of dark-eyed junco young: Do parental males benefit? *Anim. Behav.* 36: 1601–1618.

Wolf, L., E. D. Ketterson, and V. Nolan, Jr. 1990. Behavioural response of female dark-eyed juncos to experimental removal of their mates: Implications for the evolution of parental care. *Anim. Behav.* 39: 125–134.

Wood, T. K. 1976. Alarm behavior of brooding female *Umbonia crassicornis* (Homoptera: Membracidae). *Ann. Entomol. Soc. Am.* 69: 340–344.

Wood, T. K. 1977. Role of parent females and attendant ants in the maturation of the treehopper, *Entylia bactriana* (Homoptera: Membracidae). *Sociobiology* 2: 257–272.

Woodhead, A. D. 1979. Senescence in fishes. *Symp. Zool. Soc. Lond.* 44: 179–205.

Woodruff, D. S. 1977. Male postmating brooding behavior in three Australian *Pseudophryne* (Anura: Leptodactylidae). *Herpetologica* 33: 296–303.

Woolf, A., and J. D. Harder. 1979. Population dynamics of a captive white-tailed deer herd with emphasis on reproduction and mortality. *Wildl. Monogr.* 67: 1–53.

Woombs, M., and J. Laybourne-Parry. 1984. Growth, reproduction and longevity in nematodes from sewage treatment plants. *Oecologia* 64: 168–172.

Wootton, R. J. 1979. Energy costs of egg production and environmental determinants of fecundity in teleost fishes. *Symp. Zool. Soc. Lond.* 44: 133–159.

Wootton, R. J. 1984. *A Functional Biology of Sticklebacks*. Croom Helm, London.

Worlein, J. M., G. G. Eaton, D. F. Johnson, and B. B. Glick. 1988. Mating season effects on mother-infant conflict in Japanese macaques *Macaca fuscata*. *Anim. Behav.* 36: 1472–1481.

Wourms, J. P. 1977. Reproduction and development in chondrichthyan fishes. *Amer. Zool.* 17: 379–410.

Wourms, J. P. 1981. Viviparity: The maternal foetal relationship in fishes. *Amer. Zool.* 21: 473–575.

Wrangham, R. W. 1979. On the evolution of ape social systems. *Soc. Sci. Inform.* 18: 335–368.

Wrangham, R. W. 1980. An ecological model of female-bonded primate groups. *Behaviour* 75: 262–300.

Wright, J., and I. Cuthill. 1989. Manipulation of sex differences in parental care. *Behav. Ecol. Sociobiol.* 25: 171–181.

Wright, S. 1931. Evolution in Mendelian populations. *Genetics* 16, 97–159.

Wright, S. 1978. Evolution and the genetics of populations, vol. 4, *Variability within and among Natural Populations*. University of Chicago Press, Chicago.

Wright, S. L., C. B. Crawford, and J. L. Anderson. 1988. Allocation of reproductive effort in *Mus domesticus*: Responses of offspring sex ratio and quality to social density and food availability. *Behav. Ecol. Sociobiol.* 23: 357–365.

Wyatt, T. D. 1986. How a subsocial intertidal beetle, *Bledius spectabilis*, prevents flooding and anoxia in its burrow. *Behav. Ecol. Sociobiol.* 19: 323–331.

Yamamoto, I. 1987. Male parental care in the raccoon dog *Nyctereutes procyonoides* during the early rearing period. In Y. Ito, J. L. Brown, and J. Kikkawa, eds., *Animal Societies*, pp. 189–196. Japan Societies Scientific Press, Tokyo.

Yamamura, N. 1987. Biparental defence in a subsocial spider mite. *Trends in Ecology and Evolution* 2: 261–262.

Yaron, Z. 1985. Reptilian placentation and gestation: Structure, function and endocrine control. In C. Gans and F. Billett, eds., *Biology of the Reptilia*, vol. 15. Wiley, New York.

Yasukawa, K., J. L. McClure, R. A. Boley, and J. Zanocco. In press. Provisioning of nestlings by male and female red-winged blackbirds, *Agelaius phoeniceus. Anim. Behav.*

Yom-Tov, Y., and J. G. Ollason. 1976. Sexual dimorphism and sex ratios in wild birds. *Oikos* 27: 81–85.

Zahavi, A. 1977. Reliability in communication systems and the evolution of altruism. In B. Stonehouse and C. Perrins, eds., *Evolutionary Ecology*, pp. 253–259. University Park Press, Baltimore, Maryland.

Zeh, D. W., and R. L. Smith. 1985. Paternal investment by terrestrial arthropods. *Amer. Zool.* 25: 785–805.

Zeveloff, S. I., and M. S. Boyce. 1980. Parental investment and mating systems in mammals. *Evolution* 34: 973–982.

Zeveloff, S. I., and M. S. Boyce. 1982. Why human neonates are so altricial. *Am. Nat.* 120: 537–542.

Zwickel, F. C., and J. F. Bendall. 1967. Early mortality and the regulation of numbers in blue grouse. *Can. Zool.* 45: 817–851.

Author Index

Subject Index

Taxonomic Index